# Legal Mechanisms for Water Resources in the Third Millennium



Legal mechanisms for the management, development and protection of water resources have evolved over the years and have reached unprecedented levels of complexity and sophistication. This phenomenon is largely in response to the global community's sustainable development agenda, to the challenges and limitations imposed by climate variability, and to scientific and technological advances. Bringing together diverse experiences from across the world, this book analyzes existing water law and governance solutions, their shortcomings, as well as developments and trends in the light of changing circumstances. The legal mechanisms examined range from international treaties, agreements and arrangements on cooperation over transboundary water resources, to the onset of novel issues arising out of technological advances, and from domestic regulation of water abstraction and groundwater management, to domestic regulation of the water industry.

The chapters in this book were originally published in the journal *Water International*, following the XIV and the XV World Water Congresses of the International Water Resources Association (IWRA), which were held in 2011 and in 2015, respectively.

**Marcella Nanni** is an expert in national and international water law and administration, and related disciplines, and the Vice-Chairperson of the Executive Council of the International Association for Water Law (AIDA).

**Stefano Burchi** is an expert in comparative and international water law. He is the Chairman of AIDA's Executive Council.

**Ariella D'Andrea** is a natural resources lawyer specializing in fisheries, aquaculture and water resources law, and a member of AIDA.

**Gabriel Eckstein** is Professor of Law at Texas A&M University School of Law, specializing in water, environment, natural resources and international law, and a member of AIDA.

## Routledge Special Issues on Water Policy and Governance

https://www.routledge.com/series/WATER

Edited by:

*Cecilia Tortajada (IJWRD) – Third World Centre for Water Management, Mexico*
*James Nickum (WI) – International Water Resources Association, France*

Most of the world's water problems, and their solutions, are directly related to policies and governance, both specific to water and in general. Two of the world's leading journals in this area, the *International Journal of Water Resources Development* and *Water International* (the official journal of the International Water Resources Association), contribute to this special issues series, aimed at disseminating new knowledge on the policy and governance of water resources to a very broad and diverse readership all over the world. The series should be of direct interest to all policy makers, professionals and lay readers concerned with obtaining the latest perspectives on addressing the world's many water issues.

**Water for Food Security**
Challenges for Pakistan
*Edited by Claudia Ringler and Arif Anwar*

**Water Management and Climate Change**
Dealing with Uncertainties
*Edited by Cecilia Tortajada, Asit K. Biswas and Avinash Tyagi*

**Integrated Water Resources Management**
From concept to implementation
*Edited by Cecilia Tortajada*

**Water Infrastructure**
*Edited by Cecilia Tortajada and Asit K. Biswas*

**Frontiers of Land and Water Governance in Urban Regions**
*Edited by Thomas Hartmann and Tejo Spit*

**The Water-Energy-Food Nexus in the Middle East and North Africa**
*Edited by Martin Keulertz and Eckart Woertz*

**Sustainability in the Water-Energy-Food Nexus**
*Edited by Anik Bhaduri, Claudia Ringler, Ines Dombrowsky, Rabi Mohtar and Waltina Scheumann*

**Transboundary Water Cooperation**
Principles, Practice and Prospects for China and its Neighbours
*Edited by Patricia Wouters, Huiping and James E. Nickum*

**Water Reuse Policies for Potable Use**
*Edited by Cecilia Tortajada and Choon Nam Ong*

**Hydrosocial Territories and Water Equity**
Theory, Governance and Sites of Struggles
*Edited by Rutgerd Boelens, Ben Crow, Jaime Hoogesteger, Flora E Lu, Erik Swynedouw and Jeroen Vos*

**Energy for Water**
Regional Case Studies
*Edited by Christopher Napoli*

**Legal Mechanisms for Water Resources in the Third Millennium**
Select Papers from the IWRA XIV and XV World Water Congresses
*Edited by Marcella Nanni, Stefano Burchi, Ariella D'Andrea and Gabriel Eckstein*

# Legal Mechanisms for Water Resources in the Third Millennium

## Select Papers from the IWRA XIV and XV World Water Congresses

*Edited by*

**Marcella Nanni, Stefano Burchi, Ariella D'Andrea and Gabriel Eckstein**

Routledge
Taylor & Francis Group
LONDON AND NEW YORK

IWRA

International Association for Water Law
Asociación Internacional de Derecho de Aguas
ASSOCIATION INTERNATIONALE DU DROIT DES EAUX

First published 2018
by Routledge

2 Park Square, Milton Park, Abingdon, Oxfordshire OX14 4RN
52 Vanderbilt Avenue, New York, NY 10017

*Routledge is an imprint of the Taylor & Francis Group, an informa business*

First issued in paperback 2020



*British Library Cataloguing in Publication Data*
A catalogue record for this book is available from the British Library

ISBN 13: 978-0-8153-6363-7 (hbk)
ISBN 13: 978-0-367-58966-0 (pbk)

Typeset in Minion Pro
by RefineCatch Limited, Bungay, Suffolk

**Publisher's Note**
The publisher accepts responsibility for any inconsistencies that may have arisen during the conversion of this book from journal articles to book chapters, namely the possible inclusion of journal terminology.

# Contents

## Part 3: Multiple approaches for achieving access to water



## Part 4: Science and the law: bridging the gap



## Part 5: Cooperation on transboundary water resources: current state and future directions

# Citation Information

The chapters in this book were originally published in various issues of *Water International*. When citing this material, please use the original page numbering for each article, as follows:

**Chapter 2**
*A comparative review of contemporary water resources legislation: trends, developments and an agenda for reform*
Stefano Burchi
*Water International*, volume 37, issue 6 (October 2012), pp. 613–627

**Chapter 3**
*Legislation as a tool in support of adaptive water management in response to climate change*
Marcella Nanni
*Water International*, volume 37, issue 6 (October 2012), pp. 628–639

**Chapter 4**
*Water governance reform in the context of inequality: securing rights or legitimizing dispossession?*
Helle Munk Ravnborg
*Water International*, volume 41, issue 6 (2016), pp. 928–943

**Chapter 5**
*Compulsory licensing under South Africa's National Water Act*
Michael Kidd
*Water International*, volume 41, issue 6 (2016), pp. 916–927

**Chapter 6**
*Legal pluralism and customary water resources management in Guatemala*
Ariella D'Andrea
*Water International*, volume 37, issue 6 (October 2012), pp. 683–699

**Chapter 7**
*Why the Western United States' prior appropriation water rights system should weather climate variability*
Jonathan R. Schutz
*Water International*, volume 37, issue 6 (October 2012), pp. 700–707

**Chapter 8**
*Adapting water laws to increasing demand and a changing climate*
Eric L. Garner
*Water International*, volume 41, issue 6 (2016), pp. 883–899

**Chapter 9**
*The human right to water as a creature of global administrative law*
Owen McIntyre
*Water International*, volume 37, issue 6 (October 2012), pp. 654–669

**Chapter 10**
*The right to water in a transboundary context: emergence of seminal trends*
Christina Leb
*Water International*, volume 37, issue 6 (October 2012), pp. 640–653

**Chapter 11**
*Scottish Water: a public-sector success story*
Sarah Hendry
*Water International*, volume 41, issue 6 (2016), pp. 900–915

**Chapter 12**
*Are we killing the rain? Meditations on the water cycle and, more particularly, on bioprecipitation*
Jane Maslow Cohen
*Water International*, volume 37, issue 6 (October 2012), pp. 670–682

**Chapter 13**
*Governing water augmentation under the Watercourse Convention*
Rhett B. Larson
*Water International*, volume 41, issue 6 (2016), pp. 866–882

**Chapter 14**
*Joint infrastructure and the sharing of benefits in the Senegal and Niger watersheds*
Mara Tignino
*Water International*, volume 41, issue 6 (2016), pp. 835–850

**Chapter 15**
*Ad hoc legal mechanisms governing transboundary aquifers: current status and future prospects*
Laura Movilla Pateiro
*Water International*, volume 41, issue 6 (2016), pp. 851–865

**Chapter 16**
*Adaptation in the Tisza: innovation and tribulation at the sub-basin level*
Shanna N. McClain, Carl Bruch and Silvia Secchi
*Water International*, volume 41, issue 6 (2016), pp. 813–834

# Notes on Contributors

**Carl Bruch** is Senior Attorney and Director of International Programs at the Environmental Law Institute, Washington, USA.

**Stefano Burchi** is an expert in comparative and international water law. He is the Chairman of AIDA's (International Association for Water Law) Executive Council.

**Jane Maslow Cohen** is the Edward Clark Centennial Professor of Law at the University of Texas School of Law, USA.

**Ariella D'Andrea** is a natural resources lawyer specializing in fisheries, aquaculture and water resources law, and a member of AIDA.

**Gabriel Eckstein** is Professor of Law at Texas A&M University School of Law, specializing in water, environment, natural resources and international law, and a member of AIDA.

**Eric L. Garner** is a partner at Best Best & Krieger LLP, Los Angeles, CA, USA.

**Sarah Hendry** is Senior Lecturer at the Centre for Water Law, Policy and Science, University of Dundee, UK.

**Michael Kidd** is Professor at the School of Law, University of KwaZulu-Natal, Pietermaritzburg, South Africa.

**Rhett B. Larson** is a Morrison Fellow in Water Law and Associate Professor in the Sandra Day O'Connor College of Law, Arizona State University, USA.

**Christina Leb** is Senior Water Resources Specialist at The World Bank.

**Shanna N. McClain** is AAAS Fellow at the NASA Goddard Space Flight Center, USA.

**Owen McIntyre** is Senior Lecturer at the Faculty of Law, University College Cork, Ireland.

**Marcella Nanni** is an expert in national and international water law and administration, and related disciplines, and the Vice-Chairperson of the Executive Council of the International Association for Water Law (AIDA).

**Laura Movilla Pateiro** is Assistant Dean at the Department of Public Law, University of Vigo, Ourense, Spain.

**Helle Munk Ravnborg** is Senior Researcher in natural resources and development at the Danish Institute for International Studies, Copenhagen, Denmark.

**Jonathan R. Schutz** is Co-Chair of the American Bar Association's Water Resources Committee.

**Silvia Secchi** is Associate Professor at the Department of Geography and Environmental Resources, Southern Illinois University, USA.

**Mara Tignino** is based at the Platform for International Water Law/Geneva Water Hub, Faculty of Law, University of Geneva, Switzerland.

# Introduction

Marcella Nanni, Stefano Burchi, Ariella D'Andrea and Gabriel Eckstein

Legal mechanisms for the management and development of water resources, as well as for their protection against pollution, have evolved over the years and have reached unprecedented levels of complexity and sophistication within both domestic and transboundary contexts. This phenomenon is largely in response to the global community's sustainable development agenda, to the challenges and limitations imposed by climate variability, and to scientific and technological advances. Legal mechanisms, however, are not to be viewed as static contrivances that, once identified, work effectively to meet all local and global water challenges. Rather, they need to be moulded and remoulded in response to changing circumstances and, as a necessary preliminary step, to be analyzed from practical angles that are informed by experience on the ground. Only on these premises are they likely to be understood and adapted to current and emerging needs.

Bringing together diverse experiences from across the world, this book examines contemporary water law and governance approaches, their shortcomings, and existing and potential developments and trends. Together, they provide food for thought for those engaged in water sector reforms.

The book is structured into five parts elaborating on specific themes and covering a broad spectrum of legal mechanisms for water resources management. These range from international treaties, agreements and arrangements on cooperation over transboundary water resources, to the onset of novel issues arising out of technological advances, and from domestic regulation of water abstraction and groundwater management, to domestic regulation of the water industry.

Part 1 is meant to "set the stage" by illustrating key developments and trends in contemporary water resources legislation and inviting innovative thinking (Chapter 2). It further discusses legal reform approaches facilitating adaptation to climate variability and change (Chapter 3). Part 2 examines water resources legislation, and in particular provisions on water rights and groundwater management, in the context of equity (Chapters 4 and 5), legal pluralism (Chapter 6) and climate variability (Chapters 7 and 8). In Part 3, Chapters 9 and 10 are devoted to the right to water within domestic and transboundary contexts, respectively, while the chapter that follows tells a success story in relation to the regulation of the water industry in Scotland (Chapter 11). Science and the law are the subject of Part 4. Chapter 12 examines options for land use and water resource management through the lens of the science of bioprecipitation, and considers the extent to which relevant findings may help stimulate reforms in land use and water management policies and legislation. Chapter 13 evaluates how the UN Convention on the Law of the Non-navigational

Uses of International Watercourses can be interpreted and implemented to facilitate responsible and equitable implementation of water augmentation technologies. Finally, Part 5 deals with treaties, agreements and arrangements among states regarding cooperation over transboundary aquifers (Chapter 15) and international watercourses, covering the Niger and Senegal river basins (Chapter 14), and the Tizsa River, which is a tributary of the Danube (Chapter 16).

To sum up, the book consists of an in-depth, high-quality compilation of material on a selection of contemporary water law topics and across a broad spectrum of water-related disciplines. It is hoped that the collection will stimulate the pursuit of a policy dialogue regarding the problematique and direction of contemporary water law and associated review processes.

The articles in this book were originally published in two Special Issues of the journal *Water International*, following the XIV World Water Congress of the International Water Resources Association (IWRA) "Adaptive Water Management: Looking to the Future", held in Porto de Galinhas, Brazil, 25–29 September 2011, and the XV World Water Congress, "Global Water, A Resource for Development: Opportunities, Challenges and Constraints", held in Edinburgh, UK, 25–29 May 2015, respectively. As a contributor to the development of the two Congresses, the International Association for Water Law (known as AIDA from its Spanish acronym) took a leading role in the delivery of the water law and governance track in the two congress programmes, by soliciting papers and presentations from within and outside its membership covering a variety of water law and governance topics, and by contributing to the relevant debates.

AIDA is a network of legal, policy, and management specialists active in government, academia and civil society, with a specific interest in freshwater resources law and related disciplines. It was founded in 1967 by a group of like-minded, visionary water law and management specialists who, with remarkable anticipation, foresaw the growing significance of freshwater to the well-being of humankind and to its economic and social development, in a world of rapidly changing circumstances and increasing competition for what has come to be regarded as "blue gold". On the strength of the vision of its founders, AIDA is now a partner to a variety of governmental and non-governmental institutions to ensure the role of water law as a foundational pillar of the good governance of freshwater and related resources throughout the world.

# A comparative review of contemporary water resources legislation: trends, developments and an agenda for reform

Stefano Burchi

Key trends in water legislation include: attraction of water resources into the public domain; limitations on governmental authority to allocate water resources; controlled trading of water rights; the "greening" of water laws; capturing the land–water connection; and ensuring user participation in decision making and implementation. Hence the water law reform agenda is likely to be preoccupied with reconciling tenure security with risk and uncertainty; pursuing equitable resource allocation efficiency gains; raising the environment's profile in water resources allocation; rekindling the connection between water resources management and land use regulations; empowering users; and mapping the interface between customary and statutory water allocation.

## Introduction

Concern for the long-term sustainability of water resources development and use gained definitive prominence on the agenda of the world community at the Second World Water Forum and Ministerial Conference, held at The Hague in March 2000. In particular, the concept and goal of *water security* were loosely articulated there, by reference to "key challenges", namely, meeting basic needs; securing the food supply; protecting ecosystems; sharing water resources; managing risks; valuing water; and governing water wisely. As such, *water security* appears to go beyond just concern for the maintenance of supplies of drinking water, and to be of relevance not only to arid countries. Subsequently, in the Ministerial Declaration issuing from the International Freshwater Conference held in Bonn in December 2001,[1] preparatory to the United Nations World Summit for Sustainable Development (Johannesburg, 2002) and to the Third World Water Forum (Kyoto, 2003), *water governance*, including a supportive legal and institutional framework, was seen to play a central role in achieving water security. The policy pronouncements which have been articulated since then, and, in particular, the Ministerial Statement which issued from the latest (fifth) World Water Forum (Istanbul, 2009),[2] confirm that water security and governance, and a legal and institutional framework conducive to the attainment of the relevant goals, have definitively earned the attention of the international community, and found their place in the latter's agenda for the twenty-first century.

Based primarily on prior analyses of water resources legislation by this author (Burchi 2001, 2003, 2004, 2005b, 2006, 2007, 2008, 2010), this paper takes stock of the salient features and main trends detectable in the contemporary water resources management legislation of a wide spectrum of countries. The issues emerging, and the challenges ahead, are presented and discussed, for the inspiration of lawmakers as they chart the scope and the agenda of water law reform in support of water security and water governance policies and goals in the twenty-first century.

## Salient features and trends of contemporary domestic water resources legislation

A comparative analysis of contemporary legislation for the management and development of water resources discloses a number of discrete trends, some of which are novel compared to the consolidated direction of mainstream water legislation. The trends, known and emerging, that one can detect in the water laws enacted in the two decades astride the beginning of the twenty-first century can be summarized as follows.

### *Steady attraction of water resources into the public domain of the state, and into the scope of governmental (or judicial) allocation authority*

Groundwater in particular, and riparian rights in surface watercourses and in groundwater, have been steadily attracted into the ever-expanding sphere of 'public' domain waters. Public waters may include those waters held in the public domain (ownership) of the state, as in the water laws adopted in Italy in 1994, in Morocco in 1995 and in Zimbabwe in 2000, all of which bring groundwater resources in particular into the public domain. Alternatively, water resources have been vested in the state in trust for the public, as in South Africa's *National Water Act* (1998), or vested in the president in trust for the public (Tanzania, *Water Resources Management Act*, 2009), or the state may be vested with superior user rights, as in Uganda's 1995 *Water Statute* and in the Australian state of New South Wales's *Water Management Act* (2000). Whatever the legal underpinning, the result has been the same, i.e., to extricate a nation's water resources from the ownership or control of landowners, and to bring the resource and the relevant allocation under the scope of governmental authority.[3]

As a result of water becoming public property, individuals can only claim and obtain user rights in water. Such rights generally accrue from a grant made by the government (or by the courts, as in some western states of the United States), and are recorded in an administrative permit, licence, concession or the like instrument. Grants tend to be time-bound, are qualified by terms and conditions, and are subject to review and adjustment to reflect new circumstances. As a general rule and trend, the regulation of abstraction licensing seeks to reconcile the security of water rights tenure implicit in the government grant of an abstraction licence, permit or concession, with the flexibility which is desirable to adjust allocation patterns to the ever-changing circumstances of water availability, of evolving abstraction and use technologies and economics, and of shifting policies, priorities, and political agendas.

### *Checking the government authority to allocate and re-allocate water resources, thereby improving the quality of allocation decisions*

The discretionary authority the government enjoys in making abstraction grants and allocating water has traditionally been checked by the courts of law or through the hierarchical review (appeal) opportunities available through the granting process. These traditional review mechanisms are available *after* an allocation decision had been made and a permit granted. By contrast, a new generation of fetters tends to operate *before* such decisions and the relevant grants are made.

These are basically aimed at improving the quality of decision making, thereby minimizing opportunities for judicial or administrative review and litigation. The allocation authority of government is increasingly qualified by:

- Legislation imposing environmental impact assessment (EIA) requirements in respect of proposed water abstractions (e.g., Nicaragua's *General Law on National Waters*, 2007; Paraguay's *Law on Water Resources*, 2007). In what appears to constitute an effective extension of the standard scope of EIA provisions in relation to water resources, Tanzania's *Water Resources Management Act* (2009) mandates EIA requirements in respect of "any proposed development" in a water resource area (no definition provided, however) or watershed. Moreover, a strategic EIA is required of "major" water projects (no definition provided either, however, of what constitutes a "major" water project).
- Legislation prescribing water resources plans, especially where these have a binding effect on government decisions.
- Legislation prescribing the imposition of minimum ecological flow requirements in respect of surface watercourses, which seek to protect the ecology and fish life of watercourses and act as a limit on the government's allocation authority by barring new abstractions above established limits, and by forcing review and the downwards adjustment of abstractions in progress (see the discussion in the section below on raising the profile of the environment in the management of water resources). For instance, in Spain, the minimum environmental flow requirements of rivers provided for by the consolidated *Water Act* (2001) have priority call on available river flows under the *Law on the National Water Plan* (2001). In Chile, the minimum ecological flow requirements mandated by the 2005 reforms to the *Water Code* (1981) act as a barrier to the granting of new abstraction rights, but do not affect the abstraction rights already in existence.
- Legislation prescribing the formal reservation of water quantities or flows for a specific purpose – notably, the satisfaction of basic human needs and the protection of ecosystems – which also puts limits on the government's allocation authority insofar as the water reserve cannot be allocated except for the reserved purposes. The concept and mechanism of the "reserve" were pioneered by the *Water Act* (1998) of South Africa, and found their way into, among others, the *Water Act* (2002) of Kenya, the extensive 2004 amendments to the *Law on National Waters* (1992) of Mexico, the Kyrgyz *Water Code* (2005), the 2005 reform to the *Water Code* of Chile, the *Water Resources Management Act* of Namibia (2004) and the *Water Resources Management Act* of Tanzania (2009). The environmental reserve features also in Spain's *Law on the National Water Plan* (2001).

### *Controlled trading in water rights*

Increasingly, lawmakers have been turning to trading in water rights in the pursuit of efficiency of water allocation and use. Because trading of water rights empowers users to make allocation decisions instead of government, part of the expanding allocation authority vested in the government, which was observed earlier, is returned to the users.

Unregulated water trading is known to operate perhaps only in Chile. There, water is regarded as a commodity which can be freely traded through the sale of the relevant government grants. However, economic correctives aimed at discouraging the hoarding of water rights and speculation on them were introduced in 2005 through a legislative reform of the *Water Code*. Elsewhere, as under Mexico's *Law on National Waters* (1992, extensively amended in 2004), and under the water acts of most Australian states, regulations on water trading seek to minimize unwelcome 'third-party' effects, such as effects on the environment, on the interests of the area from which water is taken for use in another area, on cultural values, on resource availability to meet priority requirements and, generally, on marginal groups. This tends to be accomplished by subjecting trades to prior government scrutiny. Occasionally, only intra-sector trades are allowed, i.e., water under a licence or concession cannot be traded and applied to a different use. To varying degrees, this is the approach reflected in the water laws of Namibia (2004), Mauritania (2005), Nicaragua (2007), Paraguay (2007) and Tanzania (2009). In Mexico and Tanzania, wastewater disposal permits are also tradable, under the same controlled conditions as abstraction licences. There are a number of countries, however, like Morocco, Indonesia, Mauritania and Venezuela, where water has traditionally attached to the land where it is used, and where the trading of water rights separate from land rights is explicitly forbidden by the current water acts or water codes (dating, respectively, from 1995, 2004, 2005 and 2006).

### *Raising the profile of the environment in the management of water resources (the "greening" of water laws)*

The growing realization of the need to manage, develop and use freshwater resources in a more environmentally sustainable manner has raised the status and importance of water for the environment to a dignity which nowadays stands on a par with other societal needs and wants. As one author put it, "Fundamentally, the objective is to find a balance between water for human and economics-based demand and water for maintaining ecosystem integrity and environmental sustainability" (UNEP 2010). Put another way, the issue is reconciling the goals of socio-economic development and those of environmental protection and conservation. The resulting process of re-assessment of priorities and of water laws has been described as a "greening" of water laws (Burchi 2007, UNEP 2010).

A majority of the mechanisms of resource allocation reviewed above attest to the "greening" of water laws. Additional evidence of the "greening" process can be gleaned from a few other mechanisms of resource management and use, as reflected in the provisions of water laws governing:

- Water trading (and water trusts), insofar as invariably the water laws restrict trades having a 'third-party' effect on the environment (this point has been illustrated above). Moreover, under the state of Oregon (United States) *Instream Water Rights Act* (1987) that state may purchase water or lease water rights, mostly from farmers, and earmark them as trust rights for instream use for the protection of aquatic habitats.
- The extraction of groundwater, and restricting it to ensure its ecosystem support function. For instance, the New South Wales *Water Management Act* (2000) provides for an "aquifer interference activity" approval by the government, and in any event, the activity must avoid land degradation such as the decline of native vegetation, increased acidity, and soil erosion. The management plan for the relevant area where such controlled activity occurs must identify the nature of the aquifer interference having any effect, including cumulative impacts, on water sources or their dependent ecosystems, and the extent of those impacts. Plans for such controlled activity also deal with undertaking work with a view to rehabilitating the water source or its

dependent ecosystems and habitats. The Namibia *Water Resources Management Act* (2004) empowers the minister to establish the "safe yield" of aquifers when making determinations regarding its use, where "safe yield" refers to the amount and rate of abstraction which would not cause damage to the aquifer, quality of the water or the environment. An identical provision features in Tanzania's *Water Resources Management Act* (2009). In the state of Texas (United States), a cap on extractions from the Edwards Aquifer has been imposed under the *Edwards Aquifer Authority Act* (1993) with a view to, among other goals, protecting endangered and threatened species and ensuring the water quality of the aquifer and of the discharging springs. Under the EU *Water Framework Directive* (WFD) (2000), the quantitative status of groundwater is given particular importance also in view of the protection of the ecological quality of surface waters and terrestrial ecosystems associated with a groundwater body. The WFD considers that a certain amount of the annual groundwater recharge is needed to support interrelated and dependent ecosystems, such as wetlands.

- The prior appropriation system of water allocation control in the arid western states of the United States, insofar as leaving water in the stream for recreation and wildlife is now being deemed a beneficial use. In the state of Colorado, in particular, the notion that leaving some water where it is constitutes a beneficial use has been described as a rather radical departure from prevailing legal thinking in the law of prior appropriation (personal communication).
- Water permitting and licensing, to the extent that (a) environmental protection criteria, alongside other criteria, are prescribed to inform the terms and conditions of water abstraction licences and of wastewater disposal permits, as in Namibia's *Water Resources Management Act* (2004); (b) exceptionally, allocations to meet environmental water requirements take precedence over allocations for all other competing uses, as under the *Water Management Act* (2000) of New South Wales, or take qualified precedence, i.e., precedence over all other competing uses except for human consumption, as under the *Law on Water Resources* of Paraguay (2007); and, more generally, (c) ecologically sustainable development principles are incorporated in various ways in the water resources management laws, such as those adopted by most other Australian states in the last decade (Fisher 2009). Moreover, retrospective review and a downwards adjustment, without compensation, of allocations granted under water abstraction licenses in progress may be prompted by the changing circumstances of the environmental protection requirements mandated by the law, as the relevant provisions of Spain's *Water Act,* and the case law which the administration of those provisions has generated, show (Brufao Curiel 2008). And the extension of the all-purpose environmental permitting mechanism for environmental regulation to the discharging of wastes into freshwater bodies, inaugurated in England and Wales by the *Environmental Permitting Regulations* (2010), may herald a turnaround where environmental regulation turns 'blue' and attracts within its fold a significant segment of water resources management regulation.

### *Charging for the use of water resources, and regarding water as an economic good*

Charging for water abstraction (also known as the 'user pays' principle), seeks to influence the demand for water and constitutes the chief non-regulatory mechanism available to curb water abstraction and use. It is generally practised in combination with the regulatory mechanisms described earlier. The criteria governing rate setting vary, from the relative scarcity of water resources and the different kinds of use, as in Mexico, Colombia, New

South Wales (Australia) and Tanzania, to the volume, kind of use, location and source, as in France, to the recovery of the operating costs of the government water resources administration, as in England and Wales. Costa Rica, in particular, has added an environmental services component to the complex and sophisticated charging schemes for water resources use inaugurated by a decree issued in 2006. So have, at least in principle, neighbouring Nicaragua in the *General Law on National Waters* (2007) and Tanzania in the *Water Resources Management Act* (2009).

### *Curbing water pollution, in particular from 'non-point' sources*

Well-tested regulatory and economic instruments for the prevention and abatement of water pollution from point sources (notably, industrial outfalls and municipal sewers) feature in the contemporary water resources laws. Such instruments range from discharge permits linked to effluent quality standards and to quality objectives/standards for the receiving water source, to charging for discharging waste in water bodies (the well-known 'polluter pays' principle).

The more recent statutes bear evidence of a growing concern for pollution of groundwater resources from diffuse sources, such as the runoff and drainage of cropland in rural areas. In that context, the focus of regulation has shifted from the discharge itself to the land use giving rise to a diffuse discharge. Already in the early 1970s, the US *Clean Water Act* had required that states establish procedures, including local land use requirements, to control non-point (and point) sources of pollution. Although federal funding for this provision ended in 1980, its mandate of a local land use planning scheme still stands as the law (Hoang and Porter 2010). Cultivation practices, in particular, have been increasingly attracting regulatory restrictions aimed at preventing, abating or minimizing pollution from substances such as, notably, the nitrates employed in agriculture. In December 1991, for example, the European Union adopted Directive 91/676 directing member states to designate nitrate-sensitive (or nitrate-vulnerable) areas and to draw up a code or codes of "good agricultural practice". Within the designated areas, the provisions of such code or codes become mandatory for farmers. In similar fashion, changes in land cultivation, and the application and storage of pesticides and fertilizers in government-designated "water management areas", are the target of restrictions provided for by the *Water Resources Management Act* (2004) of Namibia, for the protection of water resources and of water-dependent habitats and ecosystems from depletion, contamination or eventual extinction. A similar approach is reflected in the *Water Code* of Kyrgyzstan (2005), in relation to designated "water protection areas".

### *The land–water connection*

The connection between land and its uses, and the quality of water resources, both surface and underground, is readily apparent from the analysis of the regulatory approaches to the diffuse pollution of water resources illustrated under the previous heading. The connection can be equally compelling in regard to water quantity management, and in particular in relation to flood control and the natural recharge of groundwater. Regarding the former, it is a known fact that urban development and human activities located in the flood plains can interfere with the natural flow of water and expose human life and property to the risk of damage from excess stream flows. The link between this particular aspect of the quantitative management of water resources and land use – or town and country planning –

regulation is readily apparent. Yet, evidence of it in the relevant mainstream water legislation is scanty at best. The EU *Water Framework Directive* (2000) held out some promise, to the extent that the mandatory river basin management plans it provides for could, arguably, afford an opportunity for water resources management to interplay with town and country planning policy and mechanisms. However, none of this is made explicit in the WFD. The subsequent EU *Directive on the Assessment and Management of Flood Risks* (FD) (2007) focuses on the assessment and management of flood risk and, in contrast to the WFD, it does not pursue an environmental objective but rather one of disaster prevention. The missing link between water resources/flood management and land management and regulation in the FD notwithstanding, the Flemish Region of Belgium adopted in 2007 legislation (technically, an amendment to the *Decree on Integral Water Policy* (2003)) integrating water quantity management into town and country planning policy and regulation. This is achieved through the instrument of the "water assessment", which is a legal tool aiming to prevent, avoid or minimize the harmful effects of town and country plans and of relevant permit decisions on the water systems. The concept is mirrored in the town and country planning law (the Flemish *Code on Spatial Planning*, 2009), which prescribes that a water assessment has to be taken into account before a permit is granted under that law (Herman 2010).

Natural recharge of groundwater reserves and of aquifers occurs on the land above, where surface runoff and precipitation infiltrate the ground and replenish groundwater stocks. If the connection is interfered with, or severed, as a result for instance of urban development or of deforestation and cultivation, groundwater reserves may be at risk of eventual depletion, let alone pollution from the diffuse sources mentioned under the preceding heading. The connection and interface between groundwater management regulation and land use regulation in the recharge areas of aquifers tend to be lost on the water legislation. Neither are they readily apparent in the legislation mandating the government to designate areas for the protection and conservation of groundwater reserves which are at risk of depletion, and to institute land use restrictions in such areas. The provisions for "groundwater protection areas" in Kyrgyzstan's *Water Code* are a pertinent example. Elsewhere, no explicit mention is made of land use restrictions for the purposes of groundwater protection from depletion. The relevant provisions in Kenya's *Water Act* (2002) are a case in point, although one could imply recourse to land use restrictions from the sweep of the regulatory authority in the matter of "groundwater conservation areas" which has been vested in the government. Another pertinent example is the EU WFD, under which EU member states are required to designate areas where surface water and groundwater require protection, both from a qualitative and from a quantitative standpoint, and areas where habitats and species directly depending on water require conservation. The list of standard "protected" areas (appended as Annex IV to the WFD), however, does not include, nor make reference to, the recharge areas of aquifers, nor do land use restrictions feature on the lists of measures, mandatory and optional, that member states are to take to achieve the WFD objectives (although one could imply such kind of restrictions from the generic terms used to describe the optional measures listed in Appendix VI). By contrast, Tanzania's *Water Resources Management Act* (2009) explicitly includes land "draining into" (i.e., naturally recharging) an aquifer in the scope of its "protected zones" provisions, and, in particular, in the scope of the government's authority to prohibit or restrict "any activity" in a designated zone.

In general, it appears that the water laws by and large fail to make the necessary connection with the uses of land where natural recharge occurs, and with the town and country planning regulations. They also fail to mandate the necessary coordination between the separate branches of government in charge of administering the water resources laws on the one hand, and the town and country planning and urban development laws and regulations on the other. The Indian state of Andhra Pradesh's *Water, Land and Trees Act* (2002) may point in the right direction, insofar as it explicitly links the artificial recharge of groundwater pursued through the mandatory installation of rainwater-harvesting equipment in residential, commercial and other premises to local building regulations, and, in particular, to building construction approvals by the local authorities (and to the provision of water and power connections to such buildings by the relevant utilities). The statute, however, focuses on the *artificial* recharge of groundwater, leaving natural recharge processes outside its ambit of operation. Still, the approach of functionally and operationally linking two separate sets of laws, and of applying them towards a common purpose, is noteworthy.

### *Participation of water users in the management of water resources*

The formation of groupings of water users for the development and management of sources of irrigation water is widely practiced and regulated in most of Latin America, in Spain and Italy in Europe, and in many South Asian countries. Customary practices also play a dominant role in some jurisdictions, such as the island of Bali (Indonesia), and in many oases in the Saharan and Sahel regions of Africa. Elsewhere, water users are called upon by legislation to shoulder increasing responsibilities in the management of water resources under stress, and to make up the membership structure of the internal organs of the government water administration.

The direct involvement of users in the management of water resources under stress is a regular feature in much contemporary water legislation. Often, this is in connection with groundwater resources in areas experiencing accelerated groundwater depletion and/or severe groundwater pollution. In the state of Texas (United States), user-level groundwater conservation districts are formed at the initiative of the state legislature, or of local landowners, or of the state government, with a view to better managing groundwater resources. Whereas these districts have varied powers, including issuing permits, spacing wells and setting the amount of withdrawals, most have not imposed mandatory restrictions, and have opted instead for voluntary self-restraint and educational programmes. As such, they do not interfere with the landowners' rights to pump groundwater. Under New South Wales' *Water Management Act* (2000), water users must be represented in the groundwater management committees established for the management of aquifers under stress.

In Spain, the consolidated *Water Act* (2001) provides for the compulsory formation of water users' groups from among the users of an aquifer which is, or is at risk of becoming, overexploited. These groups are to share groundwater management responsibilities with the government, in particular in the management and policing of groundwater extraction rights. The very same approach is reflected in the 2005 reform to Chile's *Water Code.* In Mexico since 1995 a number of 'groundwater technical committees' (*comités técnicos de aguas subterráneas*, COTAS) have been established under the auspices of the federal Water Authority, with a view to engaging users, together with federal, state and local agencies, in the formulation and implementation of programmes and regulations for aquifer preservation and recovery. Water user groups (WUGs) for the management of water supply points, and water user associations consisting of WUGs, are provided for and regulated by Uganda's *Water Statute* (1995). South Africa's *National Water Act* (1998) also provides for the formation of water user associations from among water users wishing to undertake water-related activities for their mutual benefit.

Water users and their interests are represented in the organs which make up the internal structure of river basin authorities and agencies. Thus, for instance, Spain's river basin authorities (*confederaciones hidrográficas*) include users' representatives in their decision-making and advisory organs. Similarly, users' representatives make up at least two-thirds of the total membership of the board of directors of France's water agencies (*agences de l'eau*). They are also represented on the agencies' advisory basin committees. Irrigators hold a minority of seats on the board of directors of Morocco's basin authorities, which have been formed pursuant to the *Water Act* (1995). In South Africa, water users and environmental interest groups are represented in the decision-making structure of the new catchment management agencies established under the *Water Act* (1998). A similar approach is reflected in the *Water Acts* of Kenya (2002) and Swaziland (2003), and in Panama's *Law on the Management, Protection and Conservation of River Basins* (2002). Under Brazil's *Federal Water Act* (1997), water users are represented in the basin committees, alongside the representatives of civil society and of the federal, state and municipal governments concerned. Under the extensive amendments adopted in 2004 to Mexico's *Law on National Waters* (1992), water users form no less than one-half the total membership of basin councils. Water users have a purely nominal representation instead in the internal structure of Mexico's new river basin agencies, which are branches of the government water administration. Of note, the functions of the Brazilian committees and those of the Mexican councils are essentially of a consensus-seeking nature. Water users are to enter the membership structure of the executive basin water boards, and of coordinating and problem-solving catchment and sub-catchment water committees, contemplated in the *Water Resources Management Act* (2009) of Tanzania.

### *Interface between statutory and customary water rights*

Customary law in many countries still plays an important role in water management, particularly at the community level. Customary water laws are rarely a single and unified body of norms, and vary widely from region to region, sometimes even between villages in the same region. Customary rules governing access to water have been documented in many countries, the best-known example being perhaps the *subak* allocation system of irrigation water and relevant water rights practised since time immemorial on the island of Bali, Indonesia. Another example of customary law is riparianism, which has been practised in a variety of forms in many common law countries.

Despite the social and economic significance of customary systems and practices, their interface with statutory law has seldom been mapped out and regulated in legislation. Namibia's *Water Management Act* (2004) reflects perhaps the most articulate effort to weave consideration of customary water rights and practices into the fabric of water abstraction and wastewater disposal permitting. Under that law, the government is under a duty to take due account of existing customary practices and rights in granting statutory water abstraction and waste disposal permits. The documented existence of such rights and practices will not only influence the grant or denial of an administrative permit; it will also allow special terms and conditions to be entered in the permit, for the specific purpose of protecting existing customary rights and practices. Other laws provide other mechanisms seeking to reconcile the customary and the 'modern' water rights, and to prevent or address conflict situations on the ground. In Chile for instance, under the *Ley Indigena* (Law on Indigenous Peoples) (1993), the government is under a duty not to grant a water abstraction concession if the water rights of designated traditional peoples are affected and no alternative source for supplying their properties with water has been provided first. In Mozambique, the *Water Licensing Regulations* (2007) not only accord the traditional and customary rights practised in the rural areas freedom from regulatory controls, they accord these rights priority of allocation of available water resources and place the government under an affirmative duty to facilitate the enjoyment of the rights by creating the necessary easements of access to the relevant water sources. The customary rights of traditional communities also have priority call on available water resources under the *Law on Water Resources* of Paraguay (2007), and statutory grants of water abstraction rights are subject to those rights.

In a vast majority of cases, however, the issue has been skirted, or touched upon only superficially, in the water laws or in other laws. Guyana's *Water and Sewerage Act* (2002) carried an explicit statutory acknowledgment and safeguard of customary rights in water in general, subject however to qualifications which have never been tested on the ground, and that have remained indeterminate as a result. A similar approach is reflected in Mali's and Russia's *Water Codes* (2002 and 2006, respectively), in Indonesia's *Law on Water Resources* (2004) and implementing *Irrigation Regulation* (2006), and in the *General Law on National Waters* of Nicaragua (2007). Tanzania's *Water Resources Management Act* (2009) seems to be on the same page. These statutes bear evidence of the awareness by lawmakers of the existence and significance of customary or traditional water rights in the field. These are dealt with, however, by basically separating them out of the mainstream 'modern' water rights regulated by statute, and by creating a separate legal space for them. For want of particulars, however, such legal space comes closer to being a legal limbo, which does not prevent the two sets of water rights from mutually interfering at some point, and from clashing eventually. This is probably why the Argentine province of Tucuman has chosen to mainstream traditional water rights in the *Provincial Water Code* (2001), through their transformation into administrative concessions on the initiative and request of the rights holders, and at the discretion of the government water administration. This also features as an option available to customary rights holders under Tanzania's *Water Resources Management Act* (2009). It should be added that opportunities for customary water rights to be reckoned with are routinely afforded, at least on paper, in the process of scrutinizing applications for a water abstraction/development licence or concession, or for a wastewater disposal permit, and of litigating relevant administrative decisions through the administrative or judicial review processes. As the rich experience in this specific matter of Canadian provinces, and in particular of British Columbia, proves, settlement of customary water rights via conflict and litigation with 'modern' water rights is painful and costly, and the outcome is unpredictable (Burchi 2005a). This course of action does not bode well for the peaceful coexistence of traditional and modern water rights after a challenge in court has run its course and a verdict has been laid down.

## Conclusions and pointers for the future

The attainment of water security and water-related governance goals, as they come into sharper focus, will require supportive legal and regulatory frameworks at the domestic level. The evolution to date and the level of sophistication attained by contemporary domestic water legislation indicate that the agenda of, and the scope for, water law reform will be driven by a few paramount issues emerging from the contemporary trends highlighted in this paper, namely the following.

### *Reconciling security of tenure with risk and uncertainty*

Water allocation mechanisms and relevant legal instruments are growing in complexity and sophistication, in an effort to reconcile the security of water rights tenure with changing socio-economic development policies, and changing hydrological, hydro-geological, technological, economic, environmental and political circumstances – in other words, in the effort to match risk and uncertainty with the stability of tenure and the dependability of rights sought by water-sector investors. This will call for the creative use of classic regulatory instruments, so that they are time bound and adaptable to rapidly changing circumstances with minimum conflict. This will require doing away with grants and rights of indefinite duration, fine-tuning grants to actual or predictable water availability, scaling existing grants downwards to reflect a variety of changing circumstances, including those driven by climate change, and releasing surpluses for allocation to where priorities dictate.

### *Pursuing opportunities for efficiency gains, without neglecting equity*

Where cultural and religious barriers do not oppose it, the use of market mechanisms opens up opportunities for efficiency gains in the allocation and use of water resources. This will require relaxing the rigidities built into water allocation systems based on government water abstraction licensing, where grants are tied to a specific use or to use at a specific location or on a specific piece of land. Markets for water rights, however, must be regulated to protect third-party interests and such intangibles as environmental, amenity and cultural values. Moreover, equity cannot entirely be sacrificed to economic efficiency. To a large extent, these concerns are reflected in the water laws of a vast majority of the countries which have accepted the market as a legitimate mechanism for the allocation of water resources to competing uses. Giving equity its due, however, and creating an enabling regulatory and institutional environment for the functioning of a market mechanism of water resources allocation, are challenging tasks.

### *Raising the profile of the environment in the allocation and re-allocation of water resources to competing uses*

In a fair and transparent process for the allocation and re-allocation of water resources among competing sectors and users through the available regulatory and economic instruments, the environmental requirements of freshwater bodies, and those of water-dependent habitats, will need to be granted a standing and dignity equal to, and maybe higher than, development needs. To some extent this is already happening through a number of regulatory instruments and statutory requirements which prioritize the health of surface and groundwater systems, and the survival of water-dependent habitats, in the initial allocation of water resources to competing uses, and of the subsequent review and adjustment of allocations already made. However, the implementation of the new and emerging mechanisms alluded to here can be controversial and not immune from legal challenge, particularly when implementation implies a review of existing allocations and a re-allocation of resources. Moreover, such mechanisms tend to be complex and as a result to require considerable capacity for implementation and administration.

***Rekindling the connection between water resources management regulation (and administration) and land use regulation (and administration)***

Legislation regulating the development and use of water resources in general and, in particular, of groundwater resources, and the legislation governing town and country planning and urban development, tend to go their own separate ways, and are administered by branches of government operating at different levels, under the authority of different statutes, without coordination. This is a grey area of water legislation which needs revisiting, in conjunction with a parallel revisitation of town and country planning and urban development legislation, with a view to weaving into the fabric of both sets of laws a tightly knit web of consistent substantive and procedural requirements aimed at a common purpose – that of ensuring the long-term sustainability of renewable groundwater reserves and aquifers by (a) enabling recharge to go on unimpaired by human activities on the land above, and (b) controlling land uses which are the source of diffuse pollution of groundwater reserves, as well as of streams and lakes. There is also much scope for improvement on this score in addressing the linkages between managing the harmful effects of flood waters and regulating urban development and land use in the fluvial floodplains.

***Empowering users to shoulder greater responsibilities***

With a view to minimizing conflict, critical decisions for the adaptation of water allocation patterns and of water abstraction rights to changing circumstances with respect to resources at high risk of contamination or depletion will need to be taken and monitored by, or with the participation of, the users concerned. This will require legislation enabling the formation of legally constituted groups and empowering them through the devolution or delegation of the required authority. Existing legislation is quite advanced on the former aspect, while delegation and devolution lag.

***Mapping out the interface between customary and statutory water allocation systems***

So far, customary water allocation systems have attracted, at best, the benign neglect of much contemporary water resources legislation. Yet it stands to reason that opportunities for conflict between customary water allocation and use systems and allocation systems regulated by statute should be minimized, and that the full scope of the interface between the two systems should be mapped out if coexistence is to be pursued as a serious option. This is an area of water legislation which invites innovative thinking, in the search for viable and workable mechanisms for the accommodation of traditional systems and for their coexistence alongside statutory systems of water resources allocation and use.

## Notes

1. Istanbul Ministerial Statement, 5th World Water Forum Ministerial Process, World Water Council, 22 March 2009, available from http://www.worldwatercouncil.org/fileadmin/wwc/World_Water_Forum/WWF5/Ministerial_Statement_22_3_09.pdf [accessed 5 June 2012].
2. Ministerial Declaration adopted by ministers meeting in the Ministerial Session of the International Conference on Freshwater, Bonn, 4 December 2001, available from http://www.earthsummit2002.org/ic/freshwater/Bonn%20Ministerial%20Declaration.pdf [accessed 5 June 2012]
3. All statutory references mentioned in the paper are listed in the Appendix.

## References


Brufao Curiel, P., 2008. *La revisión ambiental de las concesiones y autorizaciones de aguas*. Zaragoza: Fundación Nueva Cultura del Agua.

Burchi, S., 2001. Year-end review of comparative international water law developments. *Journal of Water Law*, 12 (6), 330–337.

Burchi, S., 2003. Year-end review of comparative international water law developments. *Journal of Water Law*, 14 (1), 7–15, and 14 (6), 281–290.

Burchi, S., 2004. Water laws for water security in the twenty-first century. *In*: J. Trottier and P. Slack, eds. *Managing water resources: past and present*. Oxford: Oxford University Press, 117–129.

Burchi, S., 2005a. *The interface between customary and statutory water rights: a statutory perspective*. Rome: FAO.

Burchi, S., 2005b. Year-end review of comparative international water law developments. *Journal of Water Law*, 16 (1), 5–13, and 16 (5), 155–166.

Burchi, S., 2006. Year-end review of comparative international water law developments. *Journal of Water Law*, 17 (6), 223–230.

Burchi, S., 2007. Year-end review of comparative international water law developments. *Journal of Water Law*, 18 (5), 149–158.

Burchi, S., 2008. Year-end review of comparative international water law developments. *Journal of Water Law*, 19 (6), 223–228.

Burchi, S., 2010. Balancing development and environmental conservation and protection of the water resources base: the "greening" of water laws. *In*: P. Cullet *et al.*, eds. *Water governance in motion: towards socially and environmentally sustainable water laws*. New Delhi: Cambridge University Press India, 333–358. Prior to publication, this paper had been posted online as FAO Legal Papers Online No. 66 (2007), available from: http://www.fao.org/legal/prs-ol/lpo66.pdf [Accessed 5 May 2012].

Fisher, D.E., 2009. Towards sustainable water resources development in Australia. *Journal of Water Law*, 20 (1), 17–27.

Herman, C., 2010. Will the Floods Directive keep our feet dry? Policies and regulations in the Flemish Region and Scotland. *Journal of Water Law*, 21 (4), 156–166.

Hoang, Y., and Porter, K., 2010. Stormwater management in the rural New York headwater areas of the Chesapeake Bay watershed. *Journal of Water Law*, 21 (1), 6–18.

UNEP (United Nations Environment Programme), 2010. *The greening of water law: managing freshwater resources for people and for the environment*. Nairobi: UNEP.


## Appendix: statutory references

*Argentina*
Tucuman Province - Water Code (2001).

*Australia*
New South Wales - Water Management Act 2000.

*Belgium (Flemish Region)*
Decree on Integral Water Policy, of 18 July 2003, as amended by Decree of 16 July 2010.
Flemish Code on Spatial Planning, of 15 May 2009.

*Brazil*
Water Policy Act No.9.433 (1997).

*Chile*
Water Code (Decree-law No.1122 of 1981), as amended by Law No. 20.017 of 2005.
Law No.19.253 of 1993 on the protection, fostering and development of indigenous peoples.

*Colombia*
Decree of the Environment Ministry No.155 of 22 January 2004, Regulating water abstraction charges.

*Costa Rica*
Decree of the President of the Republic No.32.868 of 2005, inaugurating and regulating a water charging scheme.

*European Union*
Council Directive 91/676/EEC of 12 December 1991 concerning the protection of waters against pollution caused by nitrates from agricultural sources (Nitrates Directive).
Council Directive 2000/60/EC of 23 October 2000 establishing a framework for Community action in the field of water policy (Water Framework Directive).
Council Directive 2007/60/EC of 23 October 2007 on the assessment and management of flood risks (Floods Directive).

*Guyana*
Water and Sewerage Act (2002).

*India*
Andhra Pradesh - Water Land and Trees Act 2002, Act No.10 of 2002.

*Indonesia*
Law No.7/2004 on Water Resources.
Government Regulation No. 20/2006 on Irrigation.

*Italy*
Act No.36 on Water Resources, of 1994.

*Kenya*
Water Act, 2002 (No.8 of 2002).

*Kyrgyzstan*
Water Code, Law No.8 of 2005.

*Mali*
Water Code, Law No.02-006 Of 31 January 2002.

*Mauritania*
Water Code, Law No.2005/030 of 2 February 2005.

*Mexico*
Law on National Waters, 1992, as amended by Decree of the President of the Republic of 29 April 2004.

*Morocco*
Law No.10-95 on Water Resources (1995).

*Mozambique*
Decree of the Council of Ministers No.43/2007 of 30 October 2007, carrying Regulations on Water Licences and Concessions.

*Namibia*
Water Resources Management Act, 2004 (Act No.24 of 2004).

*Nicaragua*
General Law on National Waters, Law No.620 of 29 August 2007.

*Panama*
Law No.44 of 5 August 2002 concerning the management, protection and conservation of Panama's river basins.

*Paraguay*
Law 3239/2007 of 14 June 2007, on Water Resources.

*Russian Federation*
Water Code (Federal Law No.74/FZ of 3 June 2006).

*South Africa*
National Water Act (No.36 of 1998).

*Spain*
Royal Legislative Decree No.1/2001 - Water Act (consolidated text).
Law No.10/2001 - Law of the National Hydrological Plan.

*Swaziland*
Water Act 2003 (No.7 of 2003).

*Tanzania*
Water Resources Management Act, 2009 (No.11 of 2009).

*Uganda*
Water Statute, 1995 (Statute No.9 of 1995).

*United Kingdom*
Environmental Permitting (England and Wales) Regulations 2010, S.I. No.675 of 2010.

*United States of America*
Clean Water Act, 33 U.S.Code §1251 *et seq.* (1972).
Oregon - Instream Water Rights Act, Act of 12 May 1987.
Texas - Edwards Aquifer Authority Act, Act of 30 May 1993.

*Venezuela*
Law on Waters, of 9 November 2006.

*Zimbabwe*
Water Act (Cap.20:24) (2000).

# Legislation as a tool in support of adaptive water management in response to climate change

Marcella Nanni

An attempt is made here to shed light on how water legislation could address the climate change challenge. Although climate change legislation provides a framework for the integration of climate change issues into government policies on several topics, it does not cover water resources as such. Therefore, water legislation must provide for such integration. The challenge is to find a balance between well-defined water rights, duties and administrative functions on the one hand and, on the other, the flexibility required to respond to the vagaries of a changing climate.

## Introduction

This article focuses on the extent to which water legislation is able to meet the challenges of climate change and on the issues it should address in order to facilitate mitigation and adaptation. Climate change legislation, following the United Nations Framework Convention on Climate Change (UNFCCC), signed on 9 May 1992,[1] and exemplified at the national level by the UK Climate Change Act of 2008 and Law No. 9729 of 2009 on Climate Change of the Philippines, deals with the reduction of emissions and, while containing some provisions related to water, does not provide adequate focus. In particular, it does not cover water management issues and needs as such. Thus, water legislation must provide a framework for the integration of climate change issues into government water resources policies.

In the European Union (EU), the planning tools that the Water Framework and Floods Directives required member countries to incorporate into their water legislation facilitate a response to the climate change challenge and, in general, adaptation. The European Commission has recently stressed the need for "strategies which increase the resilience to climate change of health, property and the productive functions of land, inter alia by improving the management of water resources and ecosystems" (European Commission 2009a). In other countries, however, water legislation is not always adequate (Bruch and Troell 2011).

## Does water legislation adequately address climate change-related issues?

### *Need for integration at the river basin level*

The river basin provides the best unit for managing water resources in an integrated manner, taking into account both surface water and groundwater, quantity and quality, and all water uses (Caponera 2003). There, the interdependence between water and other natural resources may be best appreciated, water resources may be quantified and water-related hazards may be best dealt with.

In Europe, Directive 2000/60/EC of 23 October, 2000, establishing a framework for community action in the field of water policy (the Water Framework Directive)[2] mandates that EU member countries manage their water resources based on river basin districts. A river basin district is defined as the area "made up of one or more neighbouring river basins together with their associated groundwaters and coastal waters, which is identified under Article 3(1) as the main unit for management of river basins" (Art. 2(10)). There is no reference to climate change in the directive, but some governments, including the UK, acknowledge that the river basin management process is well suited to take into account climate change risks (UK Environment Agency 2011). The countries were obliged to transpose the directive into their domestic legislation by 2003.

Outside the EU, in Kenya, Nigeria, Swaziland, Tanzania, Mexico, Peru and Vietnam, water legislation also requires that water resources be managed by river basin. To implement a river basin approach it is necessary to have an appropriate institutional set-up.

The Water Framework Directive requires EU legislation to designate a "competent authority" for each river basin district. In France this is the river basin coordinating prefect, and in Spain it is the river basin organization or, within regional basins, the regional authority (Caponera 2009). More typically, in countries outside the EU a number of institutions coexist and their functions overlap, even when a lead institution for river basin management is in place. In small countries the role of lead agency may also be vested in a national agency.

Newly established water institutions find it hard to compete with existing organizations, which have more experience and qualified personnel. Environmental protection ministries of some former Soviet republics encounter resistance from established ministries of "water economy" (irrigation) and from other agencies (Nanni 2003). Similarly, in Vietnam and Sri Lanka (and some Indian states) new water institutions compete with irrigation agencies without clear resolution, because water legislation there does not clearly define institutional roles or set modalities for coordination and interaction.

Thus the existence of several autonomous centres of decision making within the river basin constrains the implementation of government policies and strategies for climate change mitigation and adaptation.

### *Need for an adequate knowledge base*

In most countries, water legislation calls for the acquisition of water resources data and information with a view to the enhancement of the knowledge base. The best example is perhaps offered by EU countries, where the legislation now reflects the duty, under the Water Framework Directive, to establish water resources monitoring programmes geared to the provision of "a coherent and comprehensive overview of the water status within each river basin district" (Art. 8). There is no specific mention of climate change in the directive, but guidelines recently adopted indicate that the directive's principles underlying monitoring and assessment should remain unaltered, although climate change "has the potential to impact on virtually all quality elements included in the definition of ecological status" (European Commission 2009b). Although not designed for climate change, water resources monitoring programmes may well serve the purposes of detecting climate change impacts and monitoring progress in the implementation of mitigation and adaptation measures, provided that they are long-term and linked to meteorological monitoring. Being part of the planning cycle under the directive, monitoring programmes are subject to periodic review and updating.

The water legislation of a number of countries outside the EU tends to reflect some of the principles followed in EU countries. The South Africa National Water Act of 1998[3] requires the Minister of Water Affairs to establish monitoring systems aimed at assessing water quantity and quality, water use, compliance with quality objectives, the health of aquatic ecosystems, and atmospheric conditions. Following consultation with stakeholders, the minister is to set up mechanisms to coordinate monitoring. A further ministerial duty is to establish information systems for the purpose of disaster management. In China, the Water Law of 2002[4] makes explicit reference to the need for comprehensiveness in water resources assessments. Under the Water Law of Indonesia of 2004,[5] a water resources inventory coordinated at the basin level is to provide information on hydrological, hydrometeorological and hydrogeological conditions, the water resources potential and water needs, the environment, and socio-economic conditions. The legislation just mentioned seems to leave room for an extension of the knowledge base to the risks and uncertainties associated with climate change.

In many instances, however, the situation on the ground constrains the implementation of water legislation. On the one hand, governments often lack implementation capacity. On the other, data-related functions are scattered among several institutions, each following its own procedures and handling the collected data autonomously (Caponera 2009). There are no procedures to guide monitoring activities, so there is no guarantee that the data and information collected are at least comparable. Moreover, the legislation often fails to clearly designate a coordinating institution, with the result that there is either no coordinator or too many coordinators. In addition, there is no data exchange requirement. This situation does not facilitate climate change mitigation and adaptation.

An example of the above is offered by Ukraine, where the duty to monitor water resources is spelled out in the Water Code of 1995.[6] However, several agencies are involved in water resources monitoring, including the Ministry of Environment, the State Water Management Committee, the Hydrometeorological Service and the Ministry of Health. Monitoring activities are carried out on the basis of contradictory legal provisions, with the result that there are overlaps and duplication of efforts (Nanni 2003). Moldova experienced similar problems, but steps are now being taken to correct the situation. Following the conclusion of partnership and cooperation agreements with the EU and, more recently, the negotiation of association agreements, other former Soviet republics are making efforts to harmonize their legal systems with the *acquis communautaire* (the body of EU law).

The situation just described may also be found elsewhere. In Vietnam's 1998 Water Law,[7] for instance, water resources monitoring is a core water management function which must be performed in a "unified" manner. Several decrees and circulars define the responsibilities of various agencies at different territorial levels, but the coordinating role of the Ministry of Natural Resources and the Environment is not spelt out clearly. In practice, each ministry, agency and people's committee undertakes resource monitoring and related activities independently. Therefore, there are overlaps and gaps, and an exchange of data and information does not take place (Vietnam Ministry of Natural Resources and Environment 2006). It follows that the information on the water resources of the country, and on the risks associated with climate change, is not complete.

Climate change also requires that water uses and activities that may exert an adverse impact be kept under control. In several countries, water legislation requires that information on water use (and wastewater discharge) permits, and on permits relating to other activities with potential adverse effects (e.g. sand and gravel extraction from the beds and banks of watercourses, the construction of hydraulic structures, etc.) be entered in special registers of water rights (Caponera 2009). Registration provides elements enabling the water administration to limit certain water uses and activities, or to reallocate water, should the impacts of climate change so warrant.

In a number of countries, water legislation provides for the establishment of such registers, but in others the water administration relies solely on information provided by water users. This is, again, the case in the former Soviet republics, where data on water uses and wastewater discharges are recorded by the users in special standard forms. This system was inherited from the Soviet Union, where water users were public enterprises, institutions and organizations under state control. At present, however, it is no longer reliable, because many activities are in the hands of private individuals and entities which, fearing to have to pay high water charges, declare that they avail themselves of less than the amount of water actually used (Nanni 2003).

### *Need for mechanisms for water allocation and reallocation*

The recognition that water resources are exhaustible and that circumstances might warrant a modification of water use patterns brought about the progressive abandonment of the traditional distinction between private and public waters, which until the second half of the twentieth century was crystallized in the water legislation of civil law countries such as France, Italy and Spain and the countries under their influence. Similarly, in common law countries this recognition led to the erosion of the system of exclusive water rights vested in the land adjacent to or overlying water under the riparian doctrine (Caponera 2009).

Water legislation adopted in the past decades has brought all water resources, whether surface or underground, under state control, either by vesting them in the nation, the state or the public domain, or by asserting a superior right of the state (or the crown) to them, such as in the case of the Australian states of Victoria, Queensland and New South Wales. Thus, the state, as public trustee, has the right to allocate water resources to different users and to control the manner in which they are used. Water allocation decisions are incorporated into permits, authorizations, licences or concessions (Caponera 2009) which are issued by the public administration in accordance with water legislation and water resources plans.

The fact that water resources may not be privately owned and are the subject of use rights does not impinge on the freedom of the right holders to use water, provided that these rights are well defined and sufficiently secure. Uncertainty affecting water entitlements does not stimulate investments in the water sector. Given this consideration, most water laws contain provisions as to the duration of a permit and the circumstances under which a permit may be modified or revoked. In several countries, the laws further provide for the protection of permit holders through the recording of the permits and their subsequent modifications in water rights registers. The water administration will not authorize a new water use if the amount of water left in the water body is not likely to meet the requirements of registered users (Hodgson 2006).

If the protection afforded to water rights serves the purpose of encouraging water resources development, water scarcity, drought and emergency situations produced by climate change may require water laws to create mechanisms enabling the water administration to modify water rights, or to shift water resources from a user to another, or from a geographic area to another, when the need arises. Permits of excessive or unlimited duration, as was the case in Chile under the Water Code of 1981,[8] might constrain water reallocation (Hodgson 2006).

In a number of countries, water laws require that an order of priorities be followed while making water reallocation decisions. Normally, this order is established through water resources plans and may differ from river basin to river basin, or from area to area. It should not be crystallized in the law itself, because circumstances may dictate changes in priorities (Caponera 2009, p. 283). An example of water law indicating the order of priorities to be followed by default (i.e. in the absence of a river basin plan) is offered by the Spanish Water Law of 1985,[9] which underwent subsequent consolidation and amendments (Hodgson 2006, p. 53). Recently enacted water laws provide absolute priority to the use of water for the satisfaction of basic water needs,. In Peru, for instance, the Law on Water Resources of 2009[10] refers to the right to access water for basic needs as a fundamental human right, even in times of water scarcity.

All water use rights are ultimately uncertain to some degree, because water is not always available at the same place and in the same amount throughout the year (Hodgson 2006). If permits are modified in order to accommodate climate change considerations, care should be taken to ensure that, once the order of priorities is met, similar permits in the same area or relating to the same water source are modified equitably.

When water used under a permit is reallocated, or when the permit is modified or revoked in response to changed circumstances in accordance with water legislation or a water resources plan, the permit holder should be compensated, either in cash or with the right to use water from another source (Hodgson 2006). The water laws of some places, including New South Wales (Australia), Spain and Portugal, entitle the permit holder to compensation. Those of others, such as Victoria (Australia), enable the government to purchase the water right.

Transfers of water rights through market mechanisms are viewed as another means to facilitate water reallocation, stimulate water conservation and obtain water whenever no further licences are to be issued. Some jurisdictions do not allow them – France, Indonesia and Vietnam are cases in point – and where they are allowed, they may be subject to administrative control, since they may cause substantial impacts on other water users (especially vulnerable ones), third parties and the environment (Caponera 2009). Water is often "transferred" when land is sold. The introduction of tradable water rights requires that these rights be clearly defined. In Chile, lack of adequate definition through the Water Code of 1981 and, in particular, the unlimited duration of water rights and lack of a penalty for non-use opened the door to the hoarding of water rights by powerful enterprises and to speculation. Therefore, the law was amended in 2005 to introduce, *inter alia*, a tax for non-use of the water (Caponera 2009). Water right transfers may be temporary or permanent, and may occur within one water-using sector or between different sectors. In Australia, not only is the water right tradable, so too is the water. In New South Wales, a detailed water account (like a bank account) is set up for each licence, and water allocated to the account can be traded to other users, based on a water sharing plan.

Finally, in order to respond to drought or other natural calamities, water laws may empower the water administration to declare an emergency and ration water resources. Under the Water Law of Morocco of 1995,[11] for instance, the administration may declare a state of water scarcity within a defined area, and make regulations prohibiting certain water uses and the digging of wells, providing for water rationing and setting rules for the use of public water points. These regulations are temporary and their effect only local.

Drinking-water supplies have to be safeguarded. The vagaries of climate change may similarly require lawmakers to allow water to be shared equitably in times of shortage or in emergencies.

### *Need for adaptive groundwater management in response to climate change*

Many countries, particularly in arid and semi-arid regions, are doomed to suffer from climate change because of declines in both river flows and groundwater recharge (Kundzewicz *et al.* 2008) – hence the need for specific legal provisions to protect groundwater reserves from over-extraction and pollution.

These provisions may relate in the first place to the designation of restricted or controlled areas subject to a special legal regime (Burchi and Nanni 2003). The water law would indicate the purpose(s) for which an area may be designated, the authority responsible for such designation and the measures to be implemented within the designated area, including those in support of climate change mitigation and adaptation. The Spanish Water Law[12] empowers a basin organization to establish a control area for over-exploited aquifers and to prepare an aquifer recovery plan, which may require that individual extraction rights be turned into a collective right or that new concessions be granted only to users' associations. Pending plan approval, groundwater extraction is restricted. The basin organizations may also establish groundwater quality protection areas. In Mexico, under the Law on National Waters[13], groundwater exploitation in regulated or prohibited areas is subject to an integrated programme, and concessions and permits are granted only under certain conditions. In China, strict controls may be imposed on groundwater abstraction in intensive-use areas, and over-exploited areas may be declared groundwater control districts by local authorities. According to the New South Wales (Australia) Water Management Act of 2000,[14] the Minister may declare groundwater management areas to protect groundwater under environmental stress, and establish groundwater management committees to develop aquifer management plans. The Groundwater Management Code of Arizona (USA) of 1980[15] envisages the formulation of plans for active management areas (AMA), so designated by the director of water resources, to limit groundwater abstraction and ban new irrigation development. AMAs correspond to groundwater basins or sub-basins (Nanni 2005).

Since climate change may have a negative impact on water availability, it is important to provide, through water legislation, for the protection of groundwater used for drinking-water supply (Nanni 2005). In Portugal, the 2005 Water Act[16] states that aquifer recharge areas for drinking-water abstraction are to be delimited and shown in river basin management plans and territorial development plans. Within these areas, certain land uses may be restricted or prohibited. In Victoria (Australia), the Water Act of 1989[17] empowers the minister responsible for water resources to declare an area to be a groundwater supply protection area when the groundwater supply therein is affected by other groundwater uses. A management plan is then prepared.

The establishment of users' associations greatly helps in managing groundwater, particularly when adaptation is necessitated by water scarcity. Compulsory associations may be formed in Spain within over-exploited aquifer areas to use groundwater (and control abstraction) in cooperation with the relevant river basin organizations. Moreover, a river basin organization may require the establishment of associations for the conjunctive use of surface water and groundwater in the interest of optimum resource utilization. In Mexico, the river basin organizations rely on technical groundwater committees (*comités técnicos de aguas subterráneas* – COTAs) to implement aquifer preservation and recovery programmes and regulations (Burchi and Nanni 2003).

Finally, artificial aquifer recharge with surface water, storm-water or wastewater is an important tool in support of the augmentation of groundwater reserves. In order to ensure water quality and that operators meet certain qualifications, artificial aquifer recharge will be subject to a permit, and the operators must register with the administration (Nanni *et al.* 2006).

### *Need for economic incentives to facilitate adaptation*

Adaptive water management in response to climate change becomes easier if water users are induced to invest in water-saving technologies, effluent treatment and water conservation. Under the Water Law of Algeria of 2005,[18] subsidies and other benefits may be granted by the state to those introducing water and soil conservation techniques. The Water Law of Spain entitles those embarking on the development, introduction or modification of technologies, processes and equipment aimed at reducing water consumption or contaminant loads for wastewater discharges to apply for state financial aid. Similarly, the Law on Water Resources of Peru recognizes a right to economic incentives to those investing in water saving, pollution control and the protection of water-related ecosystems. They may deduct investment costs from the sums due to the government on account of water charges or other fees and may be issued efficiency certificates by the National Water Authority.

### *Need for comprehensive flood risk management*

Due to climate change, floods are likely to increase. Hence the need for enhanced flood risk management and for legal provisions dealing with all four key elements, or phases – prevention, preparedness, response and recovery – preferably within a river basin context. Prevention focuses on measures to prevent, or at least mitigate, the impact of a flood; preparedness, on ensuring that communities at risk are aware of potential hazards and of how to react to them; response, on measures to be taken when floods occur; and recovery, on restoring affected communities to normal conditions.

Prevention should rely on an adequate knowledge base, mechanisms to identify, map and delimit flood risk areas, and a clear-cut indication of the activities which should be prohibited or subject to prior authorization within these areas because they are likely to modify the river morphology or the flow regime, with a risk of endangering life and property. Information on flood risk areas and the activities which are prohibited or subject to authorization should be shared with the public (WMO 2006). The measures just mentioned should be provided for by water legislation, which should also make sure that the conditions to be attached to authorizations cover flood prevention requirements. If the authority issuing an authorization is not the water authority, the legislation should provide for coordination between the two (Caponera 2009).

For preparedness, water legislation should provide for the establishment and operation of flood forecasting, warning and alarm systems, and response and evacuation plans. Response measures may include the strengthening of dykes and other retention structures, the diversion of flood flows, the mobilization of volunteers and the requisitioning of equipment and material. In this connection, the legislation should indicate institutional responsibilities, as well as the fate of the equipment and material requisitioned once the flood event is over. Recovery refers to reconstruction, the recovery of damaged crops and the restoration of public utilities. For this phase, water legislation should enable the water administration to carry out an evaluation of the manner in which the three phases of prevention, preparedness and response were dealt with (WMO 2006).

EU Directive 2007/60/EC of 23 October 2007, on the assessment and management of flood risks (the Floods Directive),[19] requires EU member countries to carry out, by 2011, a preliminary assessment with a view to identifying the river basins and associated coastal areas at risk of flooding. For these areas, the countries must develop flood risk maps by 2013 and, by 2015, prepare flood risk management plans focusing on prevention, protection and preparedness, including flood forecasting and early-warning systems. Both the preliminary assessment and the plans must take into consideration the impact of climate change on the occurrence of floods (European Commission 2009b). Flood risk management plans are to be coordinated with the river basin management plans and programmes of measures developed under the Water Framework Directive. Since planning is organized into cycles, there is scope for periodic review, evaluation and adjustments. A strong emphasis is placed on stakeholder participation in plan preparation.

Elsewhere, the approach to flood risk management is not as sound as in the EU; water legislation deals with preparedness and response, but neglects prevention, with the result that people construct dwellings and carry out economic activities within areas at risk of flood. When a flood occurs, they apply for compensation by the state. This was the case in Ukraine, for instance, when it experienced severe floods (NATO 2002).

***Need for integration of climate change issues into water resources plans***

It is now widely recognized that water sharing in times of uncertainty is best handled by means of water resources plans, which become not only the basis for water allocation and reallocation, but also for water conservation, water quality protection, the prevention and mitigation of the harmful effects of natural calamities and the maintenance of environmental flows.

Recently enacted water laws tend to place emphasis on river basin management plans (Hodgson 2006). These plans are increasingly being conceived as flexible tools facilitating the management of risks and uncertainties, including those associated with climate change, even if there is no reference to climate change in the laws. In some countries, due to the size of the territory and the need to make decisions affecting more than one river basin, water legislation also calls for national plans. In Spain, the national hydrologic plan is meant to provide for the transfer of water from one river basin to another, among other things. Similarly, in Morocco the national plan provides for inter-basin water transfers. In other countries, the water law also provides for aquifer management plans. Algeria is a case in point.

Water resources plans normally acquire binding force upon publication in the official gazette, but they are limited as to their duration and are subject to periodic review. Hence, they may accommodate the needs associated with climate change.

The river basin management plans produced in EU countries in 2009 in line with the Water Framework Directive are geared to the achievement of good status for water bodies, as defined by the directive. Climate change issues are outside of the scope of the directive, but guidelines adopted at the EU level in 2009 (European Commission 2009b) call for their integration into the plans, at least on the occasion of the next planning cycle. This may require considerable efforts by EU countries, since they met with difficulties during the first planning cycle due to the nature of the analyses to be carried out, which covered river basin characteristics, the impact of human activities and water use from an economic viewpoint. Little had to be done on that occasion in France, where the existing basin planning tools – the SDAGE (general water management plans) and SAGE (local, detailed plans) – had to be adjusted content-wise to reflect the requirements of the directive (Caponera 2009). The same applies to Spain. However, the integration of climate change issues will require new analyses, particularly in connection with the detection of impacts, the identification of "hot spots" (vulnerable water bodies), and the design of mitigation and adaptation measures. This may generate additional difficulties.

The planning cycle under the EU Water Framework Directive is six years long. In countries outside the EU the water laws may provide for a longer plan duration, tempered by the possibility of revision at shorter intervals, and even earlier under exceptional circumstances. In Morocco, river basin plans and the national plan have a 20-year duration. They may be revised every five years, or earlier if the need arises.

In Australia, water resources planning has to deal with day-to-day and year-to-year change that is difficult to quantify. Since climate change creates even more uncertainty, an adaptive approach is followed. Under the New South Wales Water Management Act, water management plans are established for designated water management areas or water sources. Their water allocation/sharing provisions *must* indicate environmental requirements, the requirements of household and other users, and allocation priorities in times of water shortage. Additional provisions *may* deal with modalities for water extraction, water supply works, the operation of water accounts for the area or source (such as the carrying over of credits from one accounting period to the next), and water sharing measures for water quality protection or the rehabilitation of water sources and their dependent ecosystems. There is a tension between adaptation needs and the need for security of water rights which is reflected in the plans. Thus, on the occasion of recent drought occurrences the plans had to be suspended. Tasmania experienced a similar situation. A difficult decision to cut existing water uses had also to be made in the case of the Gwydir (New South Wales) groundwater plan. The plan provided for proportional cuts in entitlements for all license holders. However, a group of license holders pushed for a change in the way the cuts were distributed to take more account of the level of development of entitlements. Eventually, this request was satisfied (Hamstead *et al.* 2008).

### *Need for stakeholder participation*

Since climate change may require that unpopular decisions be made by the water administration, those decisions are best understood, and are more easily implemented, if those who are affected are involved in the process leading to them (Hodgson 2006).

Stakeholders may be called to participate in water resources management at various levels, ranging from an irrigation scheme to river basin management planning. France has a long-standing tradition of user and community participation in water resources management. According to the Water Law of 1992,[20] water users participate in the adoption of the SDAGE through their one-third share in the membership structure of the basin committees, while the SAGE are adopted by a local water commission, one-fourth of its members consisting of water user representatives. In Spain, also, water users contribute to river basin planning through their participation in the organs of the river basin organizations. Moreover, the Water Law makes it mandatory for users of water from over-exploited aquifers to form groundwater users' associations, as was mentioned earlier (Burchi and Nanni 2003). Following the entry into force of the EU Water Framework Directive, public participation in river basin planning has become obligatory in all EU countries (Caponera 2009).

Under the Mexican Law on National Waters, water user representatives constitute 50% of the members of the river basin councils. The councils provide support to the river basin organizations, which are branches of the National Water Commission, by considering issues of relevance to the basin, including priorities among water uses and the river basin plans, and by disseminating information. Limitations to water use by the commission due to drought, water pollution, over-exploitation or the need to preserve vital ecosystems are subject to prior discussion with the councils. The law also calls for broader participation of the society, including NGOs, academic and scientific institutions and organizations of citizens, in water resources planning and related issues. The Law on Water Resources of Peru includes indigenous and farmer communities in the National System of Water Resources Management. These communities, together with water users, are represented in the governing council of the National Water Authority.

In South Africa, water users' groups nominate members of the governing boards of the catchment management agencies. In Algeria, organizations and people's representatives are consulted on the anti-erosion plans developed in accordance with the Water Law.

## Conclusions

Water legislation has not always been conceived as a tool in support of adaptive water management. There is, however, a trend towards acknowledging the need to provide the flexibility required for addressing climate change issues.

It is increasingly recognized that climate change issues influencing water resources management are best dealt with within a river basin context (Bruch and Troell 2011). However, in many of the countries that have incorporated a river basin management approach into their water legislation this recognition has not been followed by a redefinition of institutional roles. While the legislation indicates the authority in charge of river basin management, the responsibilities of the other ministries and agencies concerned, which are normally spelt out in separate legislation, remain uncertain. Some functions are often duplicated, while there are gaps with regard to others since nobody seems to be in charge. Hence the need for legal provisions clearly defining institutional roles and setting out mechanisms and modalities for coordination of water resources management activities and for interaction with the local authorities, water users, NGOs and others.

In devising mitigation and adaptation policies, strategies and plans it is important to have a good knowledge of the impacts of climate change and of the water uses and activities producing adverse effects. It will be a task of water legislation to require that water resources be monitored within the framework of programmes coordinated by a lead institution at the river basin level. These programmes will be subject to periodic review and updating so as to make it possible to meet new challenges. The legislation will provide for the inclusion of monitoring data and information in databases accessible to all the institutions concerned.

Climate change may produce threats and uncertainty, so it is indispensable that water legislation, while clearly defining the functions and powers of the water administration and the rights and duties of water users, be flexible enough to accommodate changing needs and facilitate adaptation. This means providing for permits of limited duration and subject to conditions as to water use in case of water scarcity, drought, floods or other natural calamities. The water administration will be empowered to suspend, alter or cancel water use rights, ration water, and shift water from one user to another or from one geographical area to another, when circumstances so warrant, based on the order of priorities established in river basin (or aquifer) management plans. These plans will be conceived by the legislation as binding instruments, yet subject to periodic review and updating, so as to provide the necessary flexibility. In parallel with the above, the legislation will allow water rights markets, subject to approval by the water administration and to the protection of vulnerable water uses – absolute priority (*in primis*) water uses to satisfy basic human needs and to ensure minimum ecological flows.

Water legislation supporting mitigation and adaptation will also contain provisions enabling the water administration to declare areas as subject to a special regime to protect groundwater from over-extraction and pollution. Further provisions may be required to facilitate the setting up of groundwater users' groups and to regulate the artificial recharge of aquifers. As far as flood risk management is concerned, water legislation will pay particular attention to prevention aspects, by requiring the identification and mapping of flood risk areas and flood management plans to be integrated into, or coordinated with, river basin plans.

Finally, since the impact of climate change may call for the introduction of measures which are likely to be misunderstood or unpopular, water legislation must provide for the active involvement of stakeholders in the relevant decision-making process. In particular, stakeholder participation in river basin planning should be mandatory.

## Notes

1. Full text available at: http://unfccc.int/resource/docs/convkp/conveng.pdf [Accessed 10 August 2012].
2. *Official Journal of the European Union*, L 327, 22 December 2000, 1.
3. Act No. 36 of 1998. Full text available at: http://www.dwaf.gov.za/IO/Docs/nwa.pdf [Accessed 10 August 2012].
4. Water Law of the People's Republic of China, promulgated by Order of the President No. 74 of 29 August 2002. Available at: http://english.gov.cn/laws/2005-10/09/content_75313.htm [Accessed 10 August 2012].
5. Law No. 7/2004 of 18 March 2004 on Water Resources. In: 2/1 *Law, Environment and Development Journal* (2006), 118. Available at: http://www.lead-journal.org/content/06118.pdf [Accessed 10 August 2012].
6. Brought into effect by Resolution of the Verkhovna Rada No. 214/95-BP of 6 June 1995. Full text available at: http://www.ten-t.org.ua/en/catalog/item/273.htm [Accessed 10 August 2012].
7. Law No. 8/1998/QH10 of 20 May 1998, on Water Resources. Full text available at: http://www.vnwp.org/vanban/Law-on-Water-Resources.pdf [Accessed 10 August 2012].
8. Decree Law No. 1.122 of 13 August 1981. Diario Oficial No. 31.102, of 29 October 1981, 4851–4865.
9. Now part of the Consolidated text of 2001. Royal Legislative Decree 1/2001 of 20 July 2001, approving the Consolidated Text of the Water Law. In: BOE No. 176 of 24 July 2001.
10. Law No. 29338 of 2009. In: *El Peruano* of 31 March 2009. NORMAS LEGALES.
11. Dahir No. 1–95–154 of 16 August 1995, promulgating Law No. 10–95 on Water. In: Bulletin officiel No. 4325 of 20 September 1995, 626–662.
12. Royal Legislative Decree 1/2001 of 20 July 2001, approving the Consolidated Text of the Water Law. In: BOE No. 176 of 24 July 2001.
13. First published in: Diario Oficial de la Federación on 1 December 1992. Current version available at: http://www.diputados.gob.mx/LeyesBiblio/pdf/16.pdf.
14. Water Management Act No. 92 of 2000. Available at: http://www.legislation.nsw.gov.au/viewtop/inforce/act+92+2000+FIRST+0+N/.
15. Ch 2, 1980 Ariz. Legisl. Serv. A-575 (West). Available online at: http://www.azleg.state.az.us/ArizonaRevisedStatutes.asp?Title=45
16. Law No. 58/2005 of 29 December 2005. In: *Diário da República*—I Série-A, No. 249, 29 December 2005, 7280–7310.
17. Available at: http://www.austlii.edu.au/au/legis/vic/consol_act/wa198983/
18. Law No. 05–12 of 4 August 2005, on Water. In: Journal Officiel de la République algérienne No. 60 of 4 September 2005.
19. *Official Journal of the European Union*, L 288, 6 November 2007, 27.
20. Law No. 92–3 of 3 January 1992, on Water. Available at: http://www.ineris.fr/aida/?q=consult_doc/consultation/2.250.190.28.8.147.

## References


Bruch, C., and Troell, J., 2011. Legalizing adaptation: water law in a changing climate. *Water International*, 36 (7), 828–845.

Burchi, S., and Nanni, M., 2003. How groundwater ownership and rights influence groundwater intensive use management. *In*: E. Custodio and R. Llamas, eds. *Groundwater intensive use: challenges and opportunities*. Lisse, the Netherlands: Balkema, 227–240.

Caponera, D.A., 2003. *National and international water law and administration: selected writings*. The Hague: Kluwer Law International.

Caponera, D.A., 2009. *Les principes du droit et de l'administration des eaux – Droit interne et droit international*. 2nd edition, revised and updated by M. Nanni. Paris: Johanet.

European Commission, 2009a. *White paper on adapting to climate change: towards a European framework for action*, COM(2009) 147 Final. Brussels: European Commission.

European Commission, 2009b. *Common implementation strategy for the Water Framework Directive (200/60/EC): river basin management in a changing climate*. Technical Report No. 24. Brussels: European Commission.

Hamstead, M., Baldwin, C., and O'Keefe, V., 2008. *Water allocation planning in Australia: current practices and lessons learned*. Waterlines Occasional Paper No. 6. Canberra: National Water Commission.

Hodgson, S., 2006. *Modern water rights: theory and practice*. FAO Legislative Study No. 92. Rome: FAO.

Kundzewicz, Z.W., *et al*., 2008. The implications of projected climate change for freshwater resources and their management. *Hydrological Sciences Journal*, 53 (8), 3–10.

Nanni, M., 2003. *Joint Rivers Management Programme: legal and institutional aspects*. Report. Moscow, Almaty, Baku, Tblisi, Kyiv: EU-TACIS.

Nanni, M., 2005. *Legal aspects of groundwater management: an overview* [online]. International Groundwater Resources Assessment Centre. Available from: http://igrac.nitg.tno.nl/pub/bscw.cgi/d68616-4/*/*/*/*/Documents/Regional%20meeting%20Americas/Papers/Legal%20Aspects%20of%20Groundwater%20Management%20An%20Overview.pdf

Nanni, M., *et al*., 2006. Sustainable utilization of groundwater resources: legal and institutional aspects. *In*: S. Foster and D.P. Loucks, eds. *Non-renewable groundwater resources: a guidebook on socially-sustainable management for water policy-makers*. Paris: UNESCO, 49–57.

NATO, 2002. *Joint NATO-Ukraine Project on Flood Preparedness and Response in the Carpathian Region: comprehensive assessment report*. Brussels: NATO.

UK Environment Agency, 2011. *Integrated river basin management planning: a co-ordinated approach to planning for water*. Available from: http://www.environment-agency.gov.uk/research/planning/109000.aspx [Accessed 18 April 2012].

Vietnam Ministry of Natural Resources and Environment, 2006. *National water resources strategy: towards the year 2020*. Hanoi: Culture-Information Publishing House.

WMO, 2006. *Legal and institutional aspects of integrated flood management*. Geneva: World Meteorological Organization.

OPEN ACCESS

# Water governance reform in the context of inequality: securing rights or legitimizing dispossession?

Helle Munk Ravnborg

**ABSTRACT**

Secure and legally sanctioned access to water is gaining significance to farmers to cushion themselves against climate change and to participate in markets that are increasingly concerned with social and environmental responsibility. Nicaragua is among the countries which recently has introduced a new water rights regime as part of its water governance reform. The article analyzes the extent to which the reform has succeeded in providing water security for all. The article argues that due to selective and partial implementation, the water governance reform could lead to the concentration of enforceable water rights in the hands of the few.

## Introduction

Secure and legally sanctioned access to water is gaining significance to farmers at large, both to cushion themselves against the effects of climate change–related unpredictability of rainfall patterns (Organisation for Economic Co-operation and Development [OECD], 2014), and to market their products in domestic and international markets that are increasingly concerned with corporate and consumer social responsibility (Daniel & Sojamo, 2012; Mason, 2013).

Water governance, including a supportive legal and institutional framework, is increasingly seen as key to water security (Burchi, 2012; UN-Water, 2013). Hence, in recent decades a wave of water governance reforms has swept across the developing world (Aagaard & Ravnborg, 2006; Burchi, 2012; Jacobi et al., 2014; Ravnborg, 2015; van Koppen, 2007). Policies, laws and administrative guidelines have been proposed, challenged, fiercely contested (not least in many Latin American countries), re-conceptualized and approved, and are currently in the process of being implemented in numerous countries (Boelens et al., 2012; Burchi, 2012; De Vos, Boelens, & Bustamante, 2006; Ravnborg, 2015).

Common to this recent wave of water reform is the attempt to replace complex, often locally negotiated and overlapping sets of water rights (Bruns, Ringler, & Meinzen-Dick, 2005; Hodgson, 2004) with a single, unified legal framework for water allocation under public, i.e. government, control. The aim is to move decisions on water access and water allocation out of the realm of things that are negotiated locally into the realm of things that are negotiated and sanctioned through statutory institutions according to predefined social, environmental and economic criteria, e.g. expressed as a clear order of priority between different types of uses of water, as well as setting the conditions which should be met to enjoy continued water use rights (Ravnborg, 2015).

Nicaragua has recently reformed its legal and regulatory water governance framework. According to the General Law on National Water (Law 620), enacted in 2007, the objective of the water governance reform, of which the water law forms part, is to ensure "the sustainable and equitable use of water and the conservation of the country's water resources in terms of quantity as well as quality",[1] through the regulation of rights to access and use water resources.[2]

However, laws and regulations on paper and laws and regulations in practice are often two different things. Despite the often sound economic, social and environmental rationales underlying water governance reforms, in many cases translated into concrete legal and administrative provisions, their fulfilment is often hampered, particularly in developing countries, by being implemented in contexts characterized not only by insufficient administrative and regulatory capacity but also by significant inequality. As noted by the UNDP, this combination is a significant challenge because "the importance of power in shaping the outcomes from legislation [tends to be] inversely related to regulatory capacity", implying that "weak regulatory capacity increases the scope for exploitation of unequal relationships" (United Nations, 2006, p. 182). Thus, though probably unintended, there is a risk that under such conditions, rather than enhancing water security for all, water governance reform may introduce a new dimension of inequality, namely the dimension of legal water security for the benefit of a limited – and privileged – segment of the total population of water users.

Drawing on interviews and archival research conducted in Nicaragua before and after the approval of the new water law in 2007, this article examines the extent to which the implementation of the water governance reform, and particularly the new water rights regime regulating access to and use of water, has contributed to water security for different segments of irrigation farmers in the country. The article is organized into six sections. Following a presentation of the empirical material upon which the article is based, the third section provides an account of Nicaragua's new legal and institutional framework for water governance and the process of its coming into being. Focusing on irrigation, the fourth section examines achievements to date with respect to implementation of the new water rights regime, while the fifth analyzes the implication of these achievements with respect to water security among farmers and agricultural enterprises. The final section draws conclusions and discusses the implications for water governance reform implementation in contexts characterized by limited administrative and regulatory capacity.

## Methods and materials

In addition to interviews, conducted over more than a decade, with key actors in water governance reform and its implementation at national, district and local levels, this article is based upon two data-sets. Ideally all administrative resolutions issued by the National Water Authority, e.g. with respect to applications for permission to develop and register water infrastructure as well as for water use rights, are published in the national gazette at the cost of the applicant. Hence, the first data-set is constructed through a review of all issues of the Nicaraguan national gazette published from the establishment of the National Water Authority in 2010 up to 29 February 2016 to identify all published administrative resolutions related to it. The data-set consists of the administrative resolutions issued until the end of 2015,[3] and as a subset of these, all published administrative resolutions announcing concessions of water use rights for irrigation. This inventory was supplemented with periodic reviews of the National Public Registry of Water Rights and tabulated as a database.[4] A total of 433 administrative resolutions were identified, of which 252 announce concessions of water use rights, of which 120 are for irrigation. The second data-set is constructed as a subset of the Fourth National Agricultural Census data-set from 2011 ($N$ = 262,546 farms – INIDE, 2011) containing those farms which according to the census employ some form of irrigation ($n$ = 11,599 farms). This data-set provides a profile of the 11,599 farms which make use of water for irrigation and thus of the context in which the implementation of the water governance reform with respect to irrigation takes place.

## Water governance reform in Nicaragua

### *The coming into being of Nicaragua's water law*

It took almost a decade, gave rise to the formation of a widely anchored and highly effective civil society movement against the privatization of water and in defence of the right to water, and involved a number of competing draft law proposals before Nicaragua in 2007 passed its first General Water Law (Barrios and Wheelock, 2005; Gentes, 2011; Gómez, Ravnborg, & Rivas Hermann, 2007; Romano, 2012). The initial impetus for the legal reform process arose through the negotiations over a major loan from the Inter-American Development Bank to modernize Nicaragua's water and sanitation services in the aftermath of Hurricane Mitch, which had hit the country in late 1998 (Romano, 2012). Subsequently, legal water reform was made part of the conditions for the negotiation of the free trade agreement (CAFTA) between Central America and the United States. Strongly influenced by the neoliberal doctrine which still prevailed in the 1990s and into the first years of the new millennium, institutions like the World Bank and the Inter-American Development Bank actively advocated and supported countries to follow the example of Chile, which a decade earlier had passed a water law that turned water into a commodity to be allocated entirely through the market (Bauer, 2004). Nicaragua was one of the countries which experienced these pressures to privatize its water and sanitation services and enable market-based allocation of water rights. However, these efforts were everywhere met with resistance, if not from governments themselves, then from civil society. Thus, until now, no other country has gone as far as Chile in instituting free market–based allocation of water rights (Burchi, 2012), and even in Chile, the 1981 Water Code was subsequently reformed in an effort to regulate the market-based allocation of water rights (Bauer, 2008; Burchi, 2012; Ravnborg, 2015).

In Nicaragua, the first signs of resistance emerged in response to plans to privatize one of the country's major hydroelectric power plants. Neighbouring residents feared that this would lead to the privatization of the water feeding the plant (Romano, 2012).

Gradually, this initial resistance grew into a national movement consisting of a large number of grass-roots organizations as well as district governments opposing the privatization of water as a resource and the privatization of water supply and sanitation utilities, and defending water as a human right (Barrios and Wheelock, 2005; Romano, 2012). Coalitions were established, marches were organized, and alternative water law proposals were formulated and consulted on across the country. Supported by 29,000 signatures, the alternative water law developed by the Alliance against Water Privatization and for the Right to Water qualified to be presented for discussion at the National Assembly in 2004.[5] Thus, in 2007, when the National Assembly passed Nicaragua's water law, the Alliance could celebrate that the new law incorporated several of its demands. These included "water is a national patrimony",[6] "it is the duty of the state to prevent [water as a natural resource] becoming subject to any type of privatization",[7] and "the drinking water service will never become subject to any type of privatization, direct or indirect".[8]

### *Nicaragua's new water law*

The objective of Nicaragua's water law is to establish the legal and institutional framework for the governance of the country's water resources – surface as well as groundwater – and thereby ensure "the sustainable and equitable use of water and the conservation of the country's water resources in terms of quantity and quality",[9] with the aim "to promote social and economic development".[10] The introduction of a water rights regime is at the core of Nicaragua's new legal and institutional water governance framework. In the context of hydrological units ranging from river basins through watersheds to catchments, this new water rights regime has as its objective "to ensure the control of the use of water from both a quantitative and a qualitative perspective, and enable the effective exercise of the rights of access to water".[11] Thus, according to the law, all water use, whether by individuals or by public or private legal persons such as water utilities and companies, has to be formally sanctioned through a concession, licence or authorization.[12] To regulate the allocation of rights to use or benefit from water resources[13] and thus to act as a custodian of Nicaragua's water resources, the water law establishes the National Water Authority as the executive body charged with the regulatory and technical responsibility for water governance in the country.[14] Moreover, the water law establishes the National Public Registry of Water Rights.[15] This registry is envisaged as separate from, yet administered by, the National Water Authority. The law requires that public access to the water rights registry be guaranteed.[16] The Nicaraguan water law thus enlists itself among the so-called 'modern' water laws which a whole range of countries have adopted in recent decades (Aagaard & Ravnborg, 2006; Burchi, 2012; Ravnborg, 2015; van Koppen, 2007).

To guide the National Water Authority in the allocation of water rights, the law establishes the order of priority between different types of water use to be followed in cases where demands for water exceed the resource available.[17] Water use for human (domestic) consumption, including for rural and urban water utilities, takes precedence over all other types of use. This and the explicit statement of the state's responsibility towards providing, facilitating and regulating the adequate supply of safe drinking water for the population bring the Nicaraguan water law in line with public demand for the recognition of water as a human right, later declared a universal human right by the United Nations (2010). In order of priority, domestic consumption is followed by the use of water for agricultural, livestock and forestry production; for environmental purposes; for energy production intended to meet public needs; industrial purposes; tourism; navigation; etc.[18] Thus, rather than leaving the allocation of water rights to the market, as in the Chilean model initially advocated e.g. by the World Bank, Nicaragua opted to follow what senior officials referred to as a "Mexican model" (personal communication, Jaime Morales, leader of the Environmental Committee of the Nicaraguan Parliament, 30 August 2004), which retains significant public control over the allocation of water rights (Garduño, 2005; Wilder, 2008).

### *The new water rights regime*

Nicaragua's new water rights regime distinguishes between three different types of use permits. With respect to irrigation, it defines two types, namely *concessions* and *authorizations*, both valid for between 5 and 30 years and both potentially subject to a water use fee, for which, however, the specific design and law are still pending. The third type of permit, the *licence*, is intended exclusively for public water utilities providing drinking water and involved in hydro-power generation. According to the law, users of water for irrigation of areas larger than 20 hectares need a *concession*, while users of water for small-scale irrigation, defined as irrigation of areas smaller than three hectares, need an *authorization*. Thus, though this is probably unintended, the law is silent with respect to how to formalize water use rights for irrigation of areas between 3 and 20 hectares. Nonetheless, this omission bears witness to the limited attention given during the legislative process to the context in which the new water rights regime was to be implemented, as well as to the implementation challenges that would ensue once the law was enacted. While concessions are to be issued by the National Water Authority, authorizations of water use in minor quantities for irrigation, but also for other purposes, may be granted by the district authorities, or in the autonomous regions by the regional councils, *provided* that a signed collaboration agreement exists between the district authority or regional council and the National Water Authority. However, apart from the issuing authority, no further differences are stipulated with respect to the legal requirements or provisions for concessions and authorizations. Thus, an application, whether for a concession or for an authorization,[19] should be accompanied by the legal identification of the applicant and the farm for which irrigation is solicited, the source and the quantity of water solicited, and, if deemed necessary, by an environmental impact study. In all cases, the water right, if granted, is conditioned on the commitment to comply with the Nicaraguan environmental law and regulations, e.g. with respect to use of pesticides or other contaminants. The decision on an application for a water rights concession[20] is to be communicated through an administrative resolution.[21]

## Water governance reform implementation in Nicaragua

### *The case of water rights for irrigation*

Although the law was passed in 2007, it was not until 2010 that the National Water Authority was formally established and became operational. This led to the elaboration of a new water law regulation (Decree 44-2010),[22] and the preparation of a regulation which made the National Public Registry of Water Rights functional (Decree 33-2011). Hence, the first applications for water use rights concessions were received in 2010 and granted and made public in 2011, by means of an administrative resolution. As a legal practice, the National Water Authority introduced a standard clause in its administrative resolutions which conditions their validity upon their publication in two national newspapers, as well as in the national gazette, at the cost of the applicant. In a systematic review of the national gazette up to 29 February 2016 and the National Public Registry of Water Rights up to April 2015, we identified 433 administrative resolutions. Of these, 366 resolutions (85%) were published in the national gazette. According to the National Water Authority's own newsletter (Rubí, 2014), however, 800 concessions, licences and authorizations have been granted by the National Water Authority. This would imply that a significant share, i.e. more than half of these, have never been published.[23] A total of 252 of the administrative resolutions were concessions[24] of water rights, of which 120 were related to irrigation. Of these, 104 (87%) were published in the national gazette. A bit more than half of the 116 concessions for irrigation (65) were granted for the irrigation of sugar cane (Figure 1).

Besides building up the newly established National Water Authority, which today counts on a technical staff of between 15 and 20 persons, including lawyers and hydrologists, efforts have been devoted to developing the procedures and formats for handling and evaluating applications, and for issuing and registering the administrative resolutions. Today, these formats are in place and available from the National Water Authority's homepage.[25] Moreover, as part of the process of developing the institutional

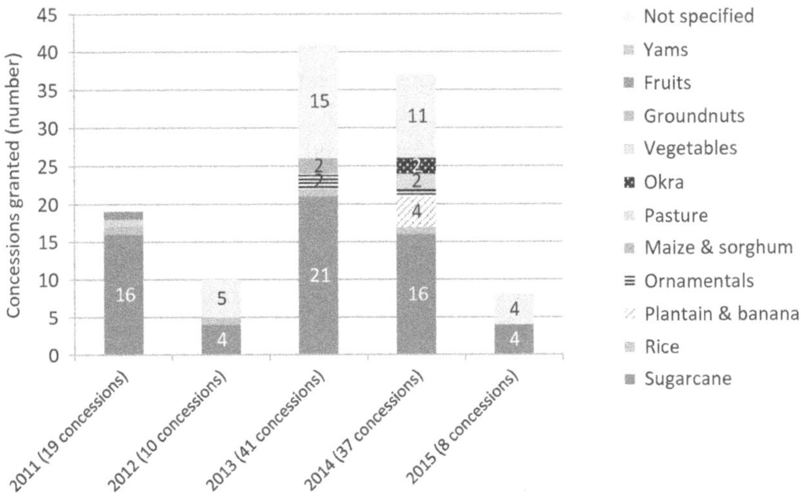


**Figure 1.** Irrigation concessions (115 total) granted by the National Water Authority, 2011–2015, by crop to be irrigated. Source: Own elaboration based on review of the national gazette, *La Gaceta – El diario oficial de Nicaragua* (August 2010 to 29 February 2016, available at http://www.lagaceta.gob.ni), and the National Public Registry of Water Rights.

Note. Whether the small number of water use rights concessions granted for irrigation in 2015 is due to a decline in 'demand' for such concessions or to failure on the part of the applicants to publish the administrative resolutions in the national gazette, is unfortunately unclear.

procedures, the National Water Authority has developed standard templates for the hydrological or hydro-geological studies[26] which are requested to accompany applications for water use concessions.[27] Of the 120 concessions granted for irrigation, 46 make reference to being supported by a hydrological or hydro-geological study. Meanwhile, no templates have been developed so far to respond to the legal requirement to consider the potential social impact of the solicited water use, including how to account for the actual use of the water solicited.[28] Mechanisms which could make up for this absence, such as seeking the endorsement of district or community authorities in the districts and communities where the water will be extracted and used, have only been applied sporadically. Thus, in Chichigalpa District, where some of Nicaragua's biggest sugar estates are located, district authority officials lamented the fact that the former practice of seeking their endorsement, e.g. before the development of new irrigation infrastructure, had been discontinued following the establishment of the National Water Authority. Instead, the sugar estate would now proceed directly to the National Water Authority to present their requests (personal communication, senior district officials, Chichigalpa, 21 January 2015). Of the 120 concessions granted for irrigation, 19 make reference to having obtained the endorsement from district authorities; none of them from Chichigalpa.

In addition to formally establishing the National Water Authority, the 2010 water law regulation (Decree 44-2010) introduced a new set of criteria for determining whether the right to use water for irrigation should be formalized through a concession or an authorization. Rather than considering the area under irrigation, which was the basis for the criterion stipulated in the water law, the new set of criteria is based on farm area, and on the destination of the produce. According to the 2010 regulation, a *concession* of water use for irrigation should be solicited from the National Water Authority when the farm where the irrigation will take place is larger than 70 hectares *or* when, irrespective of farm area, the produce is intended for the "industrial market" (*comercialización industrial*).[29] In cases where the farm in which the irrigation will take place is smaller than 70 hectares *and* where the produce is not intended for the "industrial market", water use should be formalized through an *authorization* solicited from the relevant district authority. Just like the concessions, these authorizations should subsequently be submitted to the National Water Authority for inclusion in the National Public Registry of Water Rights.[30] While thereby rectifying the ambiguity introduced by the water law with respect to how to formalize the use of water for irrigation of areas between 3 and 20 hectares, the 2010 regulation introduces a new source of ambiguity in that no definition of "industrial market" is provided. Nor is this a term or category commonly used, e.g. in national statistics.

Regardless of these legal-administrative changes in when an authorization as opposed to a concession should be solicited, limited progress has been made in institutionalizing the granting of water rights for the use of minor quantities of water by means of authorizations issued by district authorities. Much of this impasse is due to the legal and administrative ambiguity in the content and the conditions to be met to enable the National Water Authority to establish a cooperation agreement with a district authority or regional council. Neither the water law nor its regulation provide any guidance in this regard. Moreover, while recognizing the role many district authorities have played in regulating the use of water resources in order to provide for the diverse needs of their citizens for domestic, productive and recreational purposes, staff of the National Water Authority also express reservations, both with respect to whether district authorities possess the necessary technical skills to authorize water use, and with respect to the administrative capacity of the National Water Authority itself to actually handle and honour cooperation agreements with Nicaragua's 153 district authorities and two regional councils. To compensate for at least part of this impasse, the National Water Authority established a territorial delegation in Estelí in 2013, intended to cater to the five northern departments of Estelí, Matagalpa, Jinotega, Madríz and Somoto. The delegation counts on two technical staff members. The main task of the delegation is to receive and examine applications and serve as the liaison to district authorities as well as to other national agencies. Thus, efforts have recently been made to develop cooperation agreements with district authorities, and by June 2015, draft cooperation agreements, e.g. with Estelí District, had been developed, and the hope was expressed that the agreement with Estelí District could be celebrated before the end of 2015 as the first cooperation agreement between the National Water Authority and a district authority. According to our information, however, no cooperation agreement had been signed by the end of 2015, and therefore no authorizations of the right to use water in minor quantities for irrigation had been issued. Thus, what may have been envisaged as a legal option to provide equal opportunities for formalizing water use rights for irrigation, irrespective of the scale of water use, namely the district authorizations, so far appears to be a legal cul-de-sac.

## Implications for water security

The ambition presented in the water law was that all water use should be formalized, and thus sanctioned and secured through a formal water right, and that existing hydraulic infrastructure should be registered within half a year of the passing of the law (Law 620, Article 137). Given that resources (human as well as operational) did not match the passing of the law, and that only limited efforts were made to assess the actual magnitude of this ambition, in the case of irrigation to some extent explained by the lack of information on the number of farmers using irrigation,[31] this ambition appears to be a somewhat far-fetched dream. The experience of Mexico (Garduño, 2005), which was a source of inspiration for the Nicaraguan water law, would have suggested a much longer regularization period. On the other hand, although it is far from meeting the ambition stipulated in the law, progress has been made.

According to the criteria originally proposed in the water law, the legal obligation to formalize the right to use water for irrigation by means of a concession applies to owners of farms where the irrigated area is larger than 20 hectares. According to the data available from the Fourth National Agricultural Census (Table 1), this is the case at 381 farms out of the 11,599 farms which are reported to have some form of irrigation. In this context, the 120 concessions granted to date would have made up a considerable share, namely 31%. However, in the context of the new criteria introduced with the 2010 water law regulation, the achievement fades. Considering farm area, the first of these two criteria, the census identifies 1166 farms larger than 70 hectares, and reported to have some form of irrigation, and thus obliged to obtain a concession to formalize their right to use water for irrigation. This reduces the proportion of water right

**Table 1.** Profile of farms with irrigation according to irrigation type and type of legal water right requirement. Figures are number of farms.

| | According to 2007 water law[b] | | | According to 2010 water law regulation | | | |
|---|---|---|---|---|---|---|---|
| | Requiring concession | Requiring authorization | *Requirement unknown* | Requiring concession | Requiring authorization | *Requirement unknown* | |
| Irrigation type[a] | Irrigated area >20 ha | Irrigated area = <3 ha | *Irrigated area > 3 ha and ≤ 20 ha* | Farm area > 70 ha | Farm area ≤ 70 ha and produce intended for own consumption | *Farm area ≤ 70 ha but significance of "industrial market" is unclear*[c] | Total |
| Gravity | 260 | 3,773 | *740* | 622 | 1,286 | *3,072* | 4,980 |
| Sprinkler | 127 | 2,258 | *368* | 450 | 1,075 | *2,512* | 4,037 |
| Drip | 244 | 3,519 | *706* | 674 | 1,581 | *3,519* | 5,774 |
| Manual | 119 | 4,427 | *428* | 434 | 1,599 | *3,011* | 5,044 |
| Other | 9 | 183 | *37* | 26 | 84 | *126* | 236 |
| All | 381 | 8,700 | *1180* | 1166 | 3,522 | *6,911* | 11,599 |

Data source: Fourth National Agricultural Census (INIDE, 2011).

[a] Each farm with irrigation may have more than one type of irrigation system.

[b] The area under irrigation is unknown for 1338 of the 11,599 farms reported to have irrigation. Therefore, only 10,261 farms may be categorized with respect to water right type required by the 2007 water law.

[c] The census records whether the produce from a farm is primarily for own consumption or intended for the national market or for export. Assuming that farms smaller than 70 hectares and from which the produce is primarily for own consumption will be required to apply for an authorization rather than a concession, the degree of water right formalization achieved to date ranges between 1% and 10% in terms of number of farms.

formalization to 10%. Considering in addition the second criterion, whether the produce is intended for the "industrial market", the proportion may drop further – but by exactly how much is unknown, since we have no definition of "industrial market".

Yet, in the context of the amount of water estimated to be withdrawn annually for irrigation in Nicaragua, the degree of formalization of water use for irrigation achieved since the putting in place of the new water rights regime is significantly higher. Already by the end of 2011, water rights corresponding to 104 million m$^3$ annually were given in concession for irrigation, primarily for sugar cane, and entered in the National Public Registry of Water Rights. This corresponds to 9% of the total amount of water (1110 million m$^3$) estimated by FAO AQUASTAT to be withdrawn annually for irrigation (2011 data, http://www.fao.org/nr/water/aquastat/data/query/). To date, water rights for irrigation granted in concession correspond to 370 million m$^3$, or one-third of the amount estimated to be withdrawn annually for irrigation. In addition, as pointed out by Rubí and Murillo (2014) – both National Water Authority staff members – important insights with respect to the water resources and the uses they are put to are gained by the National Water Authority through the water rights allocation process itself.

The vast majority of the concessions of water rights for irrigation, namely 97 (81%) out of the 120 concessions to date for irrigation, have been granted to companies rather than to individual farmers. In most cases (87 of 97), these companies were represented by a lawyer during the application process. As already indicated, the sugar cane estates were the early movers with respect to formalizing their irrigation water use when this became possible in late 2010. As shown in Figure 1, sugar cane estates accounted for 16 of the 19 concessions of irrigation water rights granted in 2011. Although the sugar cane estates have now been joined by other types of agricultural companies, so that today the sugar cane sector accounts for 'only' 54% of the concessions for irrigation, it still accounts for most of water for which irrigation use rights have been granted.

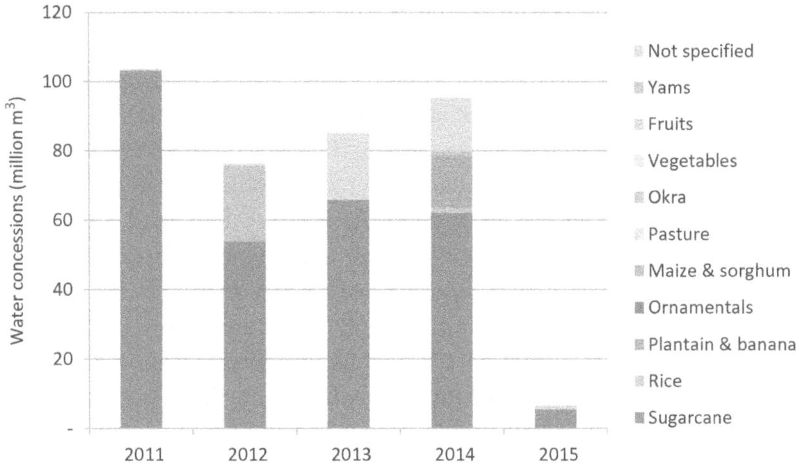


**Figure 2.** Water quantity conceded for irrigation by the National Water Authority, 2011–2015, by crop to be irrigated (115 concessions total). Source: Own elaboration based on review of the national gazette, *La Gaceta – El diario oficial de Nicaragua* (August 2010 to 29 February 2016, available at http://www.lagaceta.gob.ni), and the National Public Registry of Water Rights.

Note. Whether the small number of water use rights concessions granted for irrigation in 2015 is due to a decline in 'demand' for such concessions or to failure on the part of the applicants to publish the administrative resolutions in the national gazette, is unfortunately unclear.

Virtually all of the water given in concession for irrigation by the end of 2011 was for the irrigation of sugar cane, and combined for the period 2011–2015, 79% of the total volume of water for which irrigation water use rights had been conceded was conceded to sugar cane estates (Figure 2). In addition to a wish to comply with existing legislation, a strong incentive which allegedly has contributed to motivating the sugar cane estates to formalize existing or planned irrigation water use has been the requirement to document legal compliance, including with respect to water access, from financial institutions such as the International Finance Corporation (IFC, 2012) as well as increasingly also in commodity markets, including the biofuel market (Ponte & Daugbjerg, 2015), for which an increasing share of the sugar cane is destined (personal communication, National Water Authority officials, January 2014). This has repercussions for the standards which are gradually evolving, e.g. under the auspices of the UN Global Compact (e.g. the CEO Water Mandate[32]), within the World Economic Forum,[33] etc. (Daniel & Sojamo, 2012; Mason, 2013). Sugar cane is a commodity in expansion in Nicaragua from an annual production of 4.3 million tonnes in 2008 to 7.0 million tonnes in 2013 (FAOSTAT, http://faostat3.fao.org/). Part of this expansion has been financed by the IFC.[34] Moreover, sugar cane consumes large volumes of water. This has made the sugar cane sector particularly forthcoming with respect to soliciting the formalization of their water use and thus investing the resources necessary in terms of contracting the services of lawyers as well as of hydrologists to undertake the technical studies required to accompany the applications.

Although just starting to appear in the administrative resolutions published in the national gazette, tobacco enterprises have also begun to present their applications for irrigation water rights concessions. Whereas sugar cane is grown primarily in the plains along the Pacific coast, tobacco is primarily grown on the river banks in between the hills of the northern part of the country. Formerly, tobacco was primarily grown by cooperatives and more recently also by small-scale farmers under sharecropping agreements with the cigar manufacturers (Gómez & Ravnborg, 2011). Lately, however, enterprises, including those of the cigar manufacturers themselves, also appear to be engaging directly in the growing of tobacco (personal communication, district environmental officer, Estelí District, 22 June 2015). As this is a high-end niche market, with the tobacco destined for the production of high-quality cigars for export, demands for legal compliance and meeting social and environmental standards are gradually transmitted through the value chain. As an example, Nicaragua now has one of the first producers of organic cigars (Savona, 2013). Thus, the recognition of the rapidly growing water demands[35] – and actual withdrawals – in the northern part of the country for the irrigation of tobacco, combined with the need of the tobacco industry to formalize their irrigation water rights, contributed to the decision of the National Water Authority to establish its first and to date only functioning territorial delegation in Estelí, in the northern part of the country. Again, though, the institutionalization of the water rights regime through the National Water Authority in the case of the northern part of the country, reinforced through the establishment of the territorial delegation, has come at the cost of the efforts formerly undertaken by several of the district authorities in the area to regulate the use of available water resources. Using the legal provisions of the district law to issue district by-laws to encourage the prudent use of water for irrigation during the critical dry-season months, district authorities in several of the northern districts have made considerable efforts to ensure that sufficient water is available for domestic uses as well as for livestock keepers, small-scale irrigation of food crops, etc. (Ravnborg & Gómez, 2015). With the establishment of the new water rights regime, and particularly in the absence of cooperation agreements with the National Water Authority, district authorities have become legally excluded from playing this role. This has created what several district environmental officers refer to as "a water governance vacuum", as no other institutional body has been able to effectively fill this role (Gómez & Ravnborg, 2011; Ravnborg & Gómez, 2015).

Moreover, the fact that the new water governance regime has been only partially implemented has introduced a new source of water insecurity for the many farmers to whom the option of formalizing their right to irrigation water through a district-issued authorization has so far been unavailable. According to the criteria in the water law, approximately 9000 farmers use minor quantities of water for irrigation (irrigating less than 3 ha) and should therefore be entitled to formalize their water right by means of a district-issued authorization (Table 1). The number of farmers for whom, according to the more recent (2010) water law regulation, a district-issued authorization would suffice, is less clear due to the ambiguity in the term "industrial market". Depending on how "industrial market" is defined, the estimated number of farmers for whom a district-issued authorization would suffice ranges from 3522 to more than 10,000 (Table 1). At any rate, in terms of irrigated area as well as volumes of water withdrawn, these farmers count for only a fraction of the irrigated area (10–20%, depending on the criterion used – Table 2). Moreover, due to their relying on irrigation techniques regarded as water saving (drip and manual irrigation) to a higher extent than farmers required to solicit a concession (Table 1), this fraction is likely to be even smaller.

**Table 2.** Profile of farms with irrigation according to irrigation type and type of legal water right requirement. Figures are area under irrigation (hectares).

| | According to 2007 water law | | | According to 2010 water law regulation | | | |
|---|---|---|---|---|---|---|---|
| | Requiring concession | Requiring authorization | *Requirement unknown* | Requiring concession | Requiring authorization | *Requirement unknown* | |
| Irrigation type[a] | Irrigated area > 20 ha | Irrigated area ≤ 3 ha | *Irrigated area > 3 ha and ≤ 20 ha* | Farm area > 70 ha | Farm area ≤ 70 ha and produce intended for own consumption | *Farm area ≤ 70 ha but significance of "industrial market" is unclear*[b] | Total |
| Gravity[c] | 38,646 | 3,753 | *4,447* | 46,531 | 1,428 | *8,703* | 56,663 |
| Sprinkler[d] | 21,540 | 1,987 | *2,026* | 24,504 | 605 | *3,695* | 28,804 |
| Drip[e] | 1,413 | 187 | *198* | 3,928 | 430 | *1,276* | 5,634 |
| Manual[f] | 512 | 1,518 | *597* | 629 | 595 | *1,430* | 2,654 |
| Other[g] | 1,190 | 147 | *240* | 5,302 | 108 | *222* | 5,632 |
| All[h] | 63,302 | 7,592 | *7,507* | 61,903 | 2,775 | *13,723* | 78,401 |

Own elaboration on the basis of the Fourth National Agricultural Census (INIDE, 2011).

[a] Each farm with irrigation may have more than one type of irrigation system.

[b] The census records whether the produce from a farm is primarily for own consumption or intended for the national market or for export. Assuming that farms smaller than 70 hectares and from which the produce is primarily for own consumption will be required to apply for an authorization rather than a concession, the degree of water right formalization achieved to date ranges between 1% and 10% in terms of number of farms.

[c] The area under gravity irrigation is unknown for 5 of the 4980 farms reported to have gravity irrigation.

[d] The area under sprinkler irrigation is unknown for 1272 of the 4037 farms reported to have sprinkler irrigation.

[e] The area under drip irrigation is unknown for 4190 of the 5774 farms reported to have drip irrigation.

[f] The area under manual irrigation is unknown for 2340 of the 5044 farms reported to have manual irrigation.

[g] The area under other type of irrigation is unknown for 5 of the 236 farms reported to have other type of irrigation.

[h] The area under irrigation is unknown for 1338 of the 11,599 farms reported to have irrigation.

Nevertheless, water security is no less important to small-scale irrigation farmers, in terms of providing security for investment, economic outcome, and gaining or maintaining access to increasingly demanding commodity markets, than it is to farmers required to apply for a concession. Accentuating this, in catchments where the two categories of farmers compete for access to the same water, their different situations with respect to water security, produced by the partial implementation of the new water rights regime, may actually put the continued access to water and associated economic opportunities at risk for those who cannot document their formally sanctioned right to water, thereby reinforcing existing inequalities.

## Conclusion: the importance of implementation in securing rights or legitimizing dispossession

From the perspective of bringing irrigation water use under statutory jurisdiction, progress has been made as the implementation of the Nicaraguan water law gained pace following the establishment of the National Water Authority in 2010. With a relatively small staff and an even smaller operational budget, one-third of the volume of water which FAO estimates is withdrawn annually for irrigation in Nicaragua has been brought under statutory jurisdiction through the concession of water use rights. Yet, at the same time, precisely because of these budget constraints, which e.g. prevent the National Water Authority from undertaking independent technical studies and thereby make it rely on studies commissioned by the applicants, the National Water Authority has de facto become what could be described as a legalization service provider. Rather than being the *custodian* of Nicaragua's water resources, the regulatory role which the water rights regime was intended to enable, the National Water Authority has primarily acted in response to water users wishing to formalize their water right.

Moreover, the apparent hesitation on the part of the National Water Authority to conclude and formalize cooperation agreements with district authorities has significantly hampered the accomplishment of the intention of the water governance reform to ensure "sustainable and equitable use" of Nicaragua's water resources. First of all, it has discouraged, if not put a halt to, previous efforts undertaken by district authorities to regulate water use, primarily in order to safeguard water security for domestic water users and thus uphold the order of priority among different types of water use as stipulated in the water law. As nobody has taken over this role, this has created in many districts a water governance vacuum.

Second, the partial implementation of the water rights regime envisaged in the water law means that while a limited number of farmers and companies have obtained legally sanctioned rights to a significant part of the water resources estimated to be used in Nicaragua every year for irrigation, the majority of Nicaragua's irrigated farms, namely those using only minor quantities of water, are left in a legal and administrative limbo. Up to now, they have been prevented from formalizing their current water use for irrigation through legally sanctioned water rights by means of what was initially envisaged as a more accessible, district-issued water use authorization process through their district authority. While at best these thousands of small-scale farmers are not hampered in their endeavour to supply their families as well as the national and international markets with anything from beans, fruits and vegetables to coffee and tobacco, at worst this failure to regularize the existing water use, particularly of small-scale farmers, may weaken their water security to the extent of putting at risk their continued access to water, as well as to economic opportunities in the growing number of markets concerned about social and environmental responsibility.

As competition for water intensifies, this new dimension of inequality, namely that of legal water security, produced through the very choices made during the process of implementing the water governance reform, may turn this legal water insecurity into actual dispossession of water access. Thus, independently of the water legislation as such, the very process through which any given water legislation is implemented is of crucial importance in shaping its wider societal outcomes.

## Notes

1. This and subsequent quotes from the Nicaraguan water law (Law 620) are my own translations.
2. Law 620, Articles 1, 2 and 14c.
3. More administrative resolutions issued during 2015 may still appear in the national gazette, published from 1 March 2016 and onwards.
4. The last review of the National Public Registry of Water Rights was undertaken in April 2015.
5. http://www.simas.org.ni/noticias/28/2004-10-14-sociedad-civil-presenta-proyecto-de-la-ley-alternativo-de-agua/, last consulted 25 November 2015.
6. Law 620, Preamble, Article 3.
7. Law 620, Preamble.
8. Law 620, Article 4.

9. Law 620, Article 1.
10. Law 620, Preamble.
11. Law 620, Article 14.
12. Law 620, Article 41.
13. Law 620, Article 2.
14. Law 620, Articles 24–30. The 2010 regulation (Articles 13 and 17) states that the director of the National Water Authority has rank of minister.
15. Law 620, Articles 37–40.
16. Law 620, Article 40.
17. Law 620, Article 46.
18. Law 620, Article 46.
19. Law 620, Article 49.
20. The water law only specifies this as an obligation for the National Water Authority.
21. Law 620, Article 65; Decree 44-2010, Article 23.
22. Decree 44-2010 replaced the regulation elaborated shortly after the water law was passed in 2007 (Decree 106-2007).
23. It has not been possible to develop a comprehensive overview to document the number of administrative resolutions issued by the National Water Authority published in the two major Nicaraguan national newspapers, *El Nuevo Diario* and *La Prensa*. However, according to our archival reviews, at least 117 of the 433 administrative resolutions (27%) were published at least in one of them, while 3 resolutions were published in both. Thus, at least 98 of the administrative resolutions have been published both in the national gazette *and* in at least one of the national newspapers.
24. More concessions may have been granted but not published in the national gazette and/or registered in the National Public Registry of Water Rights prior to 17 August 2015, when the revisions of this article were concluded. The last consultation of the National Public Registry of Water Rights was made in April 2015.
25. www.ana.gob.ni.
26. http://ana.gob.ni/index.php/icons, last consulted 30 December 2015.
27. Neither the law nor its regulation strictly requires the application to be accompanied by the development of hydrological studies (in the case of surface water) or hydrogeological studies (in the case of groundwater). Rather, these studies are referred to as the basis for evaluating the application. In the practice which has evolved since 2011, the obligation to prepare the hydro(geo)logical studies rests with the applicant, who also contracts and pays for the study.
28. Law 620, Articles 46c and 49e.
29. Law 620, Article 71.
30. Ibid.
31. The 2001 National Agricultural Census available at the time of preparing and passing the law contained only limited information on the use of irrigation as compared to the 2011 National Agricultural Census.
32. http://ceowatermandate.org/.
33. https://www.weforum.org/projects/global-water-initiative/.
34. http://goo.gl/otlsqd.
35. According to the Nicaraguan Tobacco Association, Nicaragua had around 5000 hectares planted to tobacco in the beginning of 2015 (*La Prensa*, January 26, 2015; http://www.estanquers.cat/docs/1111.pdf).

## Acknowledgement

The author wishes to acknowledge the valuable participation of Daniel Chillon Olmos during the data collection and organization undertaken for this article. This article is based on research conducted as part of the programme Territorial Cohesion for Development, coordinated by

RIMISP (Latin American Centre for Rural Development). The author also wishes to thank the guest editors for this special issue, Stefano Burchi and Gabriel Eckstein, for their encouragement and suggestions, as well as for the comments received from two anonymous referees on an earlier draft of this article.

## Funding

This work was supported by the International Development Research Centre, Canada.

## References

Aagaard, C. A., & Ravnborg, H. M. (2006). Water reform. Implications for rural poor people's access to water. *DIIS Brief*. Copenhagen: Danish Institute for International Studies.

Barrios, M., & Wheelock, S. I. (2005). *Movilización social y gestión del agua en Nicaragua*. Managua: Nitlapan-Universidad de Centroamérica.

Bauer, C. J. (2004). *Siren song. Chilean water law as a model for international reform*. Washington, DC: Resources for the Future.

Bauer, C. J. (2008). *The experience of Chilean Water Markets*. Paper presented at Expo 2008, Zaragoza, Spain.

Boelens, R., Duarte, B., Manosalvas, R., Mena, P., Roa Avendaño, T., & Vera, J. (2012). Contested territories: Water rights and the struggles over indigenous livelihoods. *The International Indigenous Policy Journal*, *3*(3), Article 5. doi:10.18584/issn.1916-5781

Bruns, B. R., Ringler, C., & Meinzen-Dick, R. (Eds.). (2005). *Water rights reform: Lessons for institutional design*. Washington, DC: International Food Policy Research Institute.

Burchi, S. (2012). A comparative review of contemporary water resources legislation: Trends, developments and an agenda for reform. *Water International*, *37*, 613–627. doi:10.1080/02508060.2012.694800

Daniel, M. A., & Sojamo, S. (2012). From risks to shared value? Corporate strategies in building a global water accounting and disclosure regime. *Water Alternatives*, *5*, 636–657.

De Vos, H., Boelens, R., & Bustamante, R. (2006). Formal law and local water control in the Andean region: A fiercely contested field. *International Journal of Water Resources Development*, *22*(1), 37–48. doi:10.1080/07900620500405049

Garduño, H. (2005). Lessons from implementing water rights in Mexico. In B. R. Bruns, C. Ringler, & R. Meinzen-Dick (Eds.), *Water rights reform, lessons for institutional design*. Chapter 4. Washington, DC: International Food Policy Research Institute.

Gentes, I. (2011). Políticas hídricas, institucionalidad compleja y conflictos transfronterizos en Nicaragua. In R. Boelens, L. Cremers, & M. Zwarteveen (Eds.), *Justicia hídrica. Acumulación, conflicto y acción social* (pp. 255–278). Lima: Instituto de Estudios Peruanos.

Gómez, L., & Ravnborg, H. M. (2011). *Power, inequality and water governance: The role of third party involvement in water-related conflict and cooperation* (CAPRi Working Paper 101). Washington, DC: International Food Policy Research Institute

Gómez, L. I., Ravnborg, H. M., & Rivas Hermann, R. (2007). *Institucionalidad para la Gestión del Agua en Nicaragua* (Cuaderno de Investigación 30). Managua: Nitlapan.

Hodgson, S. (2004). *Land and water – the rights interface* (FAO Legislative Study 84). Rome: Food and Agriculture Organization.

IFC. (2012). *Performance standards on environmental and social sustainability*. Washington, DC: International Finance Corporation, The World Bank Group. Retrieved from http://www.ifc.org/wps/wcm/connect/115482804a0255db96fbffd1a5d13d27/PS_English_2012_Full-Document.pdf?MOD=AJPERES

INIDE. (2011). Data bases from the IV National Agricultural Census (IV Cenagro). Retrieved December 19, 2014, from http://www.inide.gob.ni/

Jacobi, P.R., De Stefano, L., López-Gunn, E., Solanes, M., Delacámara, G., Marín, G., ... Jiménez, A. (2014). Reforming water governance structures. In B.A. Willaarts, A. Garrido, & M.R. Llamas (Eds.) *Water for food and well-being in Latin America and the Caribbean. Social and environmental implications for a globalized economy* (pp. 285–315). London and New York: Routledge.

Mason, N. (2013). *Uncertain frontiers: Mapping new corporate engagement in water security* (Working Paper 363). London: Odi.

OECD. (2014). Climate change, water and agriculture. Towards resilient systems. *OECD Studies on Water*. Paris: OECD Publishing. doi:10.1787/9789264209138-en

Ponte, S., & Daugbjerg, C. (2015). Biofuel sustainability and the formation of transnational hybrid governance. *Environmental Politics*, *24*(1), 96–114. doi:10.1080/09644016.2014.954776

Ravnborg, H. M. (2015). Water competition, water governance and food security. In I. Christoplos & A. Pain (Eds.), *New challenges to food security: From climate change to fragile states* (pp. 109–122). London and New York: Routledge.

Ravnborg, H. M., & Gómez, L. I. (2015). The importance of inequality for natural resource governance: Evidence from two Nicaraguan territories. *World Development*, *73*, 72–84. doi:10.1016/j.worlddev.2014.11.001

Romano, S. T. (2012). From protest to proposal: The contentious politics of the Nicaraguan anti-water privatisation social movement. *Bulletin of Latin American Research*, *31*, 499–514. doi:10.1111/blar.2012.31.issue-4

Rubí, E. (2014). Concesiones para derechos de usos de agua. *Revista ANA*, *2*, 26–27. Retrieved from http://www.slideshare.net/ANA-Nicaragua/revista-anaedicin-ii?related=1

Rubí, E., & Murillo, G. (2014). Concesiones para derechos de usos de agua. *Revista ANA*, *1*, 4–5. Retrieved from http://www.slideshare.net/ANA-Nicaragua/revista-anamayo-2014

Savona, D. (2013). Talking Tobacco with Nestor Andrés Plasencia. *Cigar Aficionado*. Retrieved from http://www.cigaraficionado.com/webfeatures/show/id/17330

United Nations. (2006). *Human development report 2006: Beyond scarcity: Power, poverty and the global water crisis*. New York, NY: Palgrave, Macmillan.

United Nations. (2010). United Nations Resolution 64/292 The Human Right to Water and Sanitation. Retrieved March 20, 2014, from http://www.un.org/ga/search/view_doc.asp?symbol=A/RES/64/292

UN-Water. (2013). *Water security and the global water agenda. A UN-water analytical brief.* Ontario, Canada: United Nations University.

van Koppen, B. (2007). Dispossession at the interface of community-based water law and permit systems. In B. van Koppen, M. Giordano, & J. Butterworth (Eds.), *Community-based water law and water resource management reform in developing countries*. Wallingford, UK: CAB International.

Wilder, M. (2008). Equity and water in Mexico's changing institutional landscape. In J. M. Whiteley, H. Ingram, & R. W. Perry (Eds.), *Water, place, & equity* (pp. 95–116). Chapter 4. Cambridge: The MIT Press.

## Legal texts

Law 620: *Ley general de aguas nacionales*, published in *La Gaceta* no. 169, 4 September 2010.

Decree 44-2010: *Reglamento de la ley general de aguas nacionales*, published in *La Gaceta* nos. 150 and 151, 9 and 10 August 2010.

Decree 33-2011: *Reglamento del registro público nacional de derechos de agua*, published in *La Gaceta* no. 124, 5 July 2011.

# Compulsory licensing under South Africa's National Water Act

Michael Kidd

**ABSTRACT**
This article considers the statutory scheme for compulsory licensing in South Africa and examines its use in three case studies. Their evaluation reveals the complexity and time-consuming nature of the process. This, coupled with other challenges, particularly in relation to the achievement of equity (redress), is likely to result in the slow achievement of the National Water Act's objectives through compulsory licensing. There are no realistic alternatives to licensing if the constitutionally required goal of equity is to remain critical.

## Introduction

The transition from apartheid to democracy in the mid-1990s saw water management in South Africa characterized by two main thrusts: sustainable water use and improved access to water for those who previously had not had adequate access (referred to as 'equity' in this article). Because of the need to recognize existing water rights, pursuit of these twin goals is an ongoing process, still requiring significant progress, more than 15 years after the National Water Act 36 of 1998 ('the Act') was enacted. One of the main management tools available to water authorities in pursuing these goals is compulsory licensing.

This article examines the process of compulsory licensing as provided for in the Act. In order to understand the need for such a mechanism, it is necessary to sketch the context, both in terms of the country's water imperatives and in relation to socio-political factors. The statutory process will then be examined, followed by an evaluation of the three compulsory licensing processes carried out so far in South Africa. The experience thus far suggests that the carrying out of compulsory licensing processes will last for many years to come and that there are considerable challenges, particularly in relation to the pursuit of equity.

## Context

South Africa is a water-scarce country, with more rainfall in the east than in the interior and the arid west. Recent data indicate that the country is nearing a situation where the amount of water available will struggle to meet the demand, and many areas are already under water stress (Department of Water Affairs [DWA], 2010). At the time of writing, South Africa is experiencing the worst drought in decades. Most of South Africa's water is used for agriculture/irrigation (approximately 60%), while water for municipal needs (primarily domestic) amounts to 27% (DWA, 2013).

Prior to the advent of democracy and the enactment of the 1998 Act shortly thereafter, South Africa's water law provided for access by means of riparian rights. Land ownership was significantly skewed racially due to apartheid policies, with the result that, in the mid-1990s, most water rights (primarily for agriculture) were held by white farmers; up to a third of South Africans (primarily black people) had inadequate access to water as of 1997 (Department of Water Affairs and Forestry [DWAF], 1997).

When faced with drafting a new water law that would address the inequality in access to water rights, the government reached the conclusion that it would be all but impossible to replace overnight a system of riparian rights with a system of allocations based on administrative decision making. It was consequently decided that existing lawful water rights would remain legally recognized (Sections 4, 22, 32–35 in the Act). The decision may also have been influenced by concerns as to what effect removing water rights from farmers and other existing lawful users (e.g. mines) would have on the economy, particularly commercial agriculture (Movik, 2014). Existing lawful use rights would not be permanent, and the measure through which they could be removed or transferred is compulsory licensing. Concerns about expropriation appear not to have played a major role in the recognition of existing use rights, especially since the law provides for their temporary nature.

Whereas in other countries, introduction of administrative licensing systems has adversely affected customary law use rights, recognition of existing lawful use in the 1998 Act has the effect that limited customary law rights continue to be legally recognized (Van Koppen, van der Zaag, Manzungu, & Tapela, 2014). These rights would be identified as being held by black people, which would contribute to the equity objectives discussed below.

Before considering the statutory provisions relating to compulsory licensing, it is necessary also to consider current government policy in relation to water rights allocations, known as the water allocation reform process. The 2008 Water Allocation Reform Strategy (DWAF, 2008) stipulated targets of national water rights allocation of 60% in black hands and 50% to women by 2024. Progress is slow: only 1518 of the 4284 licences issued between 1998 and 2012 for new water rights were allocated to historically disadvantaged individuals (people other than whites), and the amount allocated was only 1.6% of the total water allocated in this period (Van Koppen & Schreiner, 2014, p. 553). The 2013 National Water Resource Strategy (NWRS) is based on a philosophy of "developmental water management", "which can be considered part of IWRM [integrated water resources management] principles in practice and which takes, as a central premise, the fact that water plays a critical role in equitable social and economic development, and that the developmental state has a critical role in ensuring that this takes place" (DWA, 2013, p. 14). According to Section 7 of the Act, any authority responsible for any decision making in relation to water is required to give effect to the NWRS when exercising any power or performing any duty in terms of the Act. Developmental water management thus appears to be IWRM with a special emphasis on equitable allocation of water resources. The NWRS regards equity as entailing three dimensions: equity in access to water services (including basic human needs); equity in access to water resources (for productive purposes); and equity in access to the benefits of water resource use ("water must be allocated so that it brings maximum benefit to all, whether directly or indirectly" – DWA, 2013, p. 45). The equity objective is particularly important in considering the law relating to compulsory licensing.

## Compulsory licensing in the National Water Act

It is beyond the scope of this article to consider in detail all of the provisions of the Act that are relevant in the licensing process (including compulsory licensing), so an overview will have to suffice. According to Section 22 of the Act, any water use requires a licence, unless it is permissible under Schedule 1 (a list of *de minimis* water uses, such as domestic use); or it is permissible as a continuation of an existing lawful use (discussed further below); or it is permissible in terms of a general authorization issued under Section 39 (essentially a 'blanket licence'). 'Water use' is defined in Section 21 of the Act, which includes a number of different activities to be regarded as water uses, not just water abstraction but effectively any activity that has an impact on water resources (such as diverting a water resource or putting substances such as waste or effluent into a water resource). A 'water resource' is defined in Section 1 of the Act as including essentially any source of water, both surface and underground.

Decisions under the Act are taken within an elaborate framework of classification of water resources, which requires determination of the class of water resources, determination of the resource quality objectives and determination of the Reserve. The determination of the class is provided for in Section 12, whereby all water resources in the country are required to be categorized into different classes (on the basis, essentially, of the water quality of, or the extent of human impact upon, the resource). The minister in this process is to determine, for all or part of every significant water resource, resource quality objectives (Sections 12 and 13). A pivotal concept in the Act is the Reserve. This is the "quantity and quality of water required to satisfy basic human needs and to protect aquatic ecosystems in order to secure ecologically sustainable development and use of the relevant water resource" (Section 1). All licensing decisions (whether compulsory licensing or otherwise) have to take into account these determinations (see Section 27, discussed below).

Licences are granted subject to conditions (Section 29), which, for example in relation to taking or storage of water, set out, *inter alia*, the specific quantity of water or percentage of flow which may be taken and the rate of abstraction. In terms of Section 28, licences must be reviewed at intervals of not longer than 5 years (see also Section 49), and the overall licence period may not exceed 40 years (Section 28(1)(e)).

Compulsory licensing itself is provided for in Section 43, which reads (in part):

> (1) If it is desirable that water use in respect of one or more water resources within a specific geographic area be licensed –
>
> (a) to achieve a fair allocation of water from a water resource … –
> (i) which is under water stress; or
> (ii) when it is necessary to review prevailing water use to achieve equity in allocations;
>
> (b) to promote beneficial use of water in the public interest;
> (c) to facilitate efficient management of the water resource; or
> (d) to protect water resource quality,

> the responsible authority may issue a notice requiring persons to apply for licences for one or more types of water use....

In relation to any licensing decision, Section 27 of the Act sets out the factors that must be considered in any such process:

In issuing a general authorisation or licence a responsible authority must take into account all relevant factors, including –

(a) existing lawful water uses;
(b) the need to redress the results of past racial and gender discrimination;
(c) efficient and beneficial use of water in the public interest;
(d) the socio-economic impact –
  (i) of the water use or uses if authorised; or
  (ii) of the failure to authorise the water use or uses;
(e) any catchment management strategy applicable to the relevant water resource;
(f) the likely effect of the water use to be authorised on the water resource and on other water users;
(g) the class and the resource quality objectives of the water resource;
(h) investments already made and to be made by the water user in respect of the water use in question;
(i) the strategic importance of the water use to be authorised;
(j) the quality of water in the water resource which may be required for the Reserve and for meeting international obligations; and
(k) the probable duration of any undertaking for which a water use is to be authorised.

This indicates that any licensing decision (but particularly a compulsory licensing process) requires substantial information relating to the water resource in question and its water users, much of this information having to be determined through complex and time-consuming scientific processes.

Section 27(*b*) contains the so-called equity factor. On the face of Section 27, this is merely one of several stated factors (bearing in mind that the list is non-exhaustive). In several cases, two of which have been taken on review to the High Court, the Department of Water Affairs (the licensing authority) insisted that equity is the overriding factor, to the extent that licence applications were refused because a decision to grant a licence would not further the equity objective. In other words, the decision was made on the basis that this factor was a trump. In those cases taken on review (*Guguletto Family Trust v Chief Director, Water Use, Department of Water Affairs and Forestry*, unreported case A556/10 (GNP), and *Goede Wellington Boerdery (Pty) Ltd v Department of Water Affairs and Forestry*, unreported case number WT W23/02/2009), the courts correctly decided that the department was wrong to take this approach. In *Guguletto*, the court held that the department was "wrong in law" to regard equity as the overriding factor, because the correct approach was that all relevant factors, including the 11 mentioned in Section 27, should be taken into account and then balanced "without attaching due weight to any one" (see, further, Kidd, 2012). Although the courts were correct in holding that the equity factor does not act as a trump, an interpretation of Section 27 taking into account the overall scheme and objective of the Act would suggest that equity does have a heightened role to play (Kidd, 2012). Seemingly because of these judicial setbacks, there are apparently moves to amend the Act to make it clear that equity does play a priority role (Van Koppen & Schreiner, 2014). Whether this involves merely giving it priority or whether it will act as a trump remains to be seen.

In relation to allocation of water rights by means of water licensing, the department has identified the following water allocation priorities (DWA, 2013, p. 47):

(1) The Reserve:
  - Basic human needs (25 lppd)
  - Ecological requirements
(2) International requirements
(3) Poverty eradication and greater racial and gender equity
(4) Nationally strategically important uses (such as power generation)
(5) General economic purposes.

These operate as a hierarchy and would be particularly important in a compulsory licensing process, as will become apparent when examining this in more detail below.

In deciding to initiate a compulsory licensing process, the department draws up an initial Catchment Assessment Report, which describes the catchment status and "is the basis for dialogue with and among stakeholders, including existing and potential water users" (Seetal, 2012, p. 21). In order to compile this assessment, it is necessary for the authority to carry out a process of verification of water uses, particularly in relation to existing lawful use. The process is aimed, ultimately, at determining the true extent of existing lawful use. It first involves validation, to determine whether the volume of water registered by the user is accurate. Users were required to register water uses, which entailed providing information relating to existing water use to the department (and is a different process from licensing, since it did not require any decision – it was an information-gathering exercise). Validation is carried out through a combination of remote-sensing techniques, augmented by information gathered from users in person. Verification entails ascertaining whether the volume of water registered was lawful in terms of the applicable law in 1998 (DWA, 2006). This can be a very difficult process, as observed in the Inkomati catchment by Movik (2009).

This is followed by a reconciliation process, which assesses water demand and availability in the catchment area. It examines "opportunities for beneficial water use, how to undertake the redress of inequities, reducing inefficient water use and recommends appropriate economic sector allocations" (Seetal, 2012, p. 22). These preparatory processes provide the information on which a decision to carry out compulsory licensing or not is based.

Once the authority has decided to initiate a compulsory licensing process in terms of Section 43, it must first publish a notice in the *Government Gazette* calling for applications and take whatever other steps may be necessary to inform affected persons. This includes everyone in the area using water, whether under an existing licence or other lawful use. Once all the applications have been received and considered, the authority draws up a proposed allocation schedule (Section 45), which must:

reflect the quantity of water to be –

(a) assigned to the Reserve and any relevant international obligations;
(b) assigned to meet the requirements of existing licences;
(c) allocated to each of the applicants to whom licences ought to be issued in order to redress the results of past racial and gender discrimination in accordance with the constitutional mandate for water reform;
(d) allocated to each of the applicants exercising existing lawful water uses to whom the licensing authority determines that licences should be issued;
(e) allocated to each of the applicants, taking into account the factors set out in section 27; and
(f) allocated to every other applicant by public tender or auction, subject to any regulation.

There is no obligation to allocate all available water (Section 45(3)). Written objections may be made to this schedule, and after consideration of any such objections, the authority publishes the preliminary allocation schedule (Section 46). Any unsuccessful objections to this schedule may be taken on appeal by "any interested person" (Section 148(1)(*g*)) to the Water Tribunal (established in terms of the Act).

Finally, once any appeals have been disposed of, and any necessary amendments made in response to successful appeals, the final allocation schedule is published (Section 47) and licences issued accordingly. Section 48 provides that any licence issued pursuant to a compulsory licensing process replaces any existing lawful water use entitlement of that person in respect of the water use in question.

## Case studies

As of January 2016, just three compulsory licensing processes have been completed, covering in total a very small part of the country and involving a small amount of the overall water resources. There were supposed to be four pilot processes: the three completed, plus Inkomati, which evidently became too complicated (see below). The reasons for choosing these areas are not publicly available, but it would appear that the four presented a good mix of different dynamics: Tosca Molopo and Jan Dissels both relatively small, with Tosca Molopo involving solely groundwater; Umhlathuze somewhat larger and an important commercial area; and Inkomati a large area, important from an agricultural perspective and also involving international dimensions. Of the overall 10,000 million $m^3$/y available in the country (DWA, 2013), the three completed processes have a total amount of water available of just under 277 million $m^3$/y, amounting to 2.77% of the total water in the country. The three processes, in order of completion, were the Tosca-Molopo, Jan Dissels and Umhlathuze. Table 1 indicates the dates on which the various statutory processes were completed.

### *Tosca-Molopo*

Tosca-Molopo is a largely rural area in the far north of the North West Province of South Africa, bordering on Botswana. Its population is approximately 4500, and its area is 1625 $km^2$ (unless specified otherwise, all geographical information relating to the

**Table 1.** Process dates.

| Area | First notice | Proposed allocation schedule | Preliminary allocation schedule | Final allocation schedule |
|---|---|---|---|---|
| Tosca-Molopo | 12 August 2010 | 17 December 2010 | 20 May 2011 | 22 July 2011 |
| Jan Dissels | 20 August 2010 | 28 September 2012 | 8 February 2013 | 26 April 2013 |
| Mhlathuze | 12 August 2010 | 10 August 2012 | 14 October 2014 | 25 March 2015 |

three areas is taken from Seetal, 2012). This is a rather unusual site from a South African perspective since all the water available in the area is groundwater.

The main purpose of compulsory licensing in Tosca-Molopo was "protection of the aquifer system from overexploitation and the management of unlawful water use" (Seetal, 2012, p. 21) – in other words, water conservation. The total water available in the area was 11.1 million $m^3$/y, while the total volume applied for, in 60 applications, was 14.424 million $m^3$/y. The total proposed compulsory licensing allocation was 9.960 million $m^3$/y (p. 20).

### *Jan Dissels*

The Jan Dissels area is part of the Cederberg, about 200 km north of Cape Town in Western Cape Province. The area contains the town of Clanwilliam, which is on the banks of the Olifants River. Other than the town itself, the area is largely rural, with a population of approximately 4000; its size is 197 $km^2$. The licensing process was carried out primarily in relation to surface water in order to alleviate severe water shortages in the area, primarily in the summer months (p. 21). The total surface water available in the area was 3.81 million $m^3$/y, while the total volume applied for, in 33 applications, was 5.436 million $m^3$/y. The total proposed compulsory licensing allocation was 3.920 million $m^3$/y, including groundwater (p. 20).

### *Mhlathuze*

The Mhlathuze area, in KwaZulu-Natal, about 150 km north of Durban, is by far the largest of the compulsory licensing sites, at 4209 $km^2$. The total area is largely rural, but it has significant urban and industrial nodes, including the large towns of Empangeni and Richards Bay. Richards Bay is one of South Africa's most significant harbours, handling more than 80 million tonnes annually, 60% of South Africa's seaborne cargo (Transnet National Ports Authority [TNPA], 2015). The total population in the area is approximately 525,000, and the compulsory licensing process involved mainly surface water.

The compulsory licensing process in Mhlathuze was necessary to address a situation of demand for water exceeding supply (Seetal, 2012, p. 21). The total water available in the area was 262 million $m^3$/y, while the total volume applied for, in 670 applications, was 401 million $m^3$/y. The total proposed compulsory licensing allocation was 288.088 million $m^3$/y (p. 20).

### *General observations*

Of the three processes, Tosca-Molopo and Jan Dissels were relatively similar in size compared with the much larger Mhlathuze area, which had more than 10 times the number of licence applications of the other two. It is thus not surprising that this process took the longest – five months shy of five years. It is not clear why the Jan Dissels process took at least two years longer than the Tosca-Molopo process.

None of the three processes involved any international water requirements, even though the Tosca-Molopo area is on an international border. It is also noteworthy that there were no appeals to the Water Tribunal in relation to any of the proposed allocation schedules, which would undoubtedly have extended the final resolution of the process.

## Analysis

### *Complexity of the process*

It is clear that a compulsory licensing process is not simple. Once a need for compulsory licensing is established (which itself requires consideration of a large amount of relevant information), it is not then simply a case of comparing the total of available water with the amount requested and allocating pro rata. Once the total amount of water available is known, the authority then first needs to decide (on the basis of relevant scientific information) how much water is required for the Reserve. This would be more difficult in a larger area because it may differ (depending on ecological requirements) in different parts and/or catchments in the area. The human needs aspect of the Reserve would meet the requirements of equity in access to water services but does not at all address the issue of productive use. Once the Reserve has been ascertained, then international requirements need to be factored in (although this was not a factor in any of the cases under review here).

The amounts required for the Reserve and international commitments having been subtracted, what is left is available for allocation. Of all the licence applications received, there is still a need to prioritize in accordance with the hierarchy set out above: first to address equity concerns, then to recognize national strategically important uses, and finally to consider "general economic purposes". These applications would all be for the use of water for productive purposes, since domestic and other *de minimis* uses would be catered for either by allocation to water services authorities (municipalities providing piped water to consumers) or by Schedule 1 of the Act. Schedule 1 sets out water uses that are regarded as sufficiently minor that they do not require licences (see Section 21). As indicated by the statistics pertinent to the case studies, in all cases the amount of water applied for was considerably greater than what was available, which would undoubtedly add to the complexity of determining the appropriate allocations in the light of the strategic imperatives. All decisions will have to involve taking into account all of the factors envisaged in Section 27.

Although not explicitly mentioned in Section 27, it would also be important for the authority to consider possible future demand for water and possible future impacts on the availability of water due to the effects of climate change. Section 45(3) allows the authority to provide an unallocated amount to meet these needs.

It seems clear that the complexity of the process is one of the main reasons the process of water allocation reform, through the compulsory licensing route, is taking so long. The difficulties in even the preliminary steps of validation and verification appear to have led to the Inkomati process being placed on the back burner, for example (Movik, 2009). The department has learnt lessons from these pilot studies (DWA, 2013), and these lessons are likely to facilitate future processes. But existing capacity constraints are unlikely to be alleviated, which means that progress will continue to be slow.

In addition to the scientific/technical difficulty involved, it is also worth bearing in mind that this is not just a desk-based exercise, and it is absolutely critical that affected persons have buy-in, particularly in relation to those processes where equity will be more significant than in the three completed case studies. This requires, as a minimum, ongoing and effective communication with affected persons and applicants.

### *Communication*

On average, most applicants will be receiving smaller allocations than they requested. In order to avoid numerous objections and appeals, applicants need not only to be apprised of the need for the compulsory licensing exercise but also to appreciate the cogency of that need. Seetal (2012, p. 24) explicitly recognizes this, observing that "early stakeholder engagements and buy-in are crucial, especially relating to agreement on the issues to be resolved and proposed time frames", followed by the maintenance of a "regular and concerted communications campaign with all stakeholders for the duration of the process". Judging by the fact that there were relatively few objections – only 19 out of 670 applications in the Mhlathuze case, for example (p. 22) – in the three processes considered above (and no appeals), the authorities appear to have been largely successful in achieving and maintaining the buy-in of the applicants in these cases.

The primary objective of compulsory licensing in all three of the case studies was related to water conservation, which is likely in the South African context to be far less controversial than a process initiated primarily to improve equity in allocations. It is unlikely that these processes would have been carried through with minimal objection had the main purpose been equity-related.

## Equity and alignment with land reform

As pointed out earlier, the Department of Water Affairs has set ambitious targets for allocation of water use rights to black and women applicants. Compulsory licensing will be the primary tool to achieve this. There are, however, not inconsiderable obstacles in the way of reaching these targets. Some of these challenges are obvious. The complexity of the process means that the three processes took an average of just under three years to complete. Numerous further processes will be necessary to bring about the allocation reforms targeted, and capacity constraints (Schreiner, 2013) make this unlikely.

The other main constraint is in ensuring that there are people of the appropriate race and gender who would be applicants in these processes. The equity targets appear not to have taken into account changes in land ownership or control. The pace of land reform (put bluntly, transfer of land from white ownership to black ownership) has been far slower than hoped, and most agricultural land in the country is still under white ownership. Commercial farms (mainly white-owned) occupy approximately 86% of farm land in the country, with subsistence agriculture (located mainly in the previous 'homelands') making up the remainder (Department of Agriculture, Forestry and

Fisheries [DAFF], 2013). In 2007 there were approximately 40,000 commercial farming units in the country (DAFF, 2013). Most water allocations are still in the hands of whites (DWA, 2013).

There is not necessarily a legal connection between the largely white-held agricultural land and largely white-held water allocations. Before the 1998 Act, the riparian rights policy meant that there was a legal connection between property (land) rights and water rights, and many of the riparian rights extant at the time the 1998 Act came into effect are still in force due to the existing lawful use principle. The 1998 Act, however, does not require ownership or control of land in order for water rights to be allocated. While this is not a legal requirement, from a pragmatic point of view it would be most implausible for someone to apply for water rights if he or she did not have at least possession of land on which to use such water productively. What this means is that, until farms are owned (or possessed) in South Africa by the numbers of blacks and women envisaged by the Department of Water Affairs water allocation targets, the targets are unlikely to be reached. (Productive use is not confined to agriculture – it could also be mining and industry – but it is unlikely that the targets specified by the department will be met without a substantial increase in black, and black women, farmers.) In other words, the transfer of at least possession of farms to blacks and women is a necessary precondition for equity reform in relation to water allocation rights.

Even if this were to occur much more quickly than is currently the case, there are other flaws. First, the targets suggest that people will be effectively forced into agriculture, though the overall population impetus is away from rural areas and farming (South African Local Government Association [SALGA], 2014). A second problem is that, increasingly, agricultural land is owned by, and water rights applied for by, juristic persons. It is trite that juristic persons have neither race nor gender, but it is possible that South Africa's Broad-Based Black Economic Empowerment laws (Kidd, 2012) could be used to ascribe, in effect, race to juristic persons. It is less clear how this could be done in relation to women.

A third problem arises in relation to successors in title. The Act (Section 51(2)) provides that a successor-in-title (presumably in relation to landownership) of any person to whom a licence to use water has been issued may, subject to the conditions of the relevant licence, continue with the water use, and must promptly inform the responsible authority of the succession, for the substitution of the name of the licensee, for the remainder of the term (a licence period may not exceed 40 years – Section 28(1)(*e*)). This is simply a clerical process: no administrative decision is involved in relation to successors-in-title, so there can be no control over equity considerations when this happens. The significance of this is that any progress in achieving equity targets in relation to water allocations may be undermined if black or women landowners sell to white men.

## Alternative to compulsory licensing?

Since licensing is particularly resource-intensive (particularly in relation to the South African legal framework, which is elaborate), it is time-consuming and difficult. Compulsory licensing, which involves numerous simultaneous licence applications, is even more complex. If added to other potential drawbacks, such as the opportunities for corruption (Bronstein, 2002), do the shortcomings of licensing militate in favour of a better way? While use of the market may well be conducive to balancing supply and demand, the market will not be able to address equity concerns, as has been observed in other jurisdictions, such as Chile (Bauer, 2004). The market would tend to reinforce existing inequalities rather than evening them out. As long as improved equity remains a primary objective of water management in South Africa, an administrative licensing process, including compulsory licensing where necessary, would appear to be the best way, although not without its shortcomings.

It is also not an option to suggest dropping equity as an objective. Achievement of equality is a constitutional imperative in South Africa, and this includes not only non-discrimination (which is relatively easy to achieve) but proactive measures to remove instances of inequity, such as racially skewed allocation of water rights. Although none of the three case studies involved equity as an underlying reason for the processes (they were all primarily required for purposes of water stress – balancing supply and demand), equity was taken into account during the allocation process, and it will undoubtedly play a more significant role in future processes. The 2013 NWRS makes it clear that compulsory licensing "will be prioritised in areas that are water stressed (over-allocated) and this priority will be based on the need to achieve equity through water re-allocation and to protect water resources from being over-exploited" (DWA, 2013); and the targets set out in the 2008 Water Allocation Reform Strategy, if they remain in place, require compulsory licensing to play a role in equity reform (DWAF, 2008).

## Conclusion

The three compulsory licensing processes already completed in South Africa are really just a drop in the ocean in relation to what will be needed in future to address issues of water stress and the imperative of improved equity in water allocations. The complexity of the processes and the consequent time taken to complete them suggest that South Africa will not be facing a large load of simultaneous compulsory licensing processes in the near future. Progress is likely to be somewhat painstaking, particularly if all the relevant factors are to be properly taken into account.

One of the geographical areas that has been earmarked for compulsory licensing, but which is still awaiting implementation, is the Inkomati region (DWAF, 2003, Appendix D). This area would throw up considerably more difficulties than the processes considered in this article. First, there is a substantial international dimension, with rivers in the Inkomati region being shared with both Mozambique and Swaziland (DWA 2013). Second, as observed by Movik (2009), attempts in the Inkomati region at quantifying the extent of existing lawful water use, a prerequisite for compulsory licensing, proved particularly difficult – no doubt a significant reason why the area has not yet been subject to compulsory licensing. This type of difficulty is likely to be replicated elsewhere, presenting a significant challenge to compulsory licensing in other parts of the country.

While there are currently no new compulsory licensing processes imminent (at least not according to information that is publicly available), it may well be necessary for the statutory process to be simplified in some way. Changing the National Water Act, however, will not overcome the problems of non-alignment with land ownership/land reform. There are apparently moves afoot to amend the Act, and not only in relation to compulsory licensing, but the department is playing its cards close to its chest and drafts of amendment proposals are not in the public domain. This makes it difficult to speculate on what compulsory licensing might look like in the future. With the identified equity and sustainable management imperatives not likely to change in substance, unless there is a radical change to the Act and to the concept of existing lawful use, compulsory licensing will certainly have an important role to play for many years into the future.

## References

Bauer, C. J. (2004). *Siren song: Chilean water law as a model for international reform.* Washington, DC: Resources for the Future.

Bronstein, V. (2002). Drowning in the hole of the doughnut: Regulatory overbreadth, discretionary licensing and the rule of law. *South African Law Journal, 119*, 469–483.

Department of Agriculture, Forestry and Fisheries. (2013). *Abstract of agricultural statistics.* Pretoria: South African Government.

Department of Water Affairs. (2006). *A guide to verification of water use.* Pretoria: South African Government.

Department of Water Affairs. (2010). *Integrated water resource planning for South Africa.* Pretoria: South African Government.

Department of Water Affairs. (2013). *National water resources strategy* (2nd ed.). Pretoria: South African Government.

Department of Water Affairs and Forestry. (1997). *White paper on a national water policy for South Africa.* Pretoria: South African Government.

Department of Water Affairs and Forestry. (2003). *National water resources strategy.* Pretoria: South African Government.

Department of Water Affairs and Forestry. (2008). *Water allocation reform strategy.* Pretoria: South African Government.

Kidd, M. (2012). Fairness floating down the stream? The Water Tribunal and administrative justice. *South African Journal of Environmental Law and Policy, 19*, 25–50.

Movik, S. (2009). *The dynamics and discourses of water allocation reform in South Africa* (STEPS Working Paper 21). Brighton: STEPS Centre.

Movik, S. (2014). A fair share? Perceptions of justice in South Africa's water allocation reform policy. *Geoforum, 54*, 187–195. doi:10.1016/j.geoforum.2013.03.003

South African Local Government Association. (2014). *Draft integrated urban development framework.* Pretoria: South African Government.

Schreiner, B. (2013). Why has the South African National Water Act been so difficult to implement? *Water Alternatives, 6*(2), 239–245.

Seetal, A. (2012). Compulsory licensing 'first ever' projects. *Water Sewage & Effluent, 2012*, 18–24.

Transnet National Ports Authority. (2015). Overview of the port of Richard Bay. Retrieved from http://www.transnetnationalportsauthority.net/OurPorts/RichardsBay/Pages/Overview.aspx

Van Koppen, B., van der Zaag, P., Manzungu, E., & Tapela, B. (2014). Roman water law in rural Africa: The unfinished business of colonial dispossession. *Water International, 39*(1), 49–62. doi:10.1080/02508060.2013.863636

Van Koppen, B., & Schreiner, B. (2014). Moving beyond integrated water resource management: Developmental water management in South Africa. *International Journal of Water Resources Development, 30*(3), 543–558. doi:10.1080/07900627.2014.912111

# Legal pluralism and customary water resources management in Guatemala

Ariella D'Andrea

The existing legal framework for water resources management in Guatemala is obsolete, inconsistent and not enforced. To bridge the gap, many indigenous and non-indigenous communities throughout the country successfully regulate water use through oral or written bylaws. This paper classifies the rules and practices adopted by local communities in order to define their scope and anticipate options to recognize customary water rights in future statutory legislation, as well as under the current legal regime, consistently with the public interest.

## Introduction

According to official data, half of the Guatemalan rural population, most of whom are indigenous, are not connected to a water supply system (Lentini 2010). Despite a clear mandate to provide basic services, local governments cannot guarantee full coverage in rural areas, and subsidized urban services often lack chlorination or sewer connection. Other water uses, such as irrigation schemes, mining activities and hydropower plants, often negatively impact the indigenous population in affected territories. Due to a general lack of water treatment, pollution levels in rivers, lakes and aquifers are alarming.

The vital need for water has caused a proliferation of community-based water committees. With little or no governmental intervention, these committees manage and operate water supply or irrigation infrastructure. Since the 2002 decentralization reform, most of them now operate under the Rural Development Councils system, which aims to incorporate rural communities in the national planning system. Their relations are ruled by custom and practices covering a range of provisions.

The Guatemalan national legal system is based on civil law and subordinates customary law to statutory or formal legislation. Domestic legislation does not fully recognize customary arrangements for water resources management, which are based on traditional concepts of collective property. Despite a 1985 constitutional reform nationalizing all waters, the Civil Code still de facto regulates water rights as individual property rights, without reference to collective or customary rights.

International human rights law protects indigenous peoples' rights, including their rights over natural resources pertaining to their traditional lands. Guatemala has ratified the relevant conventions; therefore, domestic legislation establishes the government's obligation to guarantee indigenous rights over natural resources, including water, and recognize their customary arrangements within the formal legal system. Over the past few years, several authors and international human rights courts have investigated the concept of legal pluralism as a way to recognize indigenous rights and ensure their social, cultural and economic survival.

Any new water law in Guatemala should take into account customary water rights and regulations to comply with international obligations and avert social conflict. How should that be done? Would the recognition of community-based water committees and their traditional water management system realize the rights of indigenous people in water resources, or would a more advanced form of legal pluralism be necessary in the statutory system? If so, would that be possible within the current legal order?

The following discussion aims to define the interface between formal and ancestral rights in water in accordance with international human rights law, while identifying the scope of local customary water regulations. A review of the compliance status of Guatemala with international standards then allows the identification of possible solutions within the existing legal system.

## Customary law and legal pluralism

Custom is historically the first source of law and determines the spontaneous creation of a right, while legislation represents its codified form (Caponera 2007). In contrast to formal or statutory law, customary law is typically oral and depends on two main elements: an objective requirement (i.e. a widespread and long-standing practice) and a subjective requirement (i.e. a sense of obligation to comply). In particular, customary water management law is defined as a "group of non-formal norms and behaviours that are accepted by the community, and that have endured over time in the society" (Iza and Stein 2009).

The traditional organization of community authorities allows customary law to evolve smoothly, as new regulations are agreed to by all community members, based on ancestral principles. Because regulations are internalized and accepted broadly, customary legal systems have a higher rate of compliance than formal ones (Tíu López and García Hierro 2002, Rovere and Iza 2007, Iza and Stein 2009). Internalization comes from the concept of custom, a reiterated practice that becomes a rule by constant application. Therefore, generalized compliance legitimates the authority of customary practices and transforms them into customary law.

May a set of customary rules be considered as a legal system per se? According to Santi Romano's institutional theory (1951, cited in Martines 1992), a legal system is a community organized for the achievement of a common goal. The Latin saying *ubi societas, ibi ius* ("where there is society, there is law") summarizes this concept. A community makes a social pact to obtain collective benefits and avoid conflict, with the goal of soundly managing resources for the common good. This pact legitimizes the customary legal system. "Institution" becomes synonymous with "legal system" (Martines 1992) because an institution always needs norms or rules to define its functions and the relations between its members.

Hans Kelsen's "normativism" (*Pure Theory of Law*, 1966, cited in Martines 1992) defines a strictly hierarchical distribution of norms in the construction of the legal system. In contrast, this branch of customary legal theory is aptly described as "institutionalism" because of the central role human societies play in the production of law. The positivist theory of normativism praises a complete separation between legal science and other disciplines, while institutionalism attempts to close the gap between law and reality by including social and other non-legal aspects in its analysis. This latter approach fits perfectly within anthropologist Leopold Pospisil's concept of legal pluralism, defined as a situation where one or more legal systems coexist in the same social field (Pospisil 1971, cited in Ochoa García 2002).

As a subset of customary law, indigenous law has been defined in Guatemalan legal literature as "the system of norms, procedures and authorities that regulate the social life of indigenous communities and people and allow them to resolve conflicts according to their values, worldview, needs and interests. These systems also include norms that establish how to legitimately change or create norms, authorities and procedure" (Yrigoyen 1999). Indigenous law is then defined as a legal system created by the constant practices of indigenous communities over time. What kind of legitimacy does this customary legal system have within the formal one?

Since the International Labour Organization (ILO) 1989 Convention concerning Indigenous and Tribal Peoples in Independent Countries (C169)[1] and other relevant instruments, international human rights law has promoted a form of legal pluralism within domestic legal orders, moving forward from the evolutionist approach of the 1957 Convention that assumed an ongoing "process of social, economic and cultural *integration* into the national community" (C107, emphasis added)[2].

The Inter-American Court of Human Rights (ICHR) has recognized the right to legal personality of indigenous peoples, thus legitimating the existence and legal force of the customary juridical system within the formal one. In *Saramaka People v. Suriname*, the court argued a right of the Saramaka tribal group from a rural area of Northern Suriname to collective entitlements in their traditional territories, including "the lands and natural resources necessary for their social, cultural and economic survival", as well as the right to "manage, distribute, and effectively control such territory, in accordance with their customary laws and traditional collective land tenure system" (ICHR 2007). The court based its decision on the right to legal personality of individuals established in Article 3 of the 1969 Pact of San José (American Convention on Human Rights),[3] which Guatemala also ratified.

As a result, the court ordered Suriname to adopt domestic measures to "delimitate, demarcate and grant collective title over the territory of the members of the Saramaka people, in accordance with their customary laws, and through previous, effective and fully informed consultations", as well as to "recognize the collective juridical capacity", in order to guarantee the actual exercise of their collective rights on land and natural resources. The government of Suriname was required to guarantee access to formal justice as a community "in accordance with their communal system, customary laws, and traditions" and to consult effectively or seek the favourable opinion of indigenous communities concerning development projects affecting their territories. As reparations, the court required the government to create a community development fund for water supply, housing, education and other development projects (ICHR 2007).

The court's reasoning demonstrates that different legal systems may coexist, if the formal system delimitates the scope of derived or subordinate legal systems. The State's sovereign authority may legitimate existing informal legal systems that have been operating outside its boundaries (Martines 1992). To demarcate the rights and obligations of Guatemalan indigenous people to water and other natural resources, a review of existing community-based arrangements is needed.

## Community-based water management in Guatemala

The establishment of rural water committees in Guatemala started in the 1970s. It is estimated that there are over 10,000 community-managed water systems in rural areas that are built, operated and maintained by local Water Committees, providing water primarily for human consumption (FANCA 2006). Some systems are linked to municipalities through the Rural Development Councils system, but most community systems are managed according to customary rules or agreed internal regulations. Occasionally, irrigation water users have adopted similar solutions, but irrigation is not as common among indigenous people, given the priority of clean water access for domestic purposes. Under different names, community-based water committees are common in Central America where the lack of public services in remote rural areas has obliged communities to act, often with the help of local governments and non-governmental organizations, to obtain drinking water.

Based on oral traditions, specific customary regulations are not always easy to identify and vary greatly even between villages in the same region (FAO 2009). Research on legal pluralism has demonstrated the complexity of customary legal systems that result from a mix of the formal legal system and specific historical relations (Durand Alcántara and Monzalvo 2000, Pinto 2011). In Guatemala, with an estimated population of over 14 million (INE 2011), the issue is more complicated because of the diverse traditional regimes found amongst the 23 recognized indigenous groups, 21 of which are of Mayan descent (Rovere and Iza 2007). Over 40% of the population identifies itself as indigenous, according to the 2002 census (INE 2003), although some sources report 60% (Stavenhagen 2003).

Fifty per cent of the country's forest coverage is found in indigenous territories. A review of primary and secondary sources reveals common principles and at least three types of norms regulating: (1) access to water and other natural resources; (2) water use and protection; and (3) indigenous authorities and their functioning. Compilations of indigenous norms may be found in a number of publications (e.g. CALAS 2002, 2003, 2006, 2007, 2008, Tíu López and García Hierro 2002, FANCA 2006, Rovere and Iza 2007, Gamboa *et al.* 2008, EcoLogic 2010).

Four principles undergird the Mayan lifestyle and its customary legal system: respect, trust, credibility and dignity. These principles, particularly respect, apply to all activities related to water and natural resources use. They also apply in the management of community water funds and in the payment of respective dues. For example, because a main duty of a Water Committee treasurer is to be trusted, honesty is a principal selection criterion for his appointment (Rovere and Iza 2007). Concerning conflict resolution, three general approaches apply: "flexibility" or mediation to reach an agreement, "dynamism" or participation of all affected parties, and "circulation" or compliance in order to avoid conflict.

### *Access regulations*

Water is considered a communal good, meaning it is for collective use, just like the land to which it pertains. No one owns the water, and the sale of water sources, e.g. a spring, is forbidden (CALAS 2002). The right of access to water derives from the traditional possession and use of the land where that water flows. According to the Mayan worldview, indigenous water use rights, as well as rights for the use of other natural resources, are interconnected and deeply rooted in the land tenure system.

Over time, customary access rights have been recognized by the different legal systems that have operated since the Spanish conquest. The formal titles held by indigenous communities vary greatly depending on the time of formalization. Some have acquired individual property rights that they exercise collectively; others have communal property rights granted under the Spanish rule. However, a great number occupy these lands without formal title.

### *Water use and protection regulations*

Indigenous regulations on water use and protection include prohibitions, obligations and sanctions, and they focus on four main water uses: domestic (washing, drinking, bathing); agricultural (irrigation and the watering of animals); construction (houses and other subsistence infrastructure); and cultural and conservation (craftwork, spiritual and environmental uses). The infrastructure needed may vary – from water tanks for harvesting rainwater or for storing water that is manually collected from the source; to communal sinks and basins for drinking or washing; to private and public wells; to basins excavated in the rock to collect running water during the wet season.

In the Mayan worldview, water is the origin of life. Mayan elders believe that "water is the blood of the earth" (EcoLogic 2010) and that "the health of mother earth is known by the greenery of its apparel" (FANCA 2006). The water cycle is understood and respected, even worshipped at the end of the dry season through rain ceremonies that differ according to the region and the specific Maya group (e.g. Rovere and Iza 2007, ASAECO 2009, López García 2010). Mountains are generally respected and protected by indigenous people because of their role in freshwater production: water, trees and mountains are a unit that controls the life cycle (EcoLogic 2010).

Customary provisions regulating the use and protection of water are often integrated with those concerning other natural resources. For instance, a sacred and therefore restricted area may be a forest protecting a water source or a lake that regulates local humidity. According to the Maya, human beings are part of nature and shall not disturb its course more than is needed. Therefore, norms typically dictate prohibitions or obligations aimed at equitable and just water use, to minimize human impact on the environment and ensure water quality for all. Occasionally, when a water-supply system exists, regulations provide for water charges to maintain and operate the infrastructure and ensure water quality. Sanctioning measures also help deal with possible violations of the agreed rules.

### *Institutional arrangements*

Water management arrangements in indigenous communities include indigenous steering committees for natural resources (*juntas directivas de los recursos naturales*), members of the water committees (*comités de agua, comités de operación y mantenimiento*), system operators (*fontaneros*), forest rangers (*guardabosques*) and water judges (*jueces de agua*), mainly for conflicts over irrigation. Other authorities exist, such as the elders (*ancianos*), community mayors (*alcaldes comunitarios*), spiritual guides (*sacerdotes*), sheriffs (*alguaciles*) and midwives (*comadronas*), but are not related specifically to water or natural resources management. Indigenous authorities apply customary law as well as new rules agreed to by the community to solve modern needs.

In indigenous authorities, work is often carried out pro bono or compensated by access to the resource. For instance, participants in the construction or maintenance of the system may be exempted from their dues for a specified time. Decision-making mechanisms may include voting by majority or consensus, and usually provide for a thorough consultation process. Mediation and dialogue precede the application of any sanction. Authorities are perceived as people who bring harmony and peace to the community (Rovere and Iza 2007).

The described norms and principles that govern community-based water management in Guatemala suggest that a customary legal system currently regulates water use and protection when formal legislation is lacking. However, the lack of regularization or formal recognition of customary water rights by statutory legislation produces legal insecurity, where ancestral users may lose their access rights if challenged by new users with a formal title. International human rights jurisprudence consistently recognizes the land rights of indigenous people based on customary law. Knowing that ancestral land and natural resources are intertwined, modern water legislation should provide legal space for their full recognition.

## Formal recognition of Guatemalan indigenous groups

> The recognition of the identity and rights of indigenous people is essential for the construction of a nation of national unity, multiethnic, pluricultural and multilingual. The respect and exercise of the political, cultural, economic and spiritual rights of all Guatemalans is the base of a new coexistence that reflects the diversity of its nation. (Peace Agreement 1996,[4] unofficial translation)

The Peace Agreement of 1996 recognizes the diversity of the Guatemalan population, which is comprised of four groups: the Mayas, the Garífunas (Afro-descendants from the Caribbean), the Xincas (of Nahua-Mexican origins) and the non-indigenous (includes *ladinos* and *mestizos*).

From the diversity of Guatemala's indigenous people, it appears that Guatemala was a pluricultural, multiethnic and multilingual country long before the Spanish conquest, which implanted a monistic, rather than pluralistic, model of the nation-state based on Roman civil law (Yrigoyen Fajardo 1999). The State sought to maintain legally and institutionally a homogenous social order through a strategy of integrating and assimilating indigenous groups into the dominant culture. This meant building a society of one people, one culture, one language and one legal system, ignoring the characteristics of the indigenous people and their ancestral law. The situation has barely changed in recent years (Ziegler 2006). Discrimination against indigenous groups still "permeates the Guatemalan society and characterizes the political and economic regime" at the legal, institutional and structural levels (Stavenhagen 2003, unofficial translation).

As in other colonized areas, European legislation was never applied uniformly and indigenous people were systematically unaccounted for (Hodgson 2004, COSPE 2005). Instead of applying legal security to existing uses, land tenure reforms and the nationalization of natural resources resulted only in the formalization of land usurpation (Martínez Peláez 1994; Knight 2010). After the 1954 coup hindered the adoption of long-awaited agrarian reform, the land redistribution and recognition process initiated under leftist president Jacobo Arbenz was interrupted (Pérez Brignoli 2000). The indigenous people's land tenure rights and rights over the natural resources in those lands remained in a status of legal uncertainty.

The rights of indigenous people on their land were formally recognized only in the 1985 Constitution,[5] along with their lifestyle, custom, traditions and social organization.

Collective land tenure received special protection in Articles 66 and 67. However, the necessary implementing legislation was never adopted. The strong drive for a pluralistic legal system led to the 1999 proposal for a referendum to reform the Constitution to recognize the indigenous customary rights and legal system, including internal conflict resolution within boundaries set by the national legal system, but the proposal was not approved (Ochoa García 2002).

Meanwhile, the 1993 reform to the Law of the Judicial Organ (Decree 2-89)[6] established that statutory legislation is the source of law in the Guatemalan legal system and may be complemented by jurisprudence. Custom may be recognized only by statutory delegation or in the absence of applicable legislation, but must be proven and may not violate moral or public order. Desuetude and *contra legem* custom are not admitted. However, this reform did not specifically refer to indigenous people.

The State's obligations to indigenous people were eventually defined in the Peace Agreements of 1996, signed after the 36-year internal conflict, during which these communities suffered atrocities defined as genocide by the Historical Clarification Commission (CEH 1999, Section 1257). In particular, the Agreements on Identity and Rights of Indigenous People (AIRP)[7] and on Socioeconomic Aspects and Agrarian Situation (ASAS)[8] provide for the rights of these peoples, in line with the Constitution and the C169, both ratified in that same year. Basic rights are spelt out, including: (1) the right to apply customary law for internal conflict resolution; (2) the right to secure their land tenure; and (3) the rights over their natural resources, including water. The AIRP formally recognizes the role of community authorities, "in accordance with their customary norms" but "within the municipal autonomy" (Section IV). Since their ratification by the Guatemalan Congress in 2005 (Decree 52-2005),[9] the Peace Agreements acquired legal force as pending State obligations.

To comply with the Peace Agreements, a decentralization reform was adopted through the "trilogy of laws": the Decentralization Law, the Municipal Code, and the Law on the Urban and Rural Development Councils.[10] The laws recognize the three indigenous groups of Guatemala and create an institutional opening to indigenous communities. According to the Municipal Code (Decree 12-2002), they now may acquire legal personality through inscription at the local municipal register (Article 20) and hence interact and coordinate with the local government through the Community and Municipal Rural Development Councils, which are part of the participatory planning mechanism, the Urban and Rural Development Councils System. Unfortunately, municipal registers are rarely in place. The Municipal Code also recognizes that the internal organization of such communities is regulated by their own norms, values and procedures and carried out by traditional authorities recognized by the State. It then legitimates the way relationships between indigenous communities are regulated, "according to traditional norms and criteria or to the dynamics they generate" (unofficial translation).

Neither the Constitution nor the Peace Agreements expressly mention the general human right to water, although the ASAS includes provisions on the development of water and sanitation infrastructure in rural areas. In 1988, Guatemala ratified the 1966 International Covenant on Economic, Social and Cultural Rights (ICESCR),[11] which expressed the general human right to water. The right to water and sanitation, explicitly recognized in the 2010 UN General Assembly Resolution (A/RES/62/292),[12] should not be confused with water rights regulating different water uses.

Water rights refer to all legitimate uses of water resources and define the terms and conditions of water abstraction and use for all purposes. The right to water defines a minimum standard for all human beings for domestic use, based on the rights to life, non-discrimination, food and health. According to the Committee on Economic, Social and Cultural Rights, water must be "sufficient, safe, acceptable, physically accessible and affordable for personal and domestic uses" (2002, Para. 2). The ratification of the ICESCR establishes a State's obligation to recognize the right to water and sanitation for the whole population and not only for indigenous people.

This article shall review only the status of compliance of obligations concerning natural resources provided for in the Peace Agreements, as well as the right to water, with particular attention to customary water rights, bearing in mind that no specific regulations have yet been adopted to implement fully any of the mentioned rights (Anaya 2010).

## Domestic legislation on water and indigenous natural resources

Responsibility for water management in Guatemala is shared among more than 20 authorities, 5 of which may be considered leading sectoral ministries: Environment, Agriculture, Health, Infrastructure and External Affairs (for shared waters). Over 50 texts of sectoral legislation affecting water resources do not effectively recognize indigenous rights (Aragón 2011). These laws were tailored to the water use they regulate and reflect the viewpoint of the ministry responsible for its implementation. The institutional fragmentation found at the central level is multiplied at the local level by 333 municipalities that carry responsibility for water supply and sanitation services, as well as for environmental and forestry services within their territory. Municipal authorities typically provide subsidized services in urban areas, but do not reach rural areas. Government efforts since 2008 to ensure institutional coordination through a Water Cabinet have led to a few joint initiatives but have not changed the general situation.

Official data indicate that in 2002, over 60% of the population were connected to a domestic water system, the rest having access to either a shared system or natural water sources. In rural areas, coverage falls to about half of the population. Only 40% of the population were connected to a sewerage system and as few as 2% in certain areas (Lentini 2010). Sixty-four per cent of the indigenous population lack access to a water supply system and 80% are not connected to a sewerage system (Stavenhagen 2003). On the other hand, only 4% of the municipal systems are reported to offer purified or treated water suitable for human consumption (Ziegler 2006). Official estimates in 2006 indicated that only 5% of collected wastewater actually received treatment before being discharged into water bodies (Lentini 2010).

Deforestation rates are higher in Guatemala than in Brazil or in Mexico, which threatens water sources in quality and quantity (IARNA 2006). Despite sporadic efforts to improve the situation, the institutional weakness of the main authorities tasked with the protection of water resources, namely the Ministry of Environment and the National Institute of Seismology, Volcanology, Meteorology and Hydrology, does not allow for the development of a proper scientific basis for sound national water management (Aragón, 2011).

Under the Peace Agreements, the rights of indigenous people to natural resources pertaining to their land, including water, comprise four rights: (1) the right to access, including the enjoyment of collective and individual property rights, as well as servitude rights to reach water springs or other resources they have traditionally used for subsistence or spiritual uses; (2) the right to participate in their use, management and conservation; (3) the right to consultation prior to authorization of development projects and to equitable compensation in case of damage; and (4) the right to environmental protection.

These rights should reflect the traditional relationship that these peoples maintain with their environment, recognizing the contribution of indigenous people to environmental preservation (Rovere and Iza 2007). Although no specific legislation has been adopted since the Peace Agreements to secure indigenous rights on natural resources or recognize the general right to water and sanitation, a few dispersed provisions do affect their exercise.

### *Access rights*

According to the 1985 Constitution, the ownership of all waters is vested in the State, introducing an important principle of modern water management and overruling the previous Civil Code regime, which allowed for private and public waters. The task of recognizing and regulating water uses according to social interests is delegated to a specific law (Articles 121 and 127). To date, none of these provisions have been implemented. Several water bills have been presented since 1985, but none have been approved due to opposition from both civil society and the private sector, the former advocating for full recognition and respect of indigenous rights, and the latter lobbying for less restrictive regulations to decrease business costs.

Due to the lack of a water law to define a modern water-rights system and recognize collective water rights in line with indigenous custom, Guatemalans still rely on the 1963 Civil Code[13] of Roman legal tradition (under which water rights are ancillary to the property rights of the landowner) to secure their access to water resources. Private waters may include water springs, rainwater, small lakes and groundwater extracted by artificial means, when these resources pertain to private land. Public waters are defined in the 1933 Civil Code,[14] part of which is still in force, and include navigable rivers, waters that delimit the country's borders and waters pertaining to public land. These waters may belong to national or municipal governments. The code guarantees general access to the population by providing that "all inhabitants" have a common use right on public domain goods, including public waters. Given the lack of institutional leadership at the central-government level, the concession system provided by the Civil Code for the use of public waters is not applied in actuality, although the Ministry of Agriculture has granted irrigation rights in the past (Aragón 2011).

The separation of water rights from land tenure promulgated by the Constitution does not take into account the traditional use of water by indigenous communities and its close relationship to land. The lack of a permit system and an operational authority generates legal insecurity for existing water rights, particularly those acquired after the Constitution. Since 1986, any purchase of land comprising a water source or the drilling of new wells in private estates may be declared unconstitutional. Nonetheless, the "purchase of water sources", referring to the land to which a given source pertains, is currently practiced in rural communities that wish to create a water supply system (Rovere and Iza 2007). To obtain credit for infrastructure, such a purchase may be the sole means of obtaining secure title to the water source and minimizing investment risks. Consequently, indigenous communal rights on water resources are either untitled (Colom de Morán 2005) or protected by private property on land and other land tenure rights, under the Civil Code regime (RIC 2007).

### *Right to participation*

The right to participation in use, management and conservation of natural resources may be exercised through a number of provisions not specifically designed for indigenous people. With a general provision, the Law of the Executive Organ (Decree 114-97)[15] distributes competencies amongst ministries and other government entities and establishes that the

Ministry of Environment shall promote and facilitate the equitable participation of men and women from indigenous and local communities in the sustainable use and management of natural resources, including water.

Among environmental legislation, the Law on Protected Areas (Decree 4-89)[16] creates the Subsystem for the Conservation of Rain Forests, under which private owners may administer protected areas. This specific programme requires a property title and would not apply to communities that occupy the land without a title or with a title issued under previous laws but not converted or recognized under the present land tenure system. In similar cases, the reforestation programme of the National Forests Institute (PINPEP – *Programa de incentivos para pequeños poseedores de tierras de vocación forestal o agroforestal*)[17] is more adapted, as it is aimed at small land possessors who do not have a property or other land tenure title. Also, the Forestry Law (Decree 101-96)[18] provides forestry concessions for the management of State-owned forests and promotes the participation of rural communities in forestry management.

***Right to consultation***

The right of indigenous people to consultation is often a highly conflictive issue when a construction or natural resources development project is to be authorized in their territories. This includes the right to fair and equitable compensation for their losses. The unresolved case of the communities affected by the construction of the Chixoy hydroelectric plant initiated in 1975 is emblematic, with 444 people reportedly killed by paramilitary and military forces, 3,500 displaced and 6,000 families having lost their land and means of subsistence (Arrojo 2009).

In line with international human rights standards, the Peace Agreements require the favourable opinion of indigenous communities before the initiation of a natural resources development project in their territories. Among sectoral legislation, the General Electricity Law (Decree 93-96) and the Mining Law (Decree 48-97)[19] mention only general obligations to use water rationally and respect the environment and the people, but do not establish any specific duty to consult affected communities before initiating the project. The latter law has been criticized for the preferential treatment it grants to mining companies by providing tax exemptions and the free use of water (De Schutter 2009). Though the 2007 regulations on environmental impact assessment (*Reglamento de evaluación, control y seguimiento ambiental*)[20] put on the applicant the burden to consult the population during the elaboration of the project documentation (Article 50), no provisions were found on how to consult affected parties or on the validity of good-faith consultations organized by indigenous communities according to their ancestral rules.

In February 2011, the Ministry of Labour issued Draft Regulations for the Consultation Process of Convention 169, but the population rejected it because they had not been involved in the drafting process and the text did not reflect their own consultation procedures. Also, the UN Special Rapporteur on Indigenous Rights reported that the draft did not comply with international standards (Anaya 2011). Regardless, the lack of regulations establishing a formal consultation process does not remove the obligation to consult (Anaya 2010). As argued by the UN Rapporteur, the good-faith consultations carried out by indigenous communities in line with the Municipal Code "are considered valid and relevant [under international law] inasmuch as they reflect the legitimate aspirations of the indigenous communities" (Anaya 2010), despite the Constitutional Court's opinion that such consultations are not binding because municipalities are not competent to deal with mining issues.

### *Right to environmental protection*

With regard to environmental protection, the Constitution makes a general statement, providing that the State, municipalities and all citizens must cooperate to prevent pollution, maintain the ecological balance and protect natural resources in general (Article 97). It imposes upon water users the duty to reforest the riverside and protect the riverbed. According to the Municipal Code, municipalities have the mandate to promote and manage parks, gardens and recreational areas, to develop tree nurseries for reforestation and to preserve watersheds, rivers and lakes.

The Law on the Protection and Improvement of Environment (Decree 68-86)[21] enacts provisions on pollution control, environmental impact assessment and the defence of a general interest to no contamination through a form of collective action (Article 30), but does not mention water use or indigenous communities. Two sets of regulations implement the Guatemalan environmental law: the mentioned regulations on environmental impact assessment and the regulation on wastewater discharge and reuse (*Reglamento para las descargas y reuso de aguas residuales*, 2006).[22] However, these texts have not yet shown their potential for pollution control, considering the high level of impunity, not only for environmental crimes, registered in the country (OHCHR 2010).

### *Right to water and sanitation*

National legislation does not provide for the general human right to water, nor does it establish priority of water use for domestic purposes over other uses. The Law of Agrarian Transformation (Decree 1551-1962)[23] addresses this matter, although the status of application of its provisions is unknown. Out of six water uses considered, the law places human consumption at number five in the priority order, after flood control, agrarian development zone, irrigation and energy production, and only preceding navigation and other uses (Article 258). However, it allows for the modification of existing water rights on public waters, in case such waters are needed for domestic use or public services (Article 256), as well as for the allocation of public waters and excess private waters to irrigation, domestic use and forestry, among others (Article 249).

The Health Code (Decree 90-97)[24] recognizes the rights to health and food, promotes universal coverage and obliges municipalities to provide communities within their jurisdiction with drinking water in accordance with the Municipal Code. The Health Code also promotes community-based management of drinking-water systems to guarantee sustainable water management (Articles 78 and 79). Communities may also benefit from the declaration of public utility of water sources that the Ministry of Health can adopt to satisfy the water demands of urban and rural populations. The Health Code is implemented by various regulations on water quality, the main competence of the concerned ministry in this field. A ministerial decree (*Acuerdo* 293-82)[25] issued in 1982 regulates the rural water-supply systems by transferring their administration, operation and maintenance to community-based water committees, but it has not been determined whether it has ever been applied (Tíu López and García Hierro 2002).

A later decree on agricultural servitudes (Decree 49-72)[26] declares the establishment of public utility servitudes for the construction of water systems for collective benefit, even beyond domestic use. Municipalities in the interior of the country have reportedly built drinking-water supply systems based on this text (Aragón 2011). Under the Municipal Code, municipal authorities may offer this service directly or grant a concession

(Article 73). Municipalities subsidize water supply in urban areas, but generally do not reach rural communities, who must manage their own water systems (Colom de Morán 2005). This de facto situation creates inequalities in the water charges applied in each community (Colom de Morán 2005), not always respecting the principle of non-discrimination and the right to water, as expressly defined by the Committee on Economic, Social and Cultural Rights.

## Solutions for a new water resources management law

The objective of water management policy is to optimize advantages that derive from the rational management of the available resources (Caponera 2007). In the formal legal system, this is achieved by appropriate legislation that regulates the different uses of water in an integrated fashion, mindful of environmental, social and cultural aspects. Indeed, peaceful cooperation and resource sharing are held as the main achievements of a good water law (FAO 2009). To prevent or minimize conflict, a law should effectively define water rights and provide for their protection.

It is important to distinguish between the right to own water and the right to use it (Caponera 2007). Modern water rights worldwide are increasingly based on the principle of public trust or public ownership, which in civil law countries means that the property of waters rests "under the scope of the government's allocative authority" (FAO 2002) rather than being regulated under private property legislation. Water laws must then provide for a permit and concession system to authorize the use of public waters and provide on the regularization of existing uses. Considering that "customary water rights are frequently rooted in customary *land* law" (Burchi 2005), new water legislation often meets resistance among indigenous communities, precisely because the state is appropriating all waters without considering the unity of their traditional territories.

Customary and formal water rights may interact on four occasions (Burchi 2005). First, during the planning and drafting process of new water legislation, water-right holders may be consulted or may participate in the process. Second, after the adoption of the law and during its transitional phase, the rights of existing users may be recognized in two main forms: *en bloc* recognition, which legitimates all existing uses at once, with or without conditions; and conversion of titles, which requires existing users to obtain recognition by applying for a permit within a limited time. An intermediate option is the so-called "functional recognition" of indigenous water rights on a case-by-case basis (Craig 2009). The third area of interaction occurs after the transitional phase of the law, before granting new titles. The law may allow existing water users to claim their rights by providing for public consultation, but also by requiring the water authority to take those users into consideration when establishing criteria for granting new titles. Inspired by the concept of minimum environmental flows, the notion of "cultural flows" has been proposed as a tool to protect customary water rights (Tran 2009). Finally, the fourth area of interaction occurs when a conflict arises after the granting of a title and during the exercise of the new formal water right. Therefore, the law should provide for effective mechanisms to deal with water disputes, giving special attention to more vulnerable groups.

To recognize existing de facto and titled uses and avoid conflict between water users, particular attention should be given to the second area of intersection. Appropriate provisions should be included in the water law that take into account the needs of the population and the actual capacity of the public administration. Among the three options described above for the recognition of existing water uses, *en bloc* recognition of indigenous water rights would be the most appropriate solution for Guatemala, given the large number of informal rights.

The two other options (i.e. conversion into formal titles and functional recognition on a case-by-case basis) present the inconvenience of overwhelming the water administration with applications. Furthermore, the option of converting customary water rights into formal titles "freezes" customary rights by codifying and integrating them in the formal legal system, thereby changing their oral and flexible nature. Functional recognition is more attentive to the specific needs of each community but might deepen inequalities among the diversity of Guatemalan indigenous people. Conversely, *en bloc* recognition implies the delimitation of existing customary water rights and the demarcation of a "legal space" for them to be played out by the relevant holders, with a degree of autonomy from the formal legal system. This would lead to a form of legal pluralism where statutory law and customary law coexist, beyond functional recognition (Craig 2009).

The law itself could grant de jure legal personality to all indigenous communities to facilitate the recognition process and avoid problems caused by the lack of municipal registers. The law should then make a general provision legitimating existing water uses and specific provisions while protecting the exercise of such rights from water development projects. Collective titles should be granted to each community, in order to allow community-based water management for domestic, agricultural and other uses. Based on reliable information concerning customary rights and uses, a special regime should be put in place to assign a public service function to indigenous communities. This would allow customary regulations to comply with domestic legislation and international human rights law without altering their flexible nature.

Collective water rights differ greatly from other property rights that derive from the Roman civil law tradition, such as the condominium that consists of individual rights over an undivided property. In this case, each co-owner is entitled to dispose of his or her share at his or her own will, whereas collective rights – as defined by indigenous custom – should be neither transferrable nor dividable and should only be exercised collectively. If a person leaves the community, he or she will also lose the right to use water and will not be able to request the settlement of that share. As a consequence, if the community disintegrates, the collective right should return to the State as vested owner of all waters (Pinto 2011).

Although the proposed solution gives a high degree of autonomy to indigenous communities in the management of water resources pertaining to their lands, the strong opposition of some indigenous groups, including the community association of the 48 cantons of the Department of Totonicapán, suggests that further discussion and consultation are needed for a successful delimitation of customary water rights in Guatemala. The historical reasons for the precarious and unequal land tenure situation of indigenous communities in Guatemala (Ziegler 2006) contributes to the difficulty of carrying out negotiations to accurately identify the limits of traditional land and natural resources.

## Conclusion

In a country where water administration is fragmented, legislation dispersed and rural areas neglected by public authorities, good water governance appears to come from the customary legal system of indigenous people. Although wastewater treatment is undeniably an issue due to the high costs involved, self-regulation of rural communities often results in the protection of forests and water sources, the provision of clean water to families for their daily use, the adoption of good practices in water use, the implementation of integrated water management principles and compliance with agreed regulations. This micromanagement however generates inequalities in the price of water-supply services, not always in line with the affordability and non-discrimination principles of the right to water. In addition, it lacks a larger watershed vision that is necessary to ensure sustainable water use at the national level, and it does not achieve the economies of scale which would allow the construction of expensive projects such as wastewater treatment plants or small hydropower projects.

Because existing water-related legislation does not effectively secure customary water rights, appropriate provisions in a future Guatemalan water law should recognize existing de facto and titled uses, in line with international human rights law and jurisprudence, to avoid conflict especially between industrial and subsistence users. The high level of autonomy and effectiveness shown by indigenous communities in natural resources management suggests that collective water rights be recognized *en bloc*, as well as traditional authorities. The new water law should include provisions defining the legal form of community-based water committees and delimitate ancestral water rights for their full recognition.

A clear distribution of competences among responsible institutions and for the amendment of the panoply of sectoral laws that affect water use is needed and deserves the undivided attention of national authorities. Given the highly conflictive context in which the water reform would be adopted, strong institutional leadership is crucial to sketch guidelines for the recognition of customary water management systems within the boundaries of domestic statutory law.

In the interim, a few options are offered by the current legal framework to initiate dialogue for the delimitation of existing water rights. Under current legislation, indigenous communities may legitimize their existence through municipal registries – where they exist – and manage their water supply systems, mainly in the rural areas where municipal service does not reach. Indigenous communities could then be considered water users' organizations that may be granted concessions over municipal waters for the provision of a public service, thus allowing indigenous communities to formalize their rights, pending the discussion of a comprehensive water law.

However, as the majority of water sources are currently private, the most common way for indigenous communities under the current regime to protect their access and use rights on natural resources is by purchasing the land. In practice, in the absence of a water law and of the recognition of collective land property rights, the right to private property becomes the preferred instrument to realize and protect the right to water and the rights on natural resources of indigenous people. Where no water system exists, women and children carrying water directly from the source or washing clothes by the riverside (a common sight in rural Guatemala) are arguably protected under the Civil Code regime by the general right to the common use of public waters.

Guatemala is a pluricultural society and international human rights law opens the door to legal pluralism within the existing legal system. The legal space for full recognition of customary water law is not yet defined and requires further investigation. Hopefully, despite the long-standing discrimination against indigenous people that has limited the exercise of ancestral rights, technical capacity and political will shall soon enough coincide so as to make these changes happen.

### Acknowledgements

This article was prepared during the implementation of an FAO legal assistance project on water management in Central America. The author would like to thank in particular Gloria Aragón for the detailed legal research carried out, as well as Elisa Colom for the selection of the subject. The author would like to extend her gratitude to Stefano Burchi, Gabriel Eckstein and Margret Vidar for the sharp comments made to an earlier version of the paper, and to Deepti Sharma for her review of the manuscript for English language and grammar.

## Notes

1. Convention concerning Indigenous and Tribal Peoples in Independent Countries, C169, Geneva, ILO Conference Session 76; Date of adoption: 27 June 1989, Date of entry into force: 5 September 1991.
2. Convention Concerning the Protection and Integration of Indigenous and Other Tribal and Semi-Tribal Populations in Independent Countries, C107, Geneva, Conference Session 40, Date of adoption: 26 June 1957, Date of entry into force: 2 June 1959.
3. American Convention on Human Rights, San José, Costa Rica, Inter-American Specialized Conference on Human Rights, Date of adoption: 22 November 1969, Date of entry into force: 18 July 1978.
4. Acuerdo de Paz firme y duradera, Guatemala City, Guatemala, 29 December 1996.
5. Constitución política de la República de Guatemala (Political Constitution of the Republic of Guatemala), 31 May 1985, Guatemala City, Guatemala, amended by Acuerdo legislativo 18-93, 17 November 1993.
6. Ley del Organismo Judicial, Decreto 2-89 del Congreso de la República (Law adopted by a Decree of the Congress), Guatemala City, Guatemala, 10 January 1989.
7. Acuerdo sobre identidad y derechos de los pueblos indígenas, Guatemala City, Guatemala, 31 March 1995.
8. Acuerdo sobre aspectos socioeconómicos y situación agrarian, Guatemala City, Guatemala, 6 May 1996.
9. Ley Marco de los Acuerdos de Paz, Decreto 52-2005 del Congreso de la República, Guatemala City, Guatemala, 3 August 2005.
10. Municipal Code (Código Municipal, Decreto 12-2002 del Congreso de la República, Guatemala, 2 April 2002), General Decentralization Law (Ley General de Descentralización, Decreto 14-2002 del Congreso de la República, Guatemala, 11 April 2002) and Law on Urban and Rural Development Councils (Ley de los Consejos de Desarrollo Urbano y Rural, Decreto 11-2002 del Congreso de la República, Guatemala, 12 March 2002).
11. International Covenant on Economic, Social and Cultural Rights, Adopted by United Nations General Assembly Resolution 2200A (XXI) of 16 December 1966, Entry into force 3 January 1976.
12. The Human Right to Water and Sanitation, Resolution 64/292 adopted by the United Nations General Assembly, Sixty-fourth session, 108th plenary meeting, 28 July 2010, General distribution: 3 August 2010.
13. Decreto-Ley 106, Guatemala City, Guatemala, 14 September 1963.
14. Decreto 1932 del Congreso de la República, Guatemala City, Guatemala, 13 May 1933.
15. Ley del Organismo Ejecutivo, Decreto 114-97 del Congreso de la República, Guatemala City, Guatemala, 13 November 1997.
16. Ley de Áreas Protegidas, Decreto 4-89 del Congreso de la República, Guatemala City, Guatemala, 10 January 1989.
17. Ley de incentivos forestales para poseedores de pequeñas extensiones tierras de vocación forestal o agroforestal – PINPEP, Decreto 51 del Congreso de la República, Guatemala City, Guatemala, 17 November 2010.
18. Ley Forestal, Decreto 101-96 del Congreso de la República, Guatemala City, Guatemala, 31 October 1996.
19. Ley General de Electricidad, Decreto 93-96 del Congreso de la República, Guatemala City, Guatemala, 16 October 1996; Ley de Minería, Decreto 48-97 del Congreso de la República, Guatemala City, Guatemala, 11 June 1997.
20. Acuerdo Gubernativo 431-2007, Guatemala City, Guatemala, 17 September 2007.
21. Ley de Protección y Mejoramiento del Medio Ambiente, Decreto 68-86 del Congreso de la República, Guatemala City, Guatemala, 28 November 1986.
22. Reglamento de las Descargas y Reuso de Aguas Residuales y de la Disposición de Lodos, Acuerdo Gubernativo 236-2006, Guatemala City, Guatemala, 5 May 2006.
23. Ley de Transformación Agraria, Decreto 1551-1962 del Congreso de la República, Guatemala City, Guatemala, 11 October 1962.
24. Código de Salud, Decreto 90-97 del Congreso de la República, Guatemala City, Guatemala, 2 October 1997.
25. Reglamento para la administración, operación y mantenimiento de los sistemas rurales de agua potable, Acuerdo 293-82 (cited in Tíu López and García Hierro 2002).
26. Decreto 49-72 del Congreso de la República de Guatemala (cited in Aragón 2011).

## References

Anaya, J., 2010. *Report of the Special Rapporteur on the situation of human rights and fundamental freedoms of indigenous people,* Addendum, *Preliminary note on the application of the principle of consultation with indigenous peoples in Guatemala and the case of the Marlin mine.* Geneva: UN General Assembly, A/HRC/15/37/Add.8.

Anaya, J., 2011. *Carta al Sr. Ricardo Cajas Mejía,* Director Ejecutivo del Consejo de Organizaciones Mayas de Guatemala, con *Comentarios del Relator Especial sobre los derechos de los pueblos indígenas en relación con el borrador preliminar de reglamento para el proceso de consulta del convenio 169.* UN Special Rapporteur correspondence.

Aragón, G., 2011. *Informe final para Guatemala, Diagnóstico politico-legal sobre la gestión de los recursos hídricos.* Guatemala: FAO, GCP/RLA/171/SPA.

Arrojo, P., 2009. *Agua, ríos y pueblos.* Málaga, Spain: Fundación Nueva Cultura del Agua.

ASAECO (Asociación de Agricultores Ecológicos La Laguna de Chikabal), 2009. *Toj Qtxu A' Chkab'al, En nuestra madre agua Chikab'al.* Quetzaltenango, Guatemala: Aporte para la Descentralización Cultural (ADESCA).

Burchi, S., 2005. *The interface between customary and statutory water rights: a statutory perspective.* Rome: FAO, Legal Paper Online #45.

CALAS (Centro de Acción Legal y Social de Guatemala), 2002. *Normas Indígenas sobre el uso del agua, el bosque y la vida Silvestre, Comunidades del área lingüística K'iche'.* Guatemala: Programa de derecho indígena ambiental.

CALAS, 2003. *Normas Indígenas sobre el uso del agua, el bosque y la vida silvestre Q'eqchi'.* Guatemala: Programa de derecho indígena ambiental.

CALAS, Área de derechos colectivos ambientales de Pueblos Indígenas (ADECOPI), 2006. *Normas Indígenas sobre el uso del agua, el bosque y la vida silvestre Comunidad del Área Lingüística Maya Ixil, San Gaspar Chajul, Quiché, Guatemala.* Guatemala: UNDP, PASOC.

CALAS, Área de derechos colectivos ambientales de Pueblos Indígenas (ADECOPI), 2007. *Normas indígenas Maya Q'anojb'al sobre el uso del agua, el bosque y la vida silvestre, el caso del terreno comunal Ojo de Agua Mirasol, San Pedro Soloma, Huehuetenango.* Guatemala: UNDP, PASOC.

CALAS, 2008. *Princípios de derecho indígena maya K'iché sobre el uso del agua, el bosque y la vida silvestre, el caso de San Vicente Buenabaj, Momostenango, Totonicapán, Guatemala.* Guatemala: IUCN.

Caponera, D.A., 2007. *Principles of water law and administration: national and international.* London: Taylor & Francis.

CEH, 1999. *Memoria del Silencio, Informe de la Comisión para el Esclarecimiento Histórico.* Guatemala: Government of Guatemala. Available from: http://shr.aaas.org/guatemala/ceh/mds/spanish/toc.html [Accessed 23 April 2012].

Colom de Morán, Elisa, 2005. *Estudio de los Cambios Legales en el Marco de la Privatización del Agua en Guatemala.* Guatemala: Brot für die Welt.

Committee on Economic, Social and Cultural Rights, 2002. The right to water: General Comment No. 15. Geneva: UN Economic and Social Council, E/C.12/2002/11.

COSPE (Co-operation for the Development of Emerging Countries), 2005. *El Derecho perfecto de las tierras indígenas Sutiabas.* Abangasca Sur-Nicaragua: COSPE.

Craig, D., 2009. The recognition of Indigenous customary law in water resources management. *Journal of Water Law,* 20 (5/6), 278–284.

De Schutter, O., 2009. *Report of the Special Rapporteur on the right to food,* Addendum, *Mission to Guatemala.* New York: UN General Assembly, A/HRC/13/33/Add.4.

Durand Alcántara, C., and Monzalvo, A.S., 2000. Los Sistemas de derecho indio en México (una propuesta de análisis). *In*: M. Castro, ed. *XII Congreso Internacional, Derecho consuetudinario y pluralismo legal: Desafíos en el tercer milenio.* Arica, Chile: Comisión Derecho Consuetudinario y Pluralismo Legal, I, 16–29.

EcoLogic, Junta Directiva de Alcaldes Comunales de los 48 Cantones de Totonicapán, 2010. *Recopilación de conocimientos ancestrales sobre el manejo de los recursos naturales en Totonicapán: la montaña, el bosque y el agua*, versión técnica. Guatemala: Sincronía.

FANCA (Freshwater Action Network Central America), 2006. *Las Juntas de agua en Centroamérica, Valoración de la gestión local del recurso hídrico*, estudio comparado. San José, Costa Rica: FANCA.

FAO, 2002. *Law and sustainable development since Rio: legal trends in agriculture and natural resources management*. Rome: FAO, Legislative Study No. 73.

FAO, 2009. *Law for water management: a guide to concepts and effective approaches*. Rome: FAO, Legislative Study No. 101.

Gamboa, G., *et al.*, 2008. *Esto no es un castigo de Dios. Gestión local del agua en Guatemala: el caso de los municipios de Chiantla y Huehuetenango*. Guatemala: FLACSO.

Hodgson, S., 2004. *Land and water: the rights interface*. Rome: FAO, Legislative Study No. 84.

IARNA (Instituto de Agricultura, Recursos Naturales y Ambiente), 2006. *Perfil Ambiental de Guatemala, Tendencias y reflexiones sobre la gestión ambiental*. Guatemala: Universidad Rafael Landivar.

ICHR (Inter-American Court of Human Rights), 2007. *Case of the Saramaka People v. Suriname* (Preliminary Objections, Merits, Reparations, and Costs), Judgment of November 28. San José, Costa Rica.

INE (Instituto Nacional de Estadística), 2003. *Censo Nacional XI de población y VI de habitación del 2002 del Instituto Nacional de Estadística*. Guatemala: Government of Guatemala.

INE, 2011. *Mapa con proyecciones de población en Guatemala 2011*. Guatemala: Government of Guatemala. Available from: http://www.ine.gob.gt/np/poblacion [Accessed 23 April 2012].

Iza, A., and Stein, R., eds., 2009. *Rule: reforming water governance*. Gland, Switzerland: IUCN.

Knight, R., 2010. *Statutory recognition of customary land rights in Africa: an investigation into best practices for lawmaking and implementation*. Rome: FAO, Legislative Study No. 105.

Lentini, E., 2010. *Servicios de agua potable y saneamiento en Guatemala: beneficios potenciales y determinantes de éxito*. Santiago, Chile: CEPAL.

López García, J., 2010. *Kumix, La Lluvia en la mitología y en el ritual Maya Ch'orti'*. Guatemala: Cholsamaj.

Martines, T., 1992. *Diritto Costituzionale*. Milan: Dott. A. Giuffrè Editore.

Martínez Peláez, S., 1994. *La Patria del criollo*. Guatemala: EEM.

Ochoa García, C., 2002. *Derecho consuetudinario y pluralismo jurídico*. Guatemala: Cholsamaj.

OHCHR, 2010. Report of the United Nations High Commissioner for Human Rights on the activities of her office in Guatemala (Annual Report of the United Nations High Commissioner for Human Rights and Reports of the Office of the High Commissioner and the Secretary-General), Human Rights Council, Thirteenth session, A/HRC/13/26/Add.1, General distribution: 3 March 2010.

Pérez Brignoli, H., 2000. *Breve historia de Centroamérica*. Madrid: Alianza Editorial.

Pinto, M., 2011. *Los dominios originario e indígena – Su reconstrucción histórico-dogmática en el régimen jurídico argentino*. Mendoza, Argentina: Universidad del Aconcagua.

RIC (Registro de Información Catastral), 2007. *Plan de Participación Indígena, Proyecto de administración de tierras, Fase II*. Guatemala: World Bank.

Rovere, M.B., and Iza, A., eds., 2007. *Prácticas ancestrales y derecho de aguas: de la tensión a la coexistencia*. Gland, Switzerland: IUCN, Serie de Política y Derecho Ambiental No. 68.

Stavenhagen, R., 2003. *Informe del Relator Especial sobre la situación de los derechos humanos y las libertades fundamentals de los indígenas*, Adición, *Misión a Guatemala*. New York: Economic and Social Council of the United Nations, E/CN.4/2003/90/Add.2.

Tíu López, R., and García Hierro, P., 2002. *Los Bosques comunales de Totonicapán*. Guatemala: FLACSO/MINUGUA/CONTIERRAS.

Tran, T., 2009. Valuing water in law: how can indigenous cultural values be reconciled with Australia's water law in order to strengthen indigenous water rights? *Journal of Water Law*, 20 (2/3), 144–151.

Yrigoyen Fajardo, R., 1999. El debate sobre el reconocimiento constitucional del derecho indígena en Guatemala. *In*: *América Indígena*, LVIII (1-2), Jan.–Jun. 1998. Mexico: Instituto Nacional Indigenista/Instituto Indigenista Interamericano, Organization of American States (OAS). Available from: http://www.alertanet.org/ryf-americaindigena.htm [Accessed 23 April 2012].

Ziegler, J., 2006. *Informe del Relator Especial sobre el derecho a la alimentación,* Adición, *Misión a Guatemala*. Geneva: UN Economic and Social Council, E/CN.4/2006/44/Add.1.

# Why the Western United States' prior appropriation water rights system should weather climate variability

Jonathan R. Schutz

For the past 160 years, the prior appropriation system of the Western United States has adapted to changing water uses and shifting societal priorities. This paper argues that prior appropriation creates sufficient certainty to allow for economic development and facilitate water markets, yet is flexible enough to adapt to changing priorities and water supplies. Because it possesses both flexibility and certainty, it is a strong enough framework for the Western United States to weather the possible effects of climate change on water supply.

## Introduction

> While one function of law is to give stability to institutions and predictability to the results of action, often the strength of law will lie not in immutability but in capacity for change and flexibility in the face of new forces.
>
> Frank J. Trelease, "Climatic change and water law", in *Climate, Climate Change, and Water Supply*, 70 (1977)

In the Western United States ("West"), the effects of climate change will be felt especially hard in the water sector, both through changes in the amount of water available for use and in its timing. Water delivery systems, which have been designed based on current climate models, may prove inadequate, and stresses on other sectors may multiply. For example, 19% of California's energy use is dedicated to water management – extraction, conveyance, treatment and distribution (California Energy Commission 2005, Krebs 2007). Yet precisely how and when climate change will affect the West is not certain, making adaptability imperative.

In addition to our water supply systems, our water management laws will be tested, including water rights systems. Water supplies will be strained as water regulators are forced to re-allocate scarce water resources. Many suggest that water be allocated to the highest value uses, through water markets (Hobbs 2002, Clyde 2008). While some question whether water markets can or ought to be an allocation option (Dellapenna 2005), they will certainly be at least one tool in adapting to changed climate conditions. For them to function, there must be "well-defined, enforceable and transferrable" water rights specifying quantity (Adler 2008–9; *see also* Hobbs 2002).

The prior appropriation water rights system in most of the states in the West provides such a system. This system developed because of the arid landscape and the need to move water away from water sources to its place of use. Although prior appropriation may not be the ideal water allocation system in every circumstance over the past 160 years, it has functioned well for the West, in "both a regulatory and market environment" providing relative stability of water rights (Tarlock 2001). This paper argues that modifying the prior appropriation system would be much more effective than trying to craft an entirely new system. Taking California as a case study, it concludes that although some improvements are needed, the prior appropriation model is resilient yet flexible enough to cope with the uncertainties of climate change and to facilitate water markets as a means of doing so.[1]

## Prior appropriation water law system

### *Appropriative water rights*

The law of prior appropriation – "first in time, first in right" – was initially developed in California to support the needs of the "49ers", who came to California prior to statehood to mine for gold. The system grew out of the practices that developed based upon the needs, experiences, customs and usage within the early mining camps. Colorado's prior appropriation system began in a similar manner.

Historically, an appropriative right was perfected either by posting a Notice of Intent or by actually diverting water for beneficial use. The date of priority "related back" to the first act taken to initiate the right. The scope of the right was defined by the intended purpose and place of use at the time of initiation and the right existed so long as it was being developed with "due diligence". The quantity of water obtained was the quantity needed to achieve the purpose(s) intended at the time of initiation.

In California, the law of prior appropriation was recognized shortly after statehood (1850) as a means to perfect a water right in California (Irwin 1855). The process by which a party perfects the right, noted above, was codified, requiring a permit from the state to appropriate water. Each Western state has adopted water codes that outline the details of that state's appropriation system and process by which to obtain a permit to use water. For example, after 1914 in California and 1904 in Utah, a party cannot divert water without following the prescribed process and obtaining a permit from a state agency charged with administering water rights.

In 1913, the basic California law of prior appropriation was modified with the enactment of the Water Commission Act that became effective in 1914. This law introduced a more formal application and permitting system. This system required anyone who wished to obtain an appropriative right to file an application with the state and proceed through a permitting process. No right to appropriate water is obtained in this system until a permit is issued although the priority of the right obtained "relates back" to the time the application was accepted. The permit issued specifies the purpose and place of use of water, the amount of water that maybe diverted, and other terms and conditions imposed by the permit. In approving the permit, the state determines whether water is available for the appropriation and whether granting the right harms other lawful users. There is also a determination whether the appropriation is in the public interest. The State Water Resources Control Board (SWRCB) is the agency within California that currently administers this water right permitting system. Many Western states use State Engineer offices for this function.

The law of prior appropriation worked well in the West because it was a satisfactory means to allocate a scarce resource through the granting of relative priorities or rights to all who claimed an interest in water. The first, or most senior, was able to have all of his or her beneficial needs for water met prior to the next in line having any claim to use water, and so forth, until all of the water within the system was exhausted. The right to use water also included the right to divert the water away from the stream (and adjacent lands) for use where needed.

An appropriator must divert water and apply it to beneficial use. The appropriator must exercise control over water by diverting it from the watercourse. (California Trout v. State Water Resources Control Board 1979, Fullerton v. State Water Resources Control Board 1979). All water rights must be put to a beneficial use or the right will be lost. In California, beneficial uses are defined through statute or regulation, although these codified uses are not exclusive. The use of water for domestic, municipal, agricultural and industrial purposes is deemed to be beneficial. The use of water for recreation, including snowmaking, is beneficial.

As under prior appropriation all water rights must be put to beneficial use, wasting water, or using more than is *reasonably* necessary, is not *beneficial* and is, therefore, not permitted under any water right. What is considered reasonable may change over time and has changed to include many more beneficial uses that reflect society's priorities and needs, such as instream fish flows (Tulare Dist. v. Lindsay-Strathmore Dist 1935). This shift in societal priorities is allowed under prior appropriation, though it comes with the economic cost of phasing out old uses through compensation to water right holders (Clyde 2008).

The central feature of the doctrine of prior appropriation is *priority* (Hobbs 2002). In times of water shortage, those with the most senior rights may divert their full supply, then the next most senior appropriator and so on down the priority line. Junior water rights holders are forced to limit or even curtail entirely their diversion of water to satisfy the senior water right holder: "first in time, first in right". Some argue that this rule is not always strictly enforced. For example, regardless of the rule, it would be politically unpalatable to deprive a city of water based on priority. In such a case, parties would likely negotiate a solution around the strict rule of priority, generally based on a preference list established by statute (Tarlock 2001).

In some respects, the appropriative right is flexible because it can be moved to a new place of use, dedicated to a new use, or transferred to a new water user to meet changed conditions. An appropriative right can be transferred to serve lands that did not benefit from the initial appropriation. To move water to a new place of use or purpose of use, however, the water right holder must apply to the regulating state agency. The agency will likely hold a hearing and allow public comment by affected parties. In these situations, a major consideration in allowing the transfer or modification is the impact on others. So long as other appropriators are not harmed, the water right may be modified to serve the new place of use or purpose of use. Typically, the amount of water that is transferred is based on historic consumptive use.

Other flexibilities exist for appropriative rights. An appropriative right may be used outside of the watershed of origin. Also, one can obtain an appropriative right to store water for use at times other than when water naturally occurs. In the West, the right to store water is crucial to the entire water rights system. One can obtain a right to appropriate foreign waters from outside a watercourse, which allows trans-basin transfers.

An appropriator has a right to recapture and reuse water upon lands that were originally intended to be benefited by a diversion. This right extends to the use of water conserved or developed through more efficient methods of diversion, application and use. A major consideration in allowing an appropriator to recapture or reuse water is whether it will harm other appropriators. As long as it will not harm other water users, it should be permitted.

Importantly, all water is owned by the people of the state in which it flows. Because of this, water law typically follows – and is modified to adapt to – societal needs and priorities (Hobbs 2002). The water right issued to a water user is to the use of water. Nonetheless, once a water right is perfected, it is a protected property interest of the holder. The exact nature of the property right is not entirely clear, however (Tarlock 2001, Leshy 2009). Because water is fluid, it is a less well-defined property interest than a piece of ground. But it is a sufficient property interest that it can be transferred from one water right holder to another and cannot be taken from its owner without due process.

An appropriative right can be lost by abandonment. Abandonment is established through proof of non-use coupled with an intent to abandon the water. An appropriative right may also be forfeited through non-use for a statutorily prescribed time period.

### *Public trust doctrine*

The public trust doctrine is the tool states are most likely to use if they seek to constrain historic water rights to benefit the public. This doctrine provides that certain natural resources are held in trust by the state for the benefit of the public. The public trust doctrine evolved in English common law to confer sovereign ownership to the state for "all of its navigable waterways and the lands lying beneath them 'as trustee of a public trust for the benefit of the people'" (Colberg Inc. v. State of California 1967).

Upon their admission to the United States, states obtained title to its navigable waters and underlying lands to be held in trust (City of Berkeley v. Superior Court 1980, National Audubon Society v. Superior Court 1983). The California Supreme Court extended the public trust purposes to include environmental preservation and aesthetics (Marks v. Whitney 1971). The Court subsequently held that water rights are subject to the public trust doctrine. Moreover, the public trust doctrine implies a duty of continuing supervision and the state is empowered to re-analyse water right allocations (National Audubon Society v. Superior Court 1983).

Previously, California courts had applied the public trust doctrine in ways that significantly affected the state's economy and property rights. For instance, it was a public trust doctrine decision of the California Supreme Court in 1884 that helped end the California gold rush. In *Gold Run*, hydraulic miners were diverting the waters of the American River to create high-powered water cannons used to wash away entire hillsides for gold mining purposes. The tailings from these operations entered the American River and were causing several problems, including increased flooding due to the raised riverbed; impairment of navigation; and impacts to water quality rendering the water unfit for domestic consumption (People v. Gold Run Ditch & Mining Co. 1884). The *Gold Run* court found that these mining operations impaired the public trust values of the American River and, on that basis, banned hydraulic mining. The court's ruling effectively prohibited large-scale hydraulic gold mining in California.

One century later, the California Supreme Court again invoked the public trust doctrine in the context of water rights for diversions from non-navigable tributaries to Mono Lake (National Audubon Society v. Superior Court 1983). In *National Audubon*, the court held that water rights were subject to ongoing review under the public trust doctrine.

## The effects of climate change on California's water supply system

As a state that relies heavily on winter snowpack, California presents a good example of how climate change may affect the West. Global climate change is predicted to increase the average global temperature by 3.2° to 7.2°F and may increase temperatures by 3.6° to 9°F in the West (Hall *et al.* 2008). This increase has the potential to greatly affect California's water supply due to its effects on snowpack, which stores as much water as all of California's surface storage reservoirs. Furthermore, warmer temperatures increase the capacity of the atmosphere to hold moisture, leading to the possibility of decreased rainfall. Less rain would obviously adversely affect California's water supply as approximately 75% of California's water falls in the winter in Northern California and the greatest demand for water is in Southern California in the spring and summer months.

Under some models the Sierra Nevada snowpack could decrease by as much as 90%; estimates range from 30% to 90% depending on the amount of temperature increase. With warming, more of California's precipitation would fall as rain instead of snow and the snow that does fall will melt earlier than it does currently (Anderson *et al.* 2008). The reduced snowpack would cause reservoirs to fill more quickly, instead of gradually over the spring and summer months, when supplies are needed most. Reservoirs would then either be less able to cope with flood events due to decreased storage space or would have to pass more water through to preserve space for flood events, decreasing the amount of water available in the summer months when demand is highest (Jacobs and Canger 2007). Reduced spring run-off and storage capacity translate into lower deliveries into the State Water Project and the Central Valley Project and to its water users. More than 20 million people rely on these two projects (Anderson *et al.* 2008).

Moreover, climate change is expected to cause fewer cold days and nights and increase the number of hot days and nights (Ray 2007). It may also affect evapotranspiration rates, which would affect water demand. In California, evapotranspiration is the largest consumer of DWR project water, about 80% in normal years. The length of the agricultural growing season and land use patterns may also be affected (Anderson *et al.* 2008).

Under the driest modeled scenario, one study concluded that

> [o]verall, the climate scenario reduces average annual water availability by 27 percent, which results in an average annual reduction in water deliveries of 17 percent. Statewide, average agricultural areas see water deliveries 24 percent lower than demand targets and average urban areas see 1 percent less than their demand targets. There are great regional disparities as well. Urban Southern California sees almost all scarcity in urban water deliveries; urban water scarcity is almost absent north of Southern California. (Medellin *et al.* 2006)

## Prior appropriation's ability to adapt while providing a secure allocation system

The test of the prior appropriation doctrine posed by climate change will be whether it is resilient enough to provide stability to water allocations, yet flexible enough to deal with changing, likely diminished, water supplies (Trelease 1977). As discussed below, prior appropriation does manage this balance and should be able to weather climatic variability in water supply. Some argue that prior appropriation will take on a different role – more a default rule for small conflicts, a background against which parties can negotiate voluntary solutions in complex conflicts, and a risk allocation model. Whatever its role, prior appropriation will remain the West's allocation system for some time to come (Tarlock 2001).

As previously indicated, prior appropriation is an adequate, but not perfect, system to provide security to preserve water rights, while at the same time providing the flexibility to be able to shift water to where it is needed most (Tarlock 2001). One of the features of an appropriative right that makes it able to cope with change is that it may be transferred to serve water users, lands and uses that did not benefit from the initial appropriation. Furthermore, an appropriative right may be used outside of the watershed of origin. One can obtain an appropriative right to store water for use at times other than when water naturally occurs. In other words, an appropriative right may be shifted to the most beneficial use without the holder of the right losing title to the water right. This ability to shift water to new users, uses and places of use is critical to coping with decreased water supplies from climate change.

One of the overarching principles of prior appropriation, as discussed above, is that all water use must be beneficial and reasonable. What constitutes beneficial and reasonable use may change as conditions change (Tulare Dist. v. Lindsay-Strathmore Dist. 1935). *Tulare* and other cases instruct that what is considered a beneficial use at one time, may, with changed conditions, be considered a non-beneficial use at a later time. Ignoring for now the process that is involved, uses that become non-beneficial may be adjusted or phased out. Therefore, if and when global climate change alters water supply conditions, non-beneficial uses may be phased out and shifted to the most important reasonable and beneficial uses based on existing societal needs and priorities.

Prior appropriation allows for water transfers. The ability to transfer water to an area of greater need or more beneficial use without the loss of the transferor's water right is critical to provide the certainty and flexibility needed to adjust to climate variability (Trelease 1977). Though no state has a formal water market, California and Colorado and other states do have fairly robust voluntary water markets (Hanak 2002, 2003). That is to say, there are many large and small voluntary water transfers between individuals and public agencies. A formal water market would eliminate some of the inefficiencies that exist now in the bureaucratic allocation of water and facilities transferring water from one use and user to another (Trelease 1977). But even without a formal market, informal markets occur now.

In Colorado and Utah, for example, most municipalities require developers to bring sufficient water to the water provider to cover the needs of the development. This typically involves the developer either transferring the water from the developing land to the water agency or the developer purchasing water and transferring it to the water supplier. In either case, water moves from one user to another and often from one place of use to another. These water transfers, involving millions of dollars a year, are made possible by the prior appropriation's defined property rights and allowance of transfers.

The public trust doctrine also provides a potent tool for water users to adapt to decreased water supply from climate change (Rossman 2008). The public trust doctrine includes public goods that should be maintained for the benefit of all the citizens of the state (National Audubon Society v. Superior Court 1983). As discussed above, *National Audubon* determined that the public trust applies to water rights. *National Audubon* did not determine whether the Los Angeles Department of Water and Power (LADWP) diversions should be reduced, but referred it to the State Water Resources Control Board. The State Water Resources Control Board then amended LADWP's licenses to significantly reduce the amount of water that could be lawfully diverted from the streams tributary to Mono Lake. Though the full reach of the public trust doctrine is not known because it has not been applied extensively, it could possibly be asserted in the context of climate change to curb water diversions and uses that have significant adverse affects to the water supply of the state.

Prior appropriation's history has demonstrated that it can adapt to changed societal conditions. For example, California and other Western states have passed laws that encourage water conservation and reuse of treated wastewater. In California, because a water right is for a specified amount of water, any water right holder who conserves water through more efficient means and does not use his/her full water right, is able to move the surplus water to any uses s/he wishes, including through sale (Tarlock 2001). This law should be extended throughout the West, to encourage conservation and to discourage parties from using more water than necessary solely to prevent forfeiture of unused water (Clyde 2008).

A wastewater treatment plant owns the treated water and is able to move the water into recycled water uses, such as landscaping irrigation, freeing up potable water supplies. States have also passed legislation that allows water right holders to dedicate their water rights to instream uses such as fishery flows or recreation and maintain the same priority as the original water right. Each of these laws represents a shift in how in prior appropriation water rights may be shaped according to changing societal priorities and needs.

While the focus of this article is on the existing attributes that make prior appropriation stable yet flexible, there are some changes that would make the doctrine more relevant to current societal needs and priorities. Many commentators have outlined different options to improve the prior appropriation system. For example, John Leshy and Gabriel Eckstein have suggested the following reforms: strengthening the role of states over special districts that handle a majority of high-priority water rights; amending state water laws to allow for instream flows for the protection of aquatic species; adopting "show me the water laws", which require developers to supply water rights sufficient to supply a development; further removing the obstacles to water transfers such as third-party impacts; connecting national agricultural policy to water policy; and managing transboundary waters on a basinwide approach (Eckstein 2009, Leshy 2009). These are all reforms that would allow prior appropriation states to adjust to shifting societal priorities and needs and a volatile water supply.

As stated earlier, this paper does not advocate the adoption of the prior appropriation system as the ideal water allocation system in all circumstances. However, modifying the existing system as outlined above would be much less detrimental than trying to craft a new one because it is the system currently in place that Western societies are built upon, as it has been operating for the past 160 years (Tarlock 2001, Hobbs 2002, Clyde 2008). It has functioned reasonably well in both regulatory and market environments (Tarlock 2001). While there are ways that prior appropriation doctrine may need to be reformed, there is no reason to throw the baby out with the bathwater.

## Conclusion

The West has a rich history of prior appropriation water rights law that, over time, has adapted to changing water use priorities and water availability. Prior appropriation arguably provides defined property rights in water yet has remained flexible enough for water to be transferred from one water user to another and from one place of use or purpose of use to another, without a loss of the underlying water right's priority date. Prior appropriation's defined property right and ability to move the water has allowed for water markets to develop in some states and shift water to the highest use. The beneficial-use requirement and public trust doctrine should allow societies to encourage water conservation and address shifting priorities, if the societies are willing to absorb the financial costs of such a shift. In short, prior appropriation's flexibility and certainty should allow it, and in turn its delivery systems and water users and uses, to weather the effects of climate change on the West's water supply and facilitate water markets as a means of adapting.

## Note

1. California also has a riparian water right system that is not discussed here, as our focus is on the prior appropriation system.

## References

Adler, J.H., 2008–9. Warming up to water markets. Regulation. *Regulation*, 31 (4), 14–17.

Anderson J., *et al.*, 2008. Progress on incorporating climate change in management of California's water resources. *Climatic Change*, 89 (Supplement 1), 91–108.

California Energy Commission, 2008. *California's water-energy relationship*, 1, 8 (November).

California Trout Ltd. v. State Water Resources Control Board, 90 Cal.App.3d 816 (1979).

City of Berkeley v. Superior Court, 26 Cal.3d 515 (1980).

Clyde, S.E., 2008. Marketplace reallocation in the Colorado River Basin: better utilization of the West's scarce water resources. *Journal of Land Resources & Environmental Law*, 28 (1) 49–69.

Colberg Inc. v. State of California ex rel. Dept. Public Works, 67 Cal.2d 408 (1967).

Dellapenna, J., 2005. Markets for water: time to put the myth to rest? *Journal of Contemporary Water Research and Education*, 131 (1), 33–41.

Eckstein, G., 2009. Water scarcity, conflict, and security in a climate change world: challenges and opportunities for international law and policy. *Wisconsin International Law Journal*, 27 (3), 408–461.

Fullerton v. State Water Resources Control Board, 90 Cal.App.3d 590 (1979).

Hall, N.D., Stuntz, B.B., and Abrams,R.H., 2008. Climate change and freshwater resources. *Natural Resources & the Environment*, 22 (3), 30–35.

Hanak, E., 2002. *California's water market, by the numbers*. San Francisco: Public Policy Institute of California.

Hanak, E., 2003. *Who should be allowed to sell water in California? Third-party issues and the water market*. San Francisco: Public Policy Institute of California.

Hobbs, G.J. Jr., 2002. Priority: the most misunderstood stick in the bundle. *Environmental Law*, 32 (1), 37.

Irwin v. Phillips, 5 Cal. 40 (1855).

Jacobs, N. and Canger, J., 2007. Climate change and water in California. *Western Water Law and Policy Reporter*, 37.

Krebs, M., 2007. Assembly committee on water, parks and wildlife. *Water-Related Energy Use in California*, 3, 5 (20 February).

Leshy, J., 2009. Notes on a progressive national water policy. *Harvard Land & Policy Review*, 3 (1), 138.

Marks v. Whitney, 6 Cal.3d 251 (1971).

Medellin, J., *et al.*, 2006. *Climate warming and water supply management in California*. Sacramento, CA: California Climate Change Center.

National Audubon Society v. Superior Court, 33 Cal.3d 419 (1983).

People v. Gold Run Ditch & Mining Co., 66 Cal. 138 (1884).

Ray, A., 2007. *Climate change & water in the West: a look at the 21st century*. PowerPoint presentation to Utah Water Users Association (December 3rd).

Rossman, A., 2008. Bring us laws to match our rivers. *Western Water Law and Policy Reporter* (January).

Tarlock, A.D., 2001. The future of prior appropriation in the New West. *Natural Resources Journal*, 41 (1), 769–771.

Trelease, F., 1977. Climate change and water law. *In:* Panel on Water and Climate (ed.) *Climate, Climate Change and Water Supply*. Washington, DC: National Academy of Sciences, 70–84.

Tulare Dist. v. Lindsay-Strathmore Dist., 3 Cal.2d 489, 567 (1935).

# Adapting water laws to increasing demand and a changing climate

Eric L. Garner

**ABSTRACT**
Stress on water resources throughout the world is increasing, challenging the institutions and laws that govern water use. Groundwater management efforts in Southern California have had remarkable success despite more than a half-century of strained water resources. This article explores the local, self-governing management tools that have functioned within California's extremely complex water rights system and are responsible for water management success in Southern California, and suggests that these tools can be employed effectively around the world.

## Introduction

Water demands are growing, driven by a swelling global population, economic growth and changing precipitation patterns. Traditional water law regimes have struggled to adapt to these obstacles, and leading scholars have acknowledged the need for adaptive water management tools (Burchi, D'Andrea, Eckstein, & Nanni, 2012; UNESCO, 2015).

The California drought has received worldwide media attention, much of it devoted to the dramatic overdraft and subsidence issues in parts of the Central Valley, as well as California's historic lack of substantial groundwater regulation (Barringer, 2011; Belmecheri, Babst, Wahl, Stahle, & Trouet, 2015; El Nasser, 2014; Fears, 2015; Schmit, 2014). The less well known success story in California is that much of Southern California has managed its groundwater resources very well. This article presents the method and process that have been utilized in Southern California to achieve reasonable, flexible and lasting solutions to groundwater resource conflicts in the absence of a legislative structure for managing groundwater and amidst constant hydrological uncertainty, and explains why this method and process can be applied around the world.

## California water law principles

In the United States, as in much of the world, traditional water law has focused on temporal priority or land-based rights, neither of which is useful as the amount, timing and location of water resources shift. The land-based 'riparian' doctrine, which originated in English common law and took hold in the eastern United States, permits the use of the natural flow of water on land that touches a lake, river, stream or creek (*Lux v. Haggin*, 1886). The parallel 'overlying' doctrine permits the use of groundwater on land overlying the groundwater basin (*California Water Service Co.*, 1964). A second system – prior appropriation – was developed in the western United States during the mid-to-late 1800s. Prior appropriation rewards water use. Its principle tenet is 'first in time, first in right': the first to use the water has the senior right. The scope of the right is determined by the amount of water continuously put to reasonable and beneficial use. Under prior appropriation, a senior water right holder can use the entirety of its right, even in dry years and even if no water is left for more junior water right holders. Most western states continue to follow a pure prior appropriation doctrine. California, however, incorporates both prior appropriation and land-based systems (*Lux v. Haggin*, supra; California State Water Resources Control Board, 2015). In California, the rule has traditionally been that land-based right holders have priority over appropriators, although that rule is no longer absolute.

California's water law is perhaps the most complex in the world (*Environmental Defense Fund*, 1980; Littleworth & Garner, 2007; Sax, 2003; Sommer, 2015). It has two separate legal regimes for 'surface' and 'groundwater'. It recognizes both land-based rights and appropriative rights to each type of water, and the interface between these regimes is frequently uncertain. In addition, California law recognizes an environmental use for water through the public trust doctrine, a sovereign right to water for the federal government, and a separate type of water right for Native Americans. Finally, until 2014, there was no statewide regulation of groundwater. Essentially, the groundwater system in California was to pump until you were sued and a court told you to stop.

Yet, against this complex and convoluted legal backdrop, California has also developed local, self-governing management regimes that achieve a 'physical solution' to water use conflicts. These regimes arose from California's reasonable use doctrine, initially set forth in a 1928 amendment to the California Constitution and now codified as Article X, Section 2 (2016), which reads:

> It is hereby declared that because of the conditions prevailing in this State the general welfare requires that the water resources of the State be put to beneficial use to the fullest extent of which they are capable, and that the waste or unreasonable use or unreasonable method of use of water be prevented, and that the conservation of such waters is to be exercised with a view to the reasonable and beneficial use thereof in the interest of the people and for the public welfare.

The 1928 constitutional amendment was adopted by the California legislature in response to a decision of the California Supreme Court in *Herminghaus v. Southern California* (1926). Applying the rule that riparian rights are paramount to all other rights on a watercourse, the state's highest court held that a downstream riparian right holder was entitled to the entire flow of the San Joaquin River to flood pastureland, and could thereby prevent the upstream development of a power project pursuant to an appropriative water right. The decision was widely viewed as paralyzing California's growth because it would, in much of the state, effectively prevent the storage of surface water behind dams, which was needed to serve California's rapidly growing urban population. In response, the California legislature declared instead that all water uses must be limited to the amount of water put to reasonable and beneficial use.

Since the enactment of the 1928 constitutional amendment, the reasonable use doctrine has been applied broadly. In a number of cases, higher-priority existing water uses have been limited based on the consideration of the reasonableness of those uses in comparison with the needs of others that share a common supply, including environmental needs (*City of Lodi*, 1936; *Joslin v. Marin*, 1967; *Light v. State Water*, 2014; *People ex rel.*, 1976; *Tulare Irrigation District*, 1935; *United States v. State Water*, 1986). A court still must begin with priority in comparing land-based rights and competing appropriative rights; however, a court may change the priority of water rights and eliminate vested rights after considering them in relation to the reasonable use doctrine (*City of Barstow*, 2000; *El Dorado Irrigation Dist.*, 2006; *Light v. State Water*, 2014). It is important to note that what is reasonable can vary over time and by location (*Tulare Irrigation District*,supra).

## Southern California's highly variable and uncertain water supply

Uncertainty with regard to water supply is the norm in Southern California. Southern California has a near-desert climate, part maritime and part Mediterranean. This unique climate is found in only a few places in the world – others are Spain, central Chile, western and southern Australia, and the Cape region of South Africa – each characterized by similar latitude, proximity to cold ocean currents and terrain (Schiffman, 2005).

Precipitation in Southern California is not only sparse, it is highly variable by location, season and year. The average annual precipitation across the region is roughly 38 cm. However, coastal areas regularly receive half that amount, while the San Gabriel Mountains, overlooking the Los Angeles Basin from the north-east, may get twice as much (Schoenherr, 1992; Western Regional Climate Center [WRCC], 1995). The vast majority of Southern California precipitation falls during the cold months, with about half of the annual precipitation falling between December and February and 90% between October and April (California-Nevada Climate Applications Program [CNAP], Center for Western Weather and Water Extremes [$CW^3E$], Southwest Climate Science Center and California State Climatologist, Department of Water Resources, 2015). Year-to-year variability in Southern California precipitation is by far the widest of any region in the United States, amounting to half the annual average. Local surface water supplies are likewise seasonal and highly variable.

Despite these issues, people have moved into Southern California continuously for over a century. Between 1900 and 1950, Southern California's population grew by a factor of 17 (California Department of Finance, 2010). Since 1950, it has again quadrupled. Now, approximately 23.5 million people live in Southern California, representing 60% of California's population, and the population continues to grow (California Department of Finance, 2015). A very large economy also has developed in Southern California. If it were a country, California would rank as the 7th or 8th largest economy in the world, with a gross domestic product of $2.3 trillion (California Legislative Analyst's Office, 2015; World Bank, 2015). California's leading industries include the service industry, finance, insurance and real estate. Agriculture, however, dominates the

California landscape – and water use. California agriculture, which supplies nearly half of United States–grown fruit, vegetables and nuts, only accounts for approximately $46 billion in revenue and $100 billion in agriculture-related economic activity (California Department of Food & Agriculture, 2015; United States Department of Agriculture [USDA], 2015). Yet, it occupies 17 million of California's 40 million hectares. Three and a half million of those hectares are irrigated cropland, which uses approximately 80% of the water consumed in California (Thompson, 2009). Even in Los Angeles County, the most populous county in the United States, 21% of the land is farmland, more than two-thirds of which is irrigated (Johnston, 2003).

Increasingly, because of population growth, regions around the world are facing the challenge California faces of providing water for both agricultural and urban needs. Whereas in the past a region's water demand might have primarily been from one use or the other, now agricultural and urban water uses are increasingly close together, and conflicts have followed.

The impact of Southern California's population and economic explosion has been borne largely by its groundwater basins, which continue to supply 35% of the water used in Southern California. This is despite the fact that extraordinary efforts have been made to bring water to Southern California from distant sources, including the Owens Valley, the Sacramento–San Joaquin Delta in Northern California, and the Colorado River. Transporting this water has required feats of engineering, including literally making water flow uphill, created legendary conflicts, and enabled continued growth despite prolonged droughts. But, even with these elaborate projects, the imported water is expensive, and pumping groundwater is convenient and cheap. In addition, groundwater has generally been reliably available, even in dry years when surface water was scarce.

Southern California's heavy reliance upon groundwater has continued despite a lack of information about the extent of the supply. Today, groundwater basins can be extensively modelled, for example with NASA's Gravity Recovery and Climate Experiment (GRACE) satellites, but that is a recent development. And even with this technology, the state of groundwater levels is often not readily apparent to an overlying landowner drilling a new well or straining an existing one.

All of these features make Southern California's groundwater a classic common pool resource. When coupled with California's historic lack of groundwater regulation, Southern California groundwater basins were destined for the same fate as the pasture described in Garrett Hardin's classic (1968) article on the 'tragedy of the commons':

> Picture a pasture that is open to all. It is to be expected that each herdsman will try to keep as many cattle as possible on the commons. Such an arrangement may work reasonably satisfactorily for centuries.... Finally, however, comes the day of reckoning.... At this point, the inherent logic of the commons remorsefully generates tragedy.... Each man is locked into a system that compels him to increase his herd without limit – in a world that is limited. Ruin is the destination toward which all men rush, each pursuing his own best interest.

It is worth noting that the tragedy of the commons is not a new problem. Over 2000 years ago, in his *Politics*, Aristotle identified the crux of the common pool resource management issue:

> What is common to the greatest number gets the least amount of care. Men pay most attention to what is their own; they care less for what is common; or at any rate they care for it only to the extent to which each is individually concerned. Even when there is no other cause for inattention, men are more prone to neglect their duty when they think that another is attending to it.

Southern California's growing population, open access to groundwater supplies, lack of regulatory constraints, and illusion of limitlessness combined to build a thirsty oasis of farmland and residential communities with a robust economy, all dependent on groundwater. Even as the pool of groundwater users grew, individuals failed to limit their personal use because they had to assume that if they did not take it, the water would be extracted by another. Over-extraction was the logical outcome (Ostrom, 1990).

## California's solution

While Southern California's groundwater basins illustrate the logical progression of unregulated and uncertain common pool resources towards overuse and depletion, the basins also have been fertile ground for developing a process of self-governance that has effectively broken that trajectory. The process developed and put to use in California is a management tool called a 'physical solution'. California institutions created by physical solutions have been praised by Nobel prize–winning economist Elinor Ostrom. In fact, they were the subject of her dissertation, *Public Entrepreneurship: A Case Study in Ground Water Basin Management* (1964) and inspired the inquiry addressed in her seminal book, *Governing the Commons* (1990).

A physical solution resolves water disputes by creating a combination of technical, legal and institutional structures to support continued use of a basin or watercourse for all users. In California, the development of a physical solution typically is initiated as a result of litigation. However, the majority of physical solutions are implemented voluntarily by agreement of the parties. Importantly, physical solutions depend upon the participation of the majority of stakeholders, including major stakeholders, and require a commitment to an ongoing, year-after-year process.

Physical solutions are born out of the reasonable use doctrine codified in Article X, Section 2 of the California Constitution. Consistent with this doctrine, physical solutions seek to maximize reasonable and beneficial uses while ensuring that the groundwater basin is maintained so as to ensure the continuance of the water supply (*City of Lodi*, 1936). To this end, 'reasonableness' requires all existing uses to be compared and may limit existing rights.

Why and how, in the face of water users' self-interest, have physical solutions in California been so successful? In *Governing the Commons*, Ostrom asked how individuals who share a common pool resource can avoid Hardin's tragedy by organizing and governing themselves to obtain continuing joint benefits in the face of temptations to "free-ride, shirk, or otherwise act opportunistically". Ostrom (1990, p. 90, Table 3.1) identified the essential elements of institutions that have successfully done just that:

(1) Clearly defined boundaries of both the common pool resource and the participants who have rights to use it
(2) Congruence between rules and local conditions
(3) Collective-choice arrangements that enable most individuals affected by the operational rules to participate in modifying the rules over time
(4) Monitoring of the resource conditions and the participants' behaviour
(5) Graduated sanctions
(6) Conflict-resolution mechanisms
(7) The right of participants to self-govern without challenge by external governmental authorities
(8) Nested enterprises.

Ostrom observed that successful California physical solutions have met each of these criteria. The relevant areas include the Raymond Basin, West Coast Basin,[1] Central Basin and Upper Los Angeles River Area Basin, which have now been successfully governed by physical solutions for the past 71, 60, 52 and 36 years, respectively. The physical solutions all contain allocations and adaptive management provisions which are supervised by a court. An appointed committee or engineer (called a watermaster) in each basin collects ongoing data on basin conditions which are reported annually, also to a court.[2] The state government's confidence in the self-regulating agreements reached in these basins is evidenced by their exemption from the requirements of California's 2014 Sustainable Groundwater Management Act.

The success of California physical solutions at incorporating Ostrom's elements into self-governing groundwater institutions is attributable at least in part to what has become a well-established process: (1) perform a technical analysis of the groundwater basin; (2) apportion groundwater allocations at the level of the basin's safe yield on the basis of water rights and the doctrine of reasonable use, allowing gradual pumping reductions to safe yield; and (3) create a system of continuing oversight to enable ongoing adaptive management of the basin. With an established process in place, it behoves all stakeholders to be involved in the process.

### *Technical analysis*

The first step of the physical solution process is a technical analysis of the groundwater basin. During this step, a committee of experts formed by the stakeholders (usually consisting of all the hydrologists working for the stakeholders in the basin) delineates the boundaries of the basin, identifies the water producers and potential claimants, and determines the basin's sustainable supply (usually called the safe yield or sustainable yield). This step is often completed without stakeholders, lawyers or policy makers in the room and is based as much as possible on the best available science. With direct communication between experts, and by hopefully limiting posturing or advocacy, experts frequently can agree on many of these technical issues with a wide level of consensus. This process can lead to the parties arriving at a "single, authoritative 'image' of the problem they face" (Ostrom, 1990, p. 112). These technical conclusions provide the basis for the rules, including water allocations, that will be developed by the stakeholders and help identify projects that may be needed to maximize use of existing water supplies or access new water supplies.

### *Allocating water pursuant to water rights and the doctrine of reasonable use*

After the technical issues are generally agreed upon, the next step is for the basin's stakeholders to agree to a sustainable allocation of the basin's groundwater. The allocation must not exacerbate the damage done by prior and existing water use but instead maintain the groundwater basin as a sustainable source for the future. To this end, if basin pumping exceeds the sustainable supply, physical solutions gradually curtail extractions until they fall within the basin's safe yield. Generally, physical solutions limit pumping to safe yield within five to seven years, although the greater the over-pumping, the longer it takes. In California basins, the impact of limiting pumping has usually been mitigated by importing additional water from outside the basin or through projects to increase the efficiency of water use.

Starting with water rights priorities and generally allowing existing uses to continue (or creating a market to transfer an existing use for compensation) has been essential to the success of physical solutions. No overdrafted area starts with a blank slate; usually, all of a basin's water has already been allocated, particularly in dry areas. Stakeholders will not submit to a process that might immediately extinguish their current claim of right or priority, and stakeholder participation is essential to success.

The general acceptance of existing water rights priorities and uses is where, at least to some extent, 'best practices and principles' collide with the reality that people will go to great lengths to maintain a right to water. As is often said in California, "Whiskey is for drinking; water is for fighting over."[3] It is in the context of this reality that California's physical solutions neither eliminate nor cause material harm to existing water uses. Rather, physical solutions alter the amount, nature and priority of existing water uses only as necessary to carry out the mandate of reasonable use (*City of Barstow*, 2000). From a policy or economic perspective, this may not be the ideal way to allocate water – the resulting allocations may be based on prior overuse and may not reflect the allocations that would be made if the slate were blank – but it is the only realistic way to reach a timely resolution of these serious issues.

Despite the limitations acknowledged above, it is crucial to point out that the initial allocation in the physical solution is only a starting point. Physical solutions generally provide for the transferability of allocated water rights. For example, in the Raymond Basin, the physical solution allowed those whose water rights were most adversely affected to purchase rights from others, and most of the smaller right-holders in fact sold their rights to the water companies in the basin who had higher-value uses for the water (Ostrom, 1990).

Physical solutions can also be used to create 'new' water. New water can be created through the transfer of rights, reducing transmission loss, making existing uses more efficient, recycling, desalination, stormwater capture, and importation. This new water increases the amount of water that can be allocated to users and mitigates the impacts of pumping reductions. Physical solutions sometimes impose the costs for the new water on those with more substantial water rights allocations (*Rancho Santa Maria*, 1938). Thus, physical solutions can incorporate supply- and demand-driven approaches; the short-term priority is to develop sufficient supply to meet existing demand, and the long-term priority is to increase efficiency and supply or reduce demand so that use stays at a sustainable level.

### *Continuing oversight and adaptive management*

It is also very important to ensure that the physical solution provides for continued oversight and adaptive management. In California, stakeholders in physical solutions have commonly created a committee or appointed an engineer, called a watermaster, with substantial authority to monitor basin conditions and water use, to sanction use that violates the terms of the physical solution, and to resolve conflicts that may arise. The watermaster is typically subject to court supervision and makes an annual report to the court. The watermaster is often also given flexibility to make recommendations to modify the physical solution if that is necessitated by changed circumstances.

Flexibility is paramount. The Central Basin recently revised its watermaster oversight structure to provide increased flexibility, with the goal of more optimally utilizing its basin. The change was motivated by the desire of Central Basin water rights holders to store water in the basin for later recovery. The state agency that had served as watermaster for more than 50 years was replaced with a three-arm watermaster structure: (1) an administrative body for accounting and reporting; (2) a panel to enforce issues related to pumping rights; and (3) a panel to consider and approve groundwater storage efforts.

Watermasters have in fact been instrumental in the continued success of California's physical solutions. In the Chino Basin of Southern California, land subsidence due to basin compaction began to damage existing infrastructure. The watermaster established by the Chino Basin physical solution recognized the problem, conducted a technical investigation, and developed an adaptive management programme to minimize the risk of future subsidence. In sum, the watermaster provides a ready forum and process to address issues that inevitably will arise. The watermaster's continuous collection of technical data ensures that stakeholders will be well informed and more able to adapt to changed circumstances when they do arise.

## International application

California's physical solutions have been effective and lasting in the face of a limited legal framework and uncertain and overused water supplies. Today, water users around the world, particularly those in developing nations, are facing similar struggles. Internationally, the demand for water is growing. There are now 7.3 billion people on the planet, up from 2.5 billion in 1950, and each is using more water than ever before (United Nations, 2015). Per capita water use is increasing at twice the rate of population growth, fueled by migration to cities and movement into the middle class (Garner, 2014; Leb, 2015). In addition, the environment is receiving a greater allocation of water, even within fully appropriated watercourses. At the same time, climate change has made weather more extreme and less predictable. Both flooding and drought are more common, and water supply is more difficult to forecast (Getirana, 2015; Treidel, Martin-Bordes, & Gurdak, 2012; Trigo, Gouveia, & Barriopedro, 2010; UNESCO, 2015). And, despite the tension between water supply and demand, much of the world lacks meaningful regulation of water use.

Governance – both legal and institutional – must adapt to meet these challenges. But, because much of the world's water has already been allocated, significant changes to legal regimes are rare, as they often mean that existing users must face reallocation of their current water supplies (Kamran & Shivakoti, 2013). The physical solution process generally starts with current uses and water rights priorities and gradually modifies them, using the concepts of equity and reasonableness, to get to sustainability. For the following reasons, the physical solution model seems well suited to international use, including transboundary use and use in developing nations, and can be applied to any watercourse with delineable boundaries.

First, the concepts of equity and reasonableness employed in California physical solutions have already been embraced by international law as the "cardinal rule ... applying to shared water resources" (McIntyre, 2011). The United Nations calls the requirement that watercourses be used, developed and protected in an "equitable and reasonable manner" the "cornerstone" of its 1997 Convention on the Law of Non-navigational Uses of International Watercourses (1997). Much like California's doctrine of reasonable and beneficial use, the UN explains that its equitable and reasonable requirement is intended to "attain[] maximum possible benefits for all [water users] and achiev[e] the greatest possible satisfaction of all their needs, while minimizing potential detrimental impact" (Comment 5.1.2). Factors to be considered in determining equitable and reasonable use include geography, hydrology, ecology, social and economic needs, the people dependent on the watercourse, the impacts of the water use on others and the availability of alternative uses (Article 6).

The UN's doctrine of reasonable and equitable use is specifically limited by obligations to protect ecosystem needs and vital human needs, set forth in Articles 20 and 10, respectively. Therefore, much like in California's doctrine of reasonable use, existing uses may be modified to account for the other common users of a water source and the needs of the environment.

Furthermore, the UN's doctrine of equitable and reasonable use is to be implemented in conjunction with the obligation not to cause significant harm to other water users (Article 7). The UN has explained that, read together, these requirements signify that equitable and reasonable use may cause significant harm to another water user only if all appropriate measures to prevent such harm have been taken (Comment 7.1.2). Thus, Articles 5, 6 and 7 seem to provide the same assurance afforded by California law, i.e., existing uses and priorities will be honoured, except where necessary to meet the equitable and reasonable requirement. While this assurance restricts the types of solution that can be reached, it makes a solution more likely, because stakeholders should be more willing to participate.

Second, physical solutions are a flexible and transferable tool because they reflect the specific attributes of the subject watercourse. Ostrom (1990, p. 92) cites "congruence between appropriation and provision rules and local conditions" as an element indicative of a successful self-governing arrangement. This element resonates in the context of international, transboundary watercourse management, where watercourses and their users have unique geographical, cultural and legal features and constraints. What works for one watercourse will not necessarily work for another. But physical solutions are not a one-size-fits-all proposition. Rather, the physical solution process ensures that the unique technical specifications of a particular watercourse are determined and considered before any other decisions are made. International scholars agree that this technical analysis is paramount to achieving successful governance (Aureli & Eckstein, 2011; Sugg, Varady, Gerlak, & de Grenade, 2015; UNESCO, 2015). Only after there is at least some level of consensus on this information can stakeholders in the physical solution process negotiate the self-governing structure that will regulate their shared resource. And, because all stakeholders are involved in negotiating the legal and institutional structures that will be erected by a physical solution, the ultimate system of governance should reflect the cultural and legal climate of the region. Finally, recognizing that conditions change, physical solutions often institute ongoing oversight that both recognizes when changes occur and provides a roadmap for adapting governance to account for the change.

Lastly, real-world examples demonstrate that implementation of the physical solution model outside of California and the groundwater context is possible. Processes that share some or several elements with physical solutions have already instituted systems of self-governance for a variety of watercourses in diverse regions of the world (Barry, Obuobie, Andreini, Andah, & Pluquet, 2005; Lopez-Gunn, 2003; Sugg et al., 2015; Tucker, 2014). The following are notable examples.

### *Montaña Campara reserve, Honduras*

A small-scale yet significant story of collective action and self-governance comes from the Montaña Campara region of Honduras. In a 2014 case study, Catherine M. Tucker examined the decades-long process that led to the creation of a land reserve and the establishment of institutions to manage, monitor, enforce sanctions and resolve conflicts for the benefit of a communal spring. Tucker's study illustrates the importance of stakeholder commitment and demonstrates that the technical analysis and equitable allocation of water resources called for by the physical solution model can be achieved with limited resources.

The communal spring is on a mountain that lies within three *municipios* (municipalities that closely resemble a United States county) in the poorest region of Honduras, where it supplies water to roughly 2000 households in the *municipios*' indigenous, agricultural communities. At the inception of the study, Tucker considered a grassroots solution to sustain spring access unlikely for at least four reasons:

- The three *municipios* that occupy the spring area represent distinct indigenous communities, with separate religious and agricultural traditions and governance institutions, and had been entangled in boundary disputes for more than a century.
- Users of the spring were divided over whether to form a land reserve.
- Creation of the reserve was dependent upon farmers on the mountain being displaced and abandoning their land claims.

- There were no monitoring mechanisms or enforcement structures in place when negotiations began.

In addition, the *municipios* and their residents have very limited financial and technological resources. But one significant element was in place: a collection of water committees. Water users had begun forming the water committees in the 1970s to address concerns about expanding populations and water quality, each committee representing a small group of settlements. In the late 1980s, the water committees began working together to build and maintain water projects to store and transport water from the spring. At the same time, the water committees began designing and enforcing rules governing the construction and maintenance of the water systems and the payment of modest water fees. When agricultural practices on the mountain began threatening water quality and supply in the 1990s, the water committees led the effort to establish a land reserve. Water committee meetings, which were mandatory for every head of household who uses spring water, brought all of the stakeholders together and provided a forum for reserve supporters and objecting farmers to voice their concerns. Water committees reached out to international organizations – such as the Lempira Sur project of the UN's Food and Agriculture Organization and Spain's Solidaridad – for resources, including mapping of the mountain's watersheds and other technical assistance. During the course of negotiations, the water committees developed rules to promote equity, for example by conditioning water use on the fulfilment of maintenance responsibilities. Water committees also instituted monitoring and conflict-resolution mechanisms. Tucker observed that stakeholder participation in the process permitted "the rules [to evolve] to fit the local circumstances and to meet people's needs". Ultimately, a reserve was created with "clearly demarcated boundaries and well-defined user groups". While technicalities have prevented parts of the reserve from being recognized by national law, Tucker reported that the agreements between the local stakeholders have provided the water users with confidence that the reserve will be enforced.

Despite deep conflict, lack of a pre-existing regulatory system, and modest resources, the indigenous communities of Montaña Campara achieved a solution to their shared water source conflict using a process that closely resembles the California physical solution. The Montaña Campara negotiations were supported by technical data and mapping. Water resources and maintenance responsibilities were allocated on the basis of equitable considerations. And monitoring, enforcement and conflict-resolution mechanisms were developed and have been implemented. Neither a broad oversight institution nor a specific adaptive management structure was put in place. However, Tucker has observed that the agreements at the water committee and municipal levels have provided stakeholders with confidence of enforcement, and predicts that ongoing stakeholder participation through the water committees will support adaptation of the agreements as required.

### *Guarani Aquifer, South America*

The Guarani Aquifer is one of the most massive groundwater basins in the world, with a freshwater storage volume of approximately 30,000 $km^3$. It is located in a heavily populated and rapidly developing region of South America that includes portions of Brazil, Paraguay, Uruguay and Argentina (Sugg et al., 2015). In 2003, concerns over localized pollution and over-extraction led the four overlying countries to begin discussing joint management of the Guarani Aquifer and its water resources. The discussions involved stakeholders at various levels (national, local and community) and focused on (1) increasing scientific knowledge about the aquifer, (2) establishing a permanent monitoring system, (3) creating a transboundary diagnostic analysis and a strategic action plan, and (4) identifying basic guidelines for integrated management of the resources (Sugg et al., 2015). In 2010, the four overlying nations signed the Agreement on the Guarani Aquifer, establishing a cooperative-governance regime (Agreement on the Guarani Aquifer, 2010). The agreement includes the concepts of reasonable and equitable use and no harm (Articles 3, 4, 6 and 7). Some scholars have been sceptical of the agreement, calling it "bare-bones" (International Water Law Project Blog, 2010), but others have praised it for "harmoniz[ing] multinational interests" and "provid[ing] foundational principles and a means for resolving conflicts" while "allowing for locally specific institutional arrangements" (Sugg et al., 2015). To date, only Argentina and Uruguay have ratified it.

Much is left to be done to ensure that the Guarani Aquifer is sustainably managed. Yet, the agreement has laid the groundwork for a physical solution–like process by bringing stakeholders to the table, emphasizing the importance of data collection and technical analysis and cooperative use of that information (Articles 8 and 12), and establishing the doctrines of equitable and reasonable use and no harm as criteria for future water use (Articles 3, 4, 6 and 7). However, given its emphasis on the sovereign territorial control of each country over their respective portions of the Guarani Aquifer system (Articles 1, 2 and 3), delegation of authority to a single adaptive management institution may meet resistance.

### *Volta River basin, West Africa*

The Volta River basin occupies about 400,000 km$^3$ in Western Africa across six nations – Burkina Faso, Ghana, Togo, Benin, Mali and Côte d'Ivoire (Barry et al., 2005). These nations are among the poorest in the world and have had next to no water management or regulation. The region has experienced substantial and sustained population growth. Approximately 24 million people live in the basin, up from 14 million in 1990, and the population is expected to continue to grow at the high rate of 2.5%. Water sources in the basin include rainfall, streams, rivers, lakes, groundwater and reservoirs. The region's dominant activity, agriculture, has been predominantly rain-fed, and as a result has been greatly impacted by the effects of climate change. Recently, the region has experienced both flooding and droughts, and these extreme weather events are predicted to increase in severity and frequency.

Recognizing the strain that these conditions place on the Volta Basin, the basin nations have embarked on a process to develop and implement an institution to sustainably manage their shared resource. Much like the California physical solution model, the Volta Basin nations have prioritized technical analysis, adopted equitable and reasonable use and no harm standards, and established an oversight committee to, among other things, promote permanent consultation tools and ongoing adaptive management.

The process began in 2004, when Ghana and Burkina Faso signed a joint declaration acknowledging the Volta Basin's water problems. This spurred the six basin nations to begin developing a transboundary water management institution. By July of 2004, the six nations had formed the Volta Basin Technical Committee. In 2007, the nations signed a convention that, among other things, created substantive and procedural obligations to govern use of the basin's water resources. The Volta Convention includes the requirement that basin water be used in an "equitable and reasonable manner" and obligates users not to cause damage.

The Volta Convention also created the Volta Basin Authority (VBA), which is made up of the Assembly of Heads of State and Government, the Council of Ministers in charge of Water Resources, the Forum of the Parties involved in the Volta basin development, the Committee of Experts, and the Executive Directorate of the Authority. The Volta Convention (Title III, Article 6) sets forth a substantial mandate for the VBA:

- Promote permanent consultation tools among the parties for the development of the basin
- Promote the implementation of integrated water resources management and the equitable distribution of the benefits resulting from their various utilizations
- Authorize the development of infrastructure and projects planned by the stakeholders and which could have substantial impact on the water resources of the basin
- Develop joint projects and works
- Contribute to poverty alleviation and the sustainable development of the Parties in the Volta Basin, for better socioeconomic integration in the sub-region.

The VBA subsequently formed a Technical and Financial Partners Consultative Group to "be a source of inspiration and of technical and financial support to the VBA" (Volta Basin Authority, 2010b; see also 2010a). The VBA has been working to carry out its mandate by, among other things, retaining consultants to assess the state of the basin and recommend a hydrological management system, and bringing stakeholders together for workshops (GhanaWeb, 2015; HydroWorld, 2014).

## Conclusion

The physical solution process that has been used in California is a viable way to manage water resources around the world in an era of climate change and increasing demand. Obstacles to its implementation include funding and the limited rule of law in some areas. However, physical solution processes are in place in areas such as the Volta Basin in West Africa and the Montaña Campara region of Honduras, where there are very limited financial resources. They overcame funding issues through the support of international organizations such as UNESCO, which has stated that "international support for adaptation programs should take precedence in developing countries, providing incentives for capacity strengthening in groundwater management, planning, and conceptualization of adaptation programs", and has directed stakeholders to collaborate with agencies with specialized knowledge, such as NASA's GRACE (UNESCO, 2015, pp. 4–5). Ensuring oversight and adaptive management may be a challenge in developing nations that lack a strong or well-defined rule of law. While California has had limited regulation of groundwater use, it does have a strong legal system capable of enforcing a self-governance agreement. Delegation of authority to an oversight institution may meet resistance in transboundary watercourses. However, even in these circumstances, the other physical solution elements – technical analysis, stakeholder engagement and equitable allocation – need not wait for the establishment of an oversight institution and are powerful tools on their own. Thus, despite funding challenges, lack of legal structure, and transboundary politics, the physical solution may well be the best, most adaptable mechanism for managing water in a changing climate.

## Notes

1. The West Coast Basin and its physical solution are examined in depth by Ostrom (1964).
2. The most recent annual reports for each of these basins can be found online:

   Raymond Basin Management Board Annual Report 2014–2015, http://raymondbasin.org/wp-content/uploads/2011/03/2014-15-RBMB-AR.pdf;
   California Department of Water Resources, Watermaster Service in the West Coast Basin 2013–2014, http://www.water.ca.gov/watermaster/sd_documents/west_basin_2014/westcoastbasinwatermasterreport2014.pdf;
   Water Replenishment District of California and Central Basin Watermaster Administrative Body, Watermaster Service 2013–2014, http://lagroundwater.com/pdf/2014_cb_watermaster_report_6.pdf;
   ULARA Watermaster, Annual Report 2012–2013, http://ularawatermaster.com/public_resources/WY-2012-13-ULARA-WM-Rpt-12-2014.pdf.

3. Some say this quote is originally attributable to Mark Twain.

## Acknowledgment

The author would like to thank Attorney Kira Johnson at Best Best & Krieger LLP for her research and contributions to this article.

## References

Agreement on the Guarani Aquifer. (2010). [Unofficial English translation from international water law project]. Retrieved from http://www.internationalwaterlaw.org/documents/regionaldocs/Guarani_Aquifer_Agreement-English.pdf

Aureli, A., & Eckstein, G. (2011). Strengthening cooperation on transboundary groundwater resources. *Water International, 36*(5), 549–556. doi:10.1080/02508060.2011.615137

Barringer, F. (2011, February 3). Psst... Groundwater and surface water do mix. *New York Times.* Retrieved from http://green.blogs.nytimes.com/2011/02/03/psst-groundwater-and-surface-water-do-mix/?_r=0

Barry, B., Obuobie, E., Andreini, M., Andah, W., & Pluquet, M. (2005). *The Volta river basin, comprehensive assessment of water management in agriculture: Comparative study of river basin development and management* (Working paper). Retrieved from http://catalog.ihsn.org/index.php/citations/30289

Belmecheri, S., Babst, F., Wahl, E. R., Stahle, D. W., & Trouet, V. (2015). Multi-century evaluation of Sierra Nevada snowpack. *Nature Climate Change, 6*, 2–3. doi:10.1038/nclimate2809

Burchi, S., D'Andrea, A., Eckstein, G., & Nanni, M. (2012). The future of adaptive water management: Law and governance. *Water International, 37*(6), 611–612. doi:10.1080/02508060.2012.721343

California Constitution, Article X, Section 2. (2016). Retrieved from http://www.leginfo.ca.gov/.const/.article_10

California Department of Finance. (2010). Historical census population of counties and incorporated cities in California, 1850–2010. Retrieved from http://www.dof.ca.gov/Reports/Demographic_Reports/

California Department of Finance. (2015). E-1 population estimates for cities, counties, and the state – January 1, 2014 and 2015. Sacramento, California.

California Department of Food & Agriculture. (2015). CDFA history. Retrieved from https://www.cdfa.ca.gov/CDFA-History.html

California Legislative Analyst's Office. (2015). 2014 GDP: California ranks 7th or 8th in the world. Retrieved from http://www.lao.ca.gov/LAOEconTax/Article/Detail/90

California State Water Resources Control Board. (2015). Water rights frequently asked questions. Retrieved from http://www.waterboards.ca.gov/waterrights/board_info/faqs.shtml#toc178761079

*California Water Service Co. v. Edward Sidebotham & Sons*, 224 Cal.App.2d 715 (1964).

California-Nevada Climate Applications Program, Center for Western Weather and Water Extremes, Southwest Climate Science Center and California State Climatologist, Department of Water Resources. (2015). California precipitation. Retrieved from http://www.water.ca.gov/floodmgmt/hafoo/csc/docs/CA_Precipitation_2pager.pdf

*City of Barstow v. Mojave Water Agency*, 23 Cal.4th 1224 (2000).

*City of Lodi v. East Bay Municipal Utility Dist*, 7 Cal.2d 316 (1936).

Convention on the Law of Non-navigational Uses of International Watercourses ("U.N. Watercourse Convention"). (1997). Online user's guide: What does the convention say? Retrieved from http://www.unwatercoursesconvention.org/the-convention/

*El Dorado Irrigation Dist. v. State Water Resources Control Board*, 142 Cal.App.4th 937 (2006).

El Nasser, H. (2014, August 9). In dry California, water goes to those who drill the deepest. *Al Jazeera*. Retrieved from http://america.aljazeera.com/articles/2014/8/9/in-dry-californiawatergoestothosewhodrillthedeepest.html

*Environmental Defense Fund v. East Bay Municipal Utility District*, 26 Cal.3d 183 (1980).

Fears, D. (2015, September 14). Scientists say California hasn't been this dry in 500 years. *The Washington Post*. Retrieved from https://www.washingtonpost.com/news/energy-environment/wp/2015/09/14/scientists-say-its-been-500-years-since-california-was-this-dry/

Garner, E. L. (2014). A role for reasonableness and physical solution in water law in an era of climate change. *Bloomberg BNA Water Law & Policy Monitor*, 2014(9). Retrieved from http://www.bna.com/role-reasonableness-physical-p17179882516/

Getirana, A. C. V. (2015). Extreme water deficit in Brazil detected from space. *Journal of Hydrometeorology*. E-view. doi:10.1175/JHM-D-15-0096.1

GhanaWeb. (2015). Workshop on guidelines for the development of water opens. Retrieved from http://www.ghanaweb.com/GhanaHomePage/regional/Workshop-on-guidelines-for-the-development-of-water-opens-384788

Hardin, G. (1968). The tragedy of the commons. *Science, 162*, 1243–1248. doi:10.1126/science.162.3859.1243.

*Herminghaus v. Southern California Edison Co*, 200 Cal. 81 (1926).

HydroWorld. (2014). Volta basin authority seeks hydrological studies of African basin. Retrieved from http://www.hydroworld.com/articles/2014/02/volta-basin-authority-seeks-hydrological-studies-of-african-basin.html

International Water Law Project Blog. (2010). Hydraulic harmony or water whimsy? Guarani aquifer countries sign agreement. Retrieved from http://www.internationalwaterlaw.org/blog/2010/08/05/hydraulic-harmony-or-water-whimsy-guarani-aquifer-countries-sign-agreement/

Johnston, W. E. (2003). Chapter 2: Cross sections of a diverse agriculture: Profiles of California's agricultural production regions and principal commodities. In J. Sieber (Ed.), *California agriculture: Dimensions and issues*. Giannini Foundation of Agricultural Economics, University of California, Division of Agricultural and Natural Resources. Retrieved from http://giannini.ucop.edu/CalAgbook.htm

*Joslin v. Marin Municipal Water District*, 67 Cal.2d 132 (1967).

Kamran, M. A., & Shivakoti, G. P. (2013). Comparative institutional analysis of customary rights and colonial law in spate irrigation systems of Pakistani Punjab. *Water International, 38*(5), 601–619. doi:10.1080/02508060.2013.828584

Leb, C. (2015). One step at a time: International law and the duty to cooperate in the management of shared water resources. *Water International, 40*(1), 21–32. doi:10.1080/02508060.2014.978972

*Light v. State Water Resources Control Board*, 226 Cal.App.4th 1463 (2014).

Littleworth, A., & Garner, E. (2007). *California water II*. Point Arena, CA: Solano Press Books.

Lopez-Gunn, E. (2003). The role of collective action in water governance: A comparative study of groundwater user associations in La Mancha Aquifers in Spain. *Water International, 28*(3), 367–378. doi:10.1080/02508060308691711

*Lux v. Haggin*, 69 Cal. 255 (1886).

McIntyre, O. (2011). The world court's ongoing contribution to international water law: The pulp mills case between Argentina and Uruguay. *Water Alternatives, 4*(2), 124–144.

Ostrom, E. (1964). *Public entrepreneurship: A case study in ground water basin management* (Unpublished PhD dissertation). University of California, Los Angeles.

Ostrom, E. (1990). *Governing the commons: The evolution of institutions for collective action*. New York, NY: Cambridge University Press.

*People ex rel. State Water Resources Control Board v. Forni*, 54 Cal.App.3d 743 (1976).

*Rancho Santa Maria v. Vail*, 11 Cal.2d 501 (1938).

Sax, J. L. (2003). We don't do groundwater: A morsel of California legal history. *University of Denver Water Law Review, 6*, 269.

Schiffman, P. M. (2005). Chapter 2: The Los Angeles prairie. In W. Deverell & G. Hise (Eds.), *Land of sunshine: An environmental history of metropolitan Los Angeles* (pp. 38–51). Pittsburgh, PA: University of Pittsburgh Press.

Schmit, J. (2014, March 26). In California, demand for groundwater causing huge swaths of land to sink. *National Geographic*. Retrieved from http://news.nationalgeographic.com/news/2014/03/140325-california-drought-subsidence-groundwater/

Schoenherr, A. A. (1992). *A natural history of California*. Berkeley, CA: University of California Press.

Sommer, L. (2015, May 11). Will California drought force changes in historic water rights? *KQED*. Retrieved from http://ww2.kqed.org/science/2015/05/11/how-californias-water-rights-make-it-tough-to-manage-drought/

Sugg, Z. P., Varady, R. G., Gerlak, A. K., & de Grenade, R. (2015). Transboundary groundwater governance in the Guarani aquifer system: Reflections from a survey of global and regional experts. *Water International, 40*(3), 377–400. doi:10.1080/02508060.2015.1052939

Thompson, Jr., E. (2009). *Agricultural land loss and conservation*. California Department of Food & Agriculture, AgVision 2030 White Paper. Retrieved from https://www.cdfa.ca.gov/agvision/docs/Agricultural_Loss_and_Conservation.pdf

Treidel, H., Martin-Bordes, J. L., & Gurdak, J. J. (Eds.). (2012). *Climate change effects on groundwater resources: A global synthesis of findings and recommendations*. Leiden: CRC Press/Balkema.

Trigo, R. M., Gouveia, C. M., & Barriopedro, D. (2010). The intense 2007–2009 drought in the fertile crescent: Impacts and associated atmospheric circulation. *Agricultural and Forest Meteorology, 150*, 1245–1257. doi:10.1016/j.agrformet.2010.05.006

Tucker, C. M. (2014). Creating equitable water institutions on disputed land: A Honduran case study. *Water International*, *39*(2), 216–232. doi:10.1080/02508060.2014.888986

*Tulare Irrigation District v. Lindsay-Strathmore Irrigation District*, 3 Cal.2d 489 (1935).

United Nations. (2015). Population 2030: Demographic challenges and opportunities for sustainable development planning. Retrieved from http://www.un.org/en/development/desa/population/publications/pdf/trends/Population2030.pdf

United Nations Educational, Scientific and Cultural Organization. (2015). *GRAPHIC position paper and call to action. Groundwater and climate change: Mitigating the global groundwater crisis and adapting to climate change.* Paris: UNESCO International Hydrological Programme (IHP). Retrieved from http://www.graphicnetwork.net/wp-content/uploads/2015/11/GRAPHIC_pp20151.pdf

United States Department of Agriculture. (2015). California agricultural statistics 2013 crop year. Retrieved from https://www.nass.usda.gov/Statistics_by_State/California/Publications/California_Ag_Statistics/Reports/2013cas-all.pdf

*United States v. State Water Resources Control Board*, 182 Cal.App.3d 82 (1986).

Volta Basin Authority. (2010a). Strategic plan 2010–2014. Retrieved from http://gefvolta.iwlearn.org/project-resources/studies-reports/vba-strategic-plan

Volta Basin Authority. (2010b). Framework for cooperation of Technical and Financial Partners Consultative Group of the Volta Basin Authority. Retrieved from http://www.abv-volta.org:10000/abv2/financialpartners

Western Regional Climate Center. (1995). Annual average precipitation (inches), Southern California. Retrieved from http://www.wrcc.dri.edu/pcpn/ca_south.gif

World Bank. (2015). World development indicators database, gross domestic project 2014. Retrieved from http://databank.worldbank.org/data/download/GDP.pdf

# The human right to water as a creature of global administrative law

Owen McIntyre

A global administrative law perspective helps to make sense of the diversity of sources cited in support of the human right to water and sanitation and to explain how and why the concept might exert influence on a wide variety of actors other than states, including international institutions and private-sector investors and operators involved in the provision of water and sanitation services.

## Introduction

This paper suggests that one can understand the emergence in international law of a human rights–based approach to water and sanitation entitlements as an expression of universally accepted standards of global governance, rather than as an enforceable "human right" in any strict sense of that term. This approach addresses many of the difficulties which arise in relation to the precise normative status of the human right to water under international law as well as doubts concerning its enforceability by means of the traditional enforcement mechanisms existing under human rights law. Further, an analysis of the human right to water in terms of "global administrative law" assists in explaining its potential application in such disparate doctrinal areas of law as international water resources law, international environmental law, and international investment law, as well as in national public and constitutional law. Crucially, global administrative law helps to address problems which arise with the extension of human rights requirements to the actions of private corporate actors and to the problem of extraterritoriality.

Indeed, a survey of the commonly accredited sources of rules of global administrative law helps to make sense of the wide diversity of mechanisms, in addition to the formal conventional obligations and practice of states, for the generation of rules and standards which inform the procedural and substantive normative content of the human right to water. Such mechanisms include, for example, the International Organization for Standardization, various voluntary codes of corporate conduct, international investment arbitration tribunals provided for under bilateral investment treaties, national systems of administrative law, and of course the institutional machinery which elaborates upon human rights values.

Such an analysis permits conclusions to be drawn on the key elements of the human right to water, including procedural and due process requirements, the key rule of law values, the key good governance values (relating particularly to transparency, participation and accountability), and the key human rights values impacting upon the concept. This mode of analysis goes some way towards identifying the rights and obligations created by the human right to water for a variety of actors, including individuals, vulnerable communities, transnational corporations, investors in water and sanitation services, and of course state agencies.

## Global administrative law

The emerging concept of global administrative law addresses the rapidly changing realities of transnational regulation, which increasingly involves, *inter alia*, various forms of industry self-regulation, hybrid forms of private-private and public-private regulation, network governance by state officials, and governance by inter-governmental organizations with direct or indirect regulatory powers, and "begins from the twin ideas that much global governance can be understood as administration, and that such administration is often organized and shaped by principles of an administrative law character" (Kingsbury *et al.* 2005, p. 2). It is proposed that these disparate regulatory regimes, some voluntary and some mandatory, and operating at various levels (sector-specific, national, regional and global),

> together form a variegated "global administrative space" that includes international institutions and transnational networks involving both governmental and non-governmental actors, as well as domestic administrative bodies that operate within international regimes or cause transboundary regulatory effects. (Kingsbury *et al.* 2005, p. 3)

Kingsbury (2009, p. 25) deliberates further on the idea of a "global administrative space" and explains that it "marks a departure from those orthodox understandings of international law in which the international is largely inter-governmental, and there is a reasonably sharp separation of the domestic and the international", and that it reflects the practice of global governance, whereby "transnational networks of rule-generators, interpreters and appliers cause such strict barriers to break down". Indeed, in remarking on the "highly decentralized and not very systematic" nature of much of the administration of global governance, Kingsbury also observes that "some entities are given roles in global regulatory governance which they may not wish for or be particularly designed or prepared for", bringing to mind the recent decisions of International Centre for Settlement of Investment Disputes (ICSID) tribunals which would appear to tacitly support the centrality of human rights concerns to contracts in respect of the provision of water and sanitation services.[1]

Crucially, in respect of the normative content of global administrative law, and reflective of the key procedural aspects of the human right to water as articulated by the UN Committee on Economic, Social and Cultural Rights in General Comment No. 15 (2002), the leading proponents of global administrative law observe that:

> These evolving regulatory structures are each confronted with demands for transparency, consultation, participation, reasoned decisions, and review mechanisms to promote accountability. These demands, and responses to them, are increasingly framed in terms that have an administrative law character. The growing commonality of these administrative law-type principles and practices is building a unity between otherwise disparate areas of governance. (Kingsbury *et al.* 2005, p. 2)

Of course, the function of administrative law generally is to protect individuals by checking the unauthorized, excessive, arbitrary or unfair exercise of public power and, by so doing, to give direction to the practices of administrative bodies, particularly in terms of their responsiveness to broader public interests. Proponents of global administrative law argue that it can perform a similar function for global administrative structures and point out that many of the types of regulatory measures cited above have resulted from the efforts of global administrative bodies, often stimulated by external criticism, to improve internal accountability and bolster external legitimacy (Kingsbury *et al.* 2005, p. 2). One needs only to consider the establishment of accountability mechanisms by all the major multilateral development banks or the widespread inclusion of mechanisms for NGO participation and representation in the decision-making structures of regulatory bodies. In an attempt to provide a definition of the concept of global administrative law, the same leading proponents explain that it

> encompasses the legal mechanisms, principles and practices, along with supporting social understandings, that promote or otherwise affect the *accountability* of global administrative bodies, in particular by ensuring these bodies meet adequate standards of transparency, consultation, participation, rationality, and legality, and by providing effective review of the rules and decisions these bodies make. (Kingsbury *et al* 2005, p. 5)

In addition, they accompany this definition with a broad understanding of the "global administrative bodies", which generate global administrative law norms and to which such norms might apply, to include

> intergovernmental institutions, informal inter-governmental networks, national governmental agencies acting pursuant to global norms, hybrid public-private bodies engaged in transnational administration, and purely private bodies performing public roles in transnational administration.

Thus, much of the normative content of the human right to water, and in particular the procedural rights of individuals and communities contained therein, along with the policies, procedures and decisions of the disparate entities which seek to give effect to the values contained therein, can be viewed through the prism of global administrative law.

As regards the sources of global administrative law, one of the leading scholars in the field (Kingsbury 2009, p. 23), while emphasizing that "there is no single unifying rule of recognition covering all of global administrative law", includes the conventional sources of public international law, i.e. treaties, fundamental customary international law rules, and general principles of law, but also certain principles associated with "publicness" in law. He suggests that "principles relevant to publicness include the [public] entity's adherence to legality, rationality, proportionality, rule of law, and some human rights", which are manifested in "practices of judicial-type review of the acts of global governance entities, in requirements of reason-giving, and in practices concerning publicity and transparency." In an account of global administrative law which is slightly more pessimistic about the difficulty of identifying a universal set of administrative law principles, Harlow (2006, p. 187) systematically identifies and describes four potential sources as a foundation for a global administrative law system:

> first, the largely procedural principles that have emerged in national administrative law systems, notably the principle of legality and due process principles; second, the set of rule of law values, promoted by proponents of free trade and economic liberalism; third, the good governance values, and more particularly transparency, participation and accountability, promoted by the World Bank and International Monetary Fund; and finally, human rights values.

Harlow (2006, p. 188) concludes from her examination of all these sources that "there is considerable overlap between principles found in these different sources". In addition, Kingsbury (2009, p. 23) includes among the sources of global administrative law the rules, standards and safeguards developed as a result of processes of so-called "private ordering", such as the three sets of guidelines relevant to water and sanitation services adopted in 2007 by Technical Committee 224 of the International Organization for Standardization (ISO 2007a, 2007b, 2007c).

As regards the specific normative content of global administrative law, Kingsbury (2009, p. 32) identifies certain "general principles of public law [which] combine formal qualities with normative commitments in the enterprise of channelling, managing, shaping and constraining political power". In addition to certain "more detailed elements, or requirements . . . particularly review, reason-giving, and publicity/transparency", his indicative list of such general principles of public law includes:

(1) *the principle of legality* – requiring that actors within a power system are constrained to act in accordance with the rules of the system
(2) *the principle of rationality* – requiring the justification of decisions, including that decision makers give reasons and produce a factual record for decisions
(3) *the principle of proportionality* – requiring a relationship of proportionality between means and ends
(4) *rule of law* – requiring particular deliberative and decisional procedures
(5) *human rights* – requiring protection of human rights values which are intrinsic (or natural) to a modern public law system.

Kingsbury (2009, p. 34) further identifies three broad categories of public global administrative activity to which the rules and principles of global administrative law might apply, and which in turn generate practices which can give rise to such rules and principles. These include:

(1) the institutional design, and legal constitution, of the global administrative body
(2) the norms and decisions produced by that entity, including norms and decisions that have as their addressees, or otherwise materially affect:
    (a) other such public entities
    (b) states and agencies of a particular state
    (c) individuals and other private actors
(3) procedural norms for the conduct of those public entities in relation to their rules and decisions, including arrangements for review, transparency, reason-giving, participation requirements, legal accountability and liability.

While it is perfectly clear that rules and principles of global administrative law are relevant to the institutional design, and thus to the legitimate functioning, of the myriad entities involved in elaborating upon the human right to water, it is the second and third categories of administrative activity listed above which play a significant role in the development of its normative status and content. Such entities might, for example, include the UN Committee on Economic, Social and Cultural Rights (CESCR) or other global or regional bodies concerned with the interpretation or monitoring of human rights instruments; international standard-setting bodies, such as the International Organization for Standardization; and judicial and quasi-judicial organs, such as international investment arbitration tribunals established under the International Centre for the Settlement of Investment Disputes (ICSID) or the accountability mechanisms of multilateral development banks. As will be illustrated below, interpretative statements (such as General Comment No. 15), best practice guidelines (such as the 2007 ISO guidelines), and arbitral and quasi-judicial decisions (such as those of recent ICSID tribunals)[2] adopted by such entities lend much-needed support to and substantially inform the human right to water, while also illustrating the practical utility of the global administrative law concept as a means of understanding common normative approaches which converge from complex, chaotic and pluralistic origins.

While Harlow (2006, p. 188) includes human rights values as a source of global administrative law norms, she does so "only to the extent that these are procedural in character". In other words, she highlights the fact that "many international human rights texts contain due process rights of a type traditionally developed in and protected by classical administrative law systems". However, Kingsbury (2009, p. 33) appears to suggest that the substantive normative content of human rights regimes might in some instances be relevant, by suggesting that "some human rights (perhaps of *bodily integrity*, privacy, personality) are likely to be protected by public law as an intrinsic matter (without textual authority), yet without being subsumed into 'rule of law'". Like the human right to water, the human right to bodily integrity is often closely linked to, and under many human rights texts derived from, the right to health and, indeed, further connected to mutually related standards of protection of the human environment. Therefore, Kingsbury's express reference to bodily integrity implies that substantive human rights values must be relevant to the identification of global administrative law norms, and vice versa. Indeed, though General Comment No. 15 is very largely concerned with informational, participative and other procedural elements of the human right to water, it seems difficult to imagine that substantive human rights values would not be relevant to, and captured by, such general public law principles as proportionality.

Of course, there are those who have serious misgivings about the global administrative law phenomenon and highlight its hazards for democracy and traditional political processes, for developing economies, and for the coherence and predictability of applicable legal standards (Harlow 2006, pp. 207–214). The key concern is that global administrative law tends to subvert the traditional democratic processes vital to the legitimacy of law. The role of judicial, arbitral and quasi-judicial bodies in particular raises concerns over the juridification of the political process and of "government by judges" by virtue of a general empowerment of a transnational "juristocracy" (Harlow 2006, p. 213). The undermining of sovereign democratic processes and the emergence of common and universal administrative standards presents a particular risk for developing economies, which may not have had a significant role in generating the practice upon which these standards are based. Harlow (2006, pp. 207–211) suggests that administrative law is largely a "Western construct", which is protective of Western values and interests and may impact unfavourably on developing economies. Further, due to the non-systematic nature of the processes shaping global administrative law, the rules and standards invoked as inherent to the global administrative law concept may often lack clarity and certainty. As Kingsbury (2009, p. 29) points out, the difficulty in identifying universal rules and principles stems from the fact that

> "global administrative law" is not an established field of normativity and obligation in the same way as "international law". It has no great charters, no celebrated courts, no textual provisions in national constitutions giving it status in national law, no significant long-appreciated history.

Similarly, Harlow (2006, p. 208) highlights the considerable disparities of principle that exist "even within the systems in which modern administrative law [has] developed" and points out that "at least four administrative law families have been identified within the EU alone". However, as argued below, the inclusive nature of the various institutional structures and processes which have given rise to the human right to water, as well as the detailed normative guidance adopted thereunder, do much to address such concerns about sovereign legitimacy, normative clarity or Western bias, thus marking out the human right to water as an exemplar of the global administrative law phenomenon.

Therefore, rather than attempting to provide a comprehensive and coherent unifying theory of global governance arrangements, the global administrative law concept is merely an observed phenomenon which seeks to explain the growing commonality apparent among the administrative principles and practices which apply across otherwise disparate areas of governance.

## Good water governance

Recent discourse on good water governance (Asian Development Bank 1995, United Nations Development Programme 1997) similarly reflects many of the key principles of global administrative law, including accountability, participation, predictability and transparency (Grimes 2009, p. 121). One very influential Global Water Partnership (GWP) background paper (Rogers and Hall 2003) identified the key characteristics of effective water governance, which relate both to the approach that governance should take and to performance and operational requirements of governance. It suggested that effective water governance must take an approach which is open and transparent, inclusive and communicative, coherent and integrative, and equitable and ethical, while such governance should prove to be accountable, efficient, responsive and sustainable (Rogers and Hall 2003, conclusions summarized in Grimes 2009, p. 121). The role of law is recognized by all involved in this discourse as being absolutely central in achieving good water governance in accordance with the principles outlined above. The GWP explains that effective water governance requires "an enabling environment which facilitates efficient private and public sector initiatives", which is dependent upon "a coherent legal framework with a strong and autonomous regulatory regime" (Rogers and Hall 2003, p. 37). Similarly, Tropp (2007), whose model of water governance consists of four dimensions – social, economic, political and environmental – stresses that administrative systems play a part in all four and highlights the need for strong regulatory authority at the national level that "embraces new forms of governance". Clearly, the human right to water provides just such a new form of governance, which is largely administrative in nature and "can assist in defining the ultimate goal of water governance" (Grimes 2009, p. 122). At the Third World Water Forum in 2003, the Global Water Partnership (2003) emphasized the intrinsic link between the human right to water and good water governance, stating:

> Effective water governance is necessary to solve the water crisis. . . . If we are to secure access to water for all (thus complying with a recent UN human rights declaration), maintain vital ecosystems and produce economic development out of water management, effective water governance is essential.

Indeed, the importance of legally coherent rights-based approaches to achieving good governance has long been recognized by the World Bank (1992, p. 32), which has tended to stress predictability, stating that "people's knowledge of their rights helps both to limit the arbitrary behaviour of government officials and to create the climate of predictability which is associated with the rule of law". The close connection between the principles of good governance and global administrative law is self-evident.

## Human right to water

### *Uncertainty regarding legal bases and status*

Despite recent high-profile declarative support for the legal status of the human right to water from key UN bodies, including a General Assembly resolution in July 2010 and a resolution adopted unanimously by the Human Rights Council in September 2010, considerable uncertainty persists with regard to the true normative status of the human right to water under international law. Though these resolutions contain very little in terms of normative obligations, as authoritative a commentator on the topic of such "soft law" as Dinah Shelton (2011, p. 70) concludes in respect of such resolutions that:

> The not-law can be politically binding, morally binding, and expectations can be extremely strong of compliance with the norms contained in the instrument, but the difference between a legally binding instrument and one that is not appears well understood and acted upon by government negotiators.

This is also true in respect of the numerous informative reports and publications issued by the UN Special Rapporteur on the Human Right to Water and Sanitation, which are even "softer" in terms of their normative value.

In the legal systems of many states, this uncertainty may also impact upon its status in national law. Although General Comment No. 15 identifies Articles 11 and 12 of the 1966 International Covenant on Economic, Social and Cultural Rights (ICESCR) – on the right, respectively, to an adequate standard of living and to the highest attainable standard of health – as the primary legal bases for the human right to water, it should be remembered that CESCR general comments do not formally impose legal obligations on ICESCR states parties, let alone other states, and that General Comment No. 15 merely constitutes a non-binding but "highly authoritative interpretation of the Covenant" and of the legal implications which flow from key relevant covenant provisions (McCaffrey 2005, p. 94). Indeed, Shelton (2011, p. 64) points out that "using non-binding texts to give authoritative interpretation to treaty terms is particularly useful when the issues are contentious and left unresolved in the treaty itself". McCaffrey (2005, p. 103) characterizes General Comment No. 15 as being "more in the nature of a statement *de lege ferenda* [what the law should be] rather than *lex lata* [the existing law]" and cautions that the interpretation of Articles 11 and 12 contained therein "must be accepted by the States parties to the Covenant in order to be binding upon them". Also, the fact that the human right to water derives from a number of expressly articulated primary rights may lead to confusion. As Williams (2007, p. 477) puts it, "various connected rights may implicate different state obligations"; she illustrates this point by explaining that a right to water derived from the right to life, which requires the provision of drinking water, would impose lesser state obligations than a right to water derived from the right to health, which requires the provision of water for both drinking and sanitation. Even greater uncertainty persists as to the normative status and content of any human right to sanitation, either as an element of the human right to water or as an autonomous right deriving independently from the ICESCR. As the leading proponent of a right to sanitation felt bound to admit in 2010 (Smets 2010, p. 3):

> As yet, there is no single resolution of the UN General Assembly on the right to sanitation and no world wide agreement supporting this right. There is no agreed description of core obligations on sanitation and no definition of the content of the right to sanitation.

Equally, it is not at all clear from General Comment No. 15 to what extent the human right to water is expected to guarantee water for livelihoods, nor what potential ecological impacts might flow from such an approach (McIntyre, 2012, p. 225). Of course, uncertainty remains as to the true normative status and content of a number of the economic, social and cultural rights listed under the ICESCR, from which the rights to water may be derived, inevitably leading to further confusion as to the implications of the right to water (Bluemel 2004, p. 971). For example, though some commentators describe the right to food as well established (Williams 2007, p. 479), it might be argued that it raises many questions with regard to the force and extent of such welfare rights under the ICESCR.

At any rate, though proponents of an independent right to water argue that it would result in greater interpretive consistency, state compliance, enforcement and remedies for violations (Bluemel 2004, pp. 968–972, Williams 2007, p. 478), such an independent right could only arise in international law by means of a dedicated treaty instrument or customary international law (Hardberger 2005, pp. 360–362). It is clear that there does not currently exist a general treaty instrument, nor any proposal for such an instrument, by which states might bind themselves in this regard. Also, it would appear that, despite the sustained declaratory support of international conferences and UN agencies, as well as some limited legal and constitutional state recognition, there is as yet insufficient generalized state practice to establish a right to water under customary international law that would bind those states that have not actively and formally recognized the right. Therefore, Williams (2007, p. 478) concludes that "at best, this seems to give the independent right the current status of a normative ideal". Indeed, overshadowing any discussion of whether the human right to water might exist as an ancillary or independent right is the fact that the ICESCR, and any rights derived therefrom, suffer from a clear lack of immediate enforceability, with Article 2(1) merely requiring each state party "to take steps . . . to the maximum of its available resources, with a view to achieving progressively the full realization of the rights recognized in the present Covenant". As one commentator (Tully 2004, p. 103) has noted in relation to such "second-generation" rights, "the principal challenge is therefore linking the expectations of individuals as rights-holders with the duties owed by others." At the level of the practical enforceability of the obligations set out under the ICESCR, McCaffrey (2005, p. 97) points out that the language of Article 2(1) would provide a lawyer acting for a state accused of breaching its obligations with "ample bases for a defense".

Provisions of other international human rights instruments are also cited, even less convincingly, as providing support for the normative status of the human right to water concept in international law. Article 25 of the 1948 Universal Declaration of Human Rights (UDHR), which proclaims "the right to a standard of living adequate for the health and well-being of himself and his family, including food", is often cited (Gleick 1998, p. 491), even though, as a UN General Assembly Resolution, the UDHR is not binding per se. While it is generally accepted that many of the basic human rights contained therein have become part of customary international law, or at least constitute authoritative interpretations of the UN Charter's provisions on human rights, and thus bind states generally, such customary status is normally only accorded to the so-called "liberty rights" contained in the declaration, rather than the "welfare rights" (Schachter 1982, p. 340), of which the right to an adequate standard of living is one. Similarly, though Article 22 of the UDHR refers generally to the individual's entitlement "to realization . . . of the economic, social and cultural rights indispensable for his dignity and the free development of his personality", the IESCR ought, as the specific implementing instrument, to be considered the primary source of justiciable rights in this category (Hardberger 2005, p. 337). Paragraph 3 of General Comment

No. 15 also links the human right to water to the right to life set out under Article 6 of the 1966 International Covenant on Civil and Political Rights (ICCPR), even though it remains unclear whether Article 6 "merely protects against arbitrary deprivation of life by the state, or also guarantees against death from such causes as lack of water or food, exposure to the elements, or lack of medical attention" (McCaffrey 2005, p. 97). Though the Human Right Committee (HRC)[3] has since 1982 interpreted the reference to the "inherent right to life" in Article 6 to mean that the right to life includes a socioeconomic component and demands positive action by states,[4] it has traditionally been understood only to extend to arbitrary deprivations of life by the state (Williams 2007, p. 474). However, the view that "rights such as one to an appropriate means of subsistence belong within the category of economic, social and cultural rights" is more in accordance with the reality of water services provision and with the true intentions of states parties to the covenants (McCaffrey 2005, p. 98).

Particular provisions of a number of binding international conventions which apply in specific contexts are also commonly cited in support of the human right to water. These include Article 14(2) of the 1979 Convention on the Elimination of All Forms of Discrimination Against Women (CEDAW);[5] Article 24(2)(c) of the 1989 Convention on the Rights of the Child (CRC);[6] the 1949 Geneva Convention (III) on the Treatment of Prisoners of War;[7] and the 1949 Geneva Convention (IV) on the Treatment of Civilian Persons in Time of War.[8] Though Gleick (1998, p. 494) heralds Article 24 of the CRC as the first-ever example in a binding international treaty instrument having general application of "explicit recognition of the connection between resources, the health of the environment, and human health", McCaffrey (2005, p. 98) is rather more circumspect and points out that

> none of these agreements casts the corresponding entitlement in human rights terms. Instead, they place a duty on governments to ensure that water, among other things necessary to life and good health, is provided to members of groups that have been identified as requiring special protection.

He cautions that, while the clear need of vulnerable groups for access to sufficient and safe water supplies has been recognised in such instruments, "this does not necessarily mean that states have recognized a human right to water, with all its implications, either generally or in those specific instruments" (p. 107).

Of course, implicit support for the values inherent to the human right to water can be found in a very wide and diverse range of legal instruments operating at both the international and national levels and covering a variety of areas of activity. In relation to international water resources law, for example, though the UN Watercourses Convention[9] addresses the obligations of international watercourse states rather than the rights of individuals, it would appear to support the existence of a state obligation to cater for the basic needs of citizens for water by expressly providing in Article 10(2) for watercourse states to have "special regard . . . to the requirements of vital human needs" over and above all other classes of uses of shared water resources. Similar priority is given to such needs in the 1999 Protocol on Water and Health adopted under the 1992 UNECE Convention on the Protection and Use of Transboundary Watercourses and International Lakes,[10] and in the International Law Association's (ILA) 2004 Berlin Rules on Water Resources Law.[11] The "requirements of vital human needs" would appear to correspond closely with the obligations of states and the entitlements of individuals under the human right to water (Tanzi and Arcari 2001, p. 139). Thus, a learned body as influential as the ILA expressly links the human right to water to the position, widely acknowledged in international codifications and accepted in state and arbitral practice, that uses required for the satisfaction of vital human needs take priority over other, less urgent uses. Indeed, General Comment No. 15 suggests on numerous occasions the significance of the emergence of the human right to water for inter-state practice in respect of shared water resources.

### *Procedural elements of the human right to water*

Significantly, as regards a global administrative law analysis of human right to water, it is clear that any elaboration of the human right to water involves the inclusion of detailed procedural elements regarded as inherent to the concept. In addition to those provisions of global and regional human rights instruments which might be argued explicitly or implicitly to include the human right to water, it is quite clear that all such instruments would now be interpreted and applied so as to require that states generally facilitate a participative approach in respect of projects or policies that might impact on human rights, by ensuring the adoption of procedures by which interested groups or individuals or communities likely to be affected by such projects or policies can receive and access relevant information, meaningfully participate in decision making and, if necessary, have access to some appropriate means of legal recourse (Cullett and Gowlland-Gaultieri 2005). Such a participatory approach to guaranteeing human rights would equally apply to projects or policies which might impact on the availability of water resources, particularly where this might arise by virtue of environmental risk, and procedural and participative rights are a very significant element of the normative content of the human right to water as put forward in General Comment No. 15. Indeed, the requirement for states parties to the ICESCR to ensure a participatory and transparent process for the adoption and implementation of a national water strategy and plan of action is included among the non-derogable "core obligations" of states under paragraphs 37(f) of General Comment No. 15. In the *Ogoni* case, for example, the African Commission on Human Rights gave a broad participative reading to Article 24 of the African Charter on Human and Peoples' Rights, which acknowledges all peoples' right to a generally satisfactory environment, to include specific procedural guarantees concerning the carrying out of environmental and social impact assessment (Cullett and Gowlland-Gaultieri 2005, pp. 313–314). Clearly, such procedural requirements, which correlate closely with the procedural and informational requirements of the human right to water as set out under General Comment No. 15, would equally apply, under existing regional human rights instruments, to any major project or policy initiative, such as the privatization of a water utility, which threatened the quality or availability of water supply or sanitation services.

These procedural requirements appear all the more widely accepted and applied when one considers that broad informational and participatory rights are generally also included under regional and global environmental instruments. The concept of participation in international environmental law is exemplified by the 1998 UNECE Aarhus Convention,[12] and such participation requirements are also central to the carrying out of an adequate environmental impact assessment (EIA) consistent with the standards established under international law. More generally, in the field of sustainable development, all seminal instruments purport to establish participatory standards which apply not only to states but also to international organizations, including multilateral development banks (MDBs). Participatory rights are absolutely central to Chapter 18 on freshwater resources in the Agenda 21 of the United Nations Conference on Environment and Development. Therefore, the accumulated practice of regional human rights enforcement bodies strongly suggests that the CESCR's General Comment No. 15 largely involves a codification of existing state obligations under general international human rights law and general international environmental and sustainable development law, rather than an attempt at the progressive development of participatory principles applying to matters of access to water. The same might be said of the origins and normative basis of the principle of non-discrimination, which forms another essential substantive element of the human right to water as set out under General Comment No. 15 and is also included among the non-derogable "core obligations" of states. Likewise, the inclusion of special protections for indigenous peoples under General Comment No. 15 might be traced to and justified under ILO Conventions 107 and 169 on indigenous and tribal peoples.[13,14] This focus on procedural obligations arising from a wide diversity of legal sources certainly lends itself to a global administrative law analysis.

### *Extension of human right to water to private corporate actors*

The global administrative law concept helps address the difficult issue of extension of human rights norms and values to private corporate actors. The involvement of the private sector in the provision of water and sanitation services makes global administrative law a useful prism through which to view and analyze the altered regulatory obligations imposed on state authorities and private actors by the emergence of the human right to water. According to Bronwyn Morgan (2006, p. 216):

> Private-sector participation from outside national borders in the provision of basic goods makes urban water services a fascinating case study for exploring the potential ambit of what scholars have provocatively called "global administrative law".

She justifies this mode of analysis by explaining that such arrangements comprise "hybrid blends of public and private actors linked in routines of both formal and informal participation at multiple levels of governance" (p. 216).

Despite recent interest in the idea of extending the application of key international human rights norms so as to apply directly to corporations (Ratner 2001), which could make the requirements of the human right to water central to arrangements for the privatization of water and sanitation services, it is clear that such norms could only as yet dictate corporate behaviour where individual corporations have voluntarily agreed to abide by codes of conduct which explicitly or implicitly require compliance with international human rights norms (Murphy 2005). High-profile examples of such voluntary initiatives include the UN Global Compact, providing a set of 10 core principles which, while not referring explicitly to water rights or entitlements, does provide that companies should comply with international human rights norms, which might be argued to include a right to water and sanitation (Williams 2007, pp. 488–491). Similarly, the Organisation for Economic Co-operation & Development (OECD) *Guidelines for Multinational Enterprises* (1976/2000) represent another legally non-binding initiative which can operate to support application of key elements of the human right to water to private companies. They consist of recommendations providing voluntary principles and standards for responsible conduct for multinational corporations operating in or from states which adhere to the OECD declaration. The guidelines cover business conduct in such areas of relevance to the implementation of the human right to water as human rights, environment, information disclosure, the combating of bribery, and consumer interests. In addition, there is a clear trend in the declarative practice of states towards extending responsibility for respecting human rights to private companies involved in the provision of private services. For example, the Draft Declaration on the Right to Access to Essential Services, proposed by France in the context of the 2002 World Summit on Sustainable Development, relates to essential services indispensible for a dignified life, expressly including drinking water and sanitation, and would apply equally to both public and private sector providers (Morgan 2004, pp.181–182). In respect of the development of an appropriate regulatory framework for private-sector water service providers which is informed by international human rights values and widely accepted by states and leading private operators, the adoption in 2007 of three sets of guidelines by Technical Committee 224 of the International Organization for Standardization (ISO) represents a significant step (Morgan 2004, pp. 182–183, Morgan 2006, pp. 224–227). The guidelines set standards for service activities relating to the provision of drinking water supply and sewerage services, that apply equally to both public and private actors, both as service provider and user, and even attempt to deal with the role of "cost" or "price" within the standard of service. With 35 participating countries and 17 observer countries involved in their development and adoption, in liaison with a range of interested international organizations, including the World Health Organization, World Bank and International Water Association, and with leading industry interests, the ISO guidelines are likely to prove influential in determining an acceptable level of service provision where a dispute arises involving a private-sector provider.

Of potentially far greater significance in this regard is the work of the UN Special Representative of the Secretary General on the issue of human rights and transnational corporations and other business enterprises, whose guiding principles were adopted by the Human Rights Council in 2011 (Ruggie 2011). The "protect, respect and remedy" framework proposed for ensuring that private corporations operationalize human rights is described as resting on three complementary pillars:

> the State duty to protect against human rights abuses by third parties, including business, through appropriate policies, regulation, and adjudication . . . the corporate responsibility to respect human rights, which means to act with due diligence to avoid infringing on the rights of others . . . [and] greater access by victims to effective remedy, judicial and non-judicial. (Ruggie 2011, p. 4)

This approach also enjoys considerable support among leading business entities and civil society (Ruggie 2009, pp. 3–4).

Further, recent developments in international investment law, the body of rules that function to provide protection to private-sector investors operating in foreign jurisdictions against arbitrary interference with their property or business interests by the sovereign actions of host states, help to inform the requirements imposed by the human right to water on host states and private actors in cases of water services privatization. The concept would appear to have received solid, if implicit, support in a number of recent statements of ICSID international arbitration tribunals engaged in the settlement of investor-state disputes over water and sanitation service contracts.[15] In relation to a series of primarily procedural issues, ICSID tribunals have recognized that the provision of water and sanitation services inevitably involves questions of human rights. Though the precise mechanism by means of which such human rights values came to inform the reasoning of international investment arbitration tribunals was nowhere set out by the tribunals in question, it can be easily explained in terms of the application of global administrative law standards of good governance.

## Conclusion

The diversity of legal sources for the human right to water and the continuing uncertainties surrounding its legal status, along with the procedural character of many of its inherent requirements and the challenges presented by the extension of human rights values to non-state actors, make a global administrative law analysis a useful approach to better understanding the human right to water discourse. Further, the human right to water might be presented as an exemplar of the global administrative law phenomenon. Indeed, the human right to water would appear to answer most, if not all, of the concerns raised earlier about the global administrative law concept (Harlow 2006, pp. 207–214). First, in response to the allegation that global administrative law subverts traditional democratic political processes and the principle of state sovereignty, a June 2010 study by the French NGO Coalition Eau claims that 190 states have thus far declared support for the human right to water at a ministerial level. Indeed, as regards the concern that global administrative law is a "Western construct", the same study points out that such ministerial declaratory support can be attributed to 139 developing states and 51 developed states. Further, the level of developing state representation in the CESCR and ISO Technical Committee 224 should help to address such concerns in respect of the human right to water (Morgan 2004, pp. 182–183). The existence of detailed normative and technical guidance on the human right to water, contained in such documents as General Comment No. 15 and the 2007 ISO guidelines, should go some way towards addressing concerns about the lack of clarity and consistency in respect of global administrative law norms. Indeed, several of the key areas of legal practice supporting and informing the human right to water, such as human rights law and international investment law, are themselves firmly "established fields of normativity" with established institutional structures and autonomous bodies of jurisprudential thought (Kingsbury 2009, p. 29). In response to concerns about the "juridification" of political processes, the areas of human rights law and international investment law are already substantially "juridified", but the discretion of judicial and quasi-judicial decision makers is largely constrained by highly developed normative and technical guidance.

## Notes

1. See, for example, ICSID Case No. ARB/03/19, *Aguas Argentinas S.A., Suez, Sociedad General de Aguas de Barcelona S.A. and Vivendi Universal S.A. v. The Argentine Republic, Order in Response to a Petition for Transparency and Participation as Amicus Curiae* (19 May 2005); ICSID Case No. ARB/03/17, *Aguas Provinciales de Santa Fe S.A., Suez, Sociedad General de Aguas de Barcelona S.A. and InterAguas Servicios Integrales del Agua S.A. v. The Argentine Republic, Order in Response to a Petition for Participation as Amicus Curiae* (17 March 2006); ICSID Case No. ARB/03/19, *Suez, Sociedad General de Aguas de Barcelona S.A. and Vivendi Universal S.A. v. The Argentine Republic, Order in Response to a Petition by Five Non-Governmental Organizations for Permission to Make an Amicus Curiae Submission* (12 February 2007); and ICSID Case No. ARB/05/22, *Biwater Gauff (Tanzania) Ltd. V. United Republic of Tanzania*, Procedural Order No. 5 (2 February 2007).
2. See Note 1.
3. This body, commonly referred to as the Human Rights Committee, can also be more accurately referred to as the Committee on Civil and Political Rights which operates under the auspices of the Office of the High Commissioner for Human Rights and has provided detailed general Comments on the key provisions of the 1966 Covenant on Civil and Political Rights, including Article 6, see http://www.unhchr.ch/tbs/doc.nsf/(Symbol)/84ab9690ccd81fc7c12563ed0046fae3?Opendocument
4. See the Committee on Civil and Political Rights, General Comment No. 6: The right to life (Article 6), 30 April 1982, para. 5. http://www.unhchr.ch/tbs/doc.nsf/(Symbol)/84ab9690ccd81fc7c12563ed0046fae3?Opendocument

5. See CEDAW: U.N. Doc. A/34/46 (1979). Entered into force 1981. Available from: http://www.un.org/womenwatch/daw/cedaw/cedaw.htm
6. CRC: U.N. Doc. A/44/49 (1989). Entered into force 1990. Available from: http://www2.ohchr.org/english/law/crc.htm
7. 1949 Geneva Convention III: (1949) 75 UNTS 135. Entered into force 1950.
8. 1949 Geneva Convention IV: (1949) 75 UNTS 287. Entered into force 1950.
9. 1997 UN Convention on the Non-Navigational Uses of International Watercourses, (1997) 36 ILM 719 (New York, 21 May 1997, not yet in force).
10. 1999 Protocol on Water and Health to the 1992 UNECE Convention on the Protection and Use of Transboundary Watercourses and International Lakes, U.N. Doc. MP.WAT/AC.1/1991/1 (17 June 1999).
11. ILA, Berlin Rules on Water Resources Law (2004), Available at www.asil.org/ilib/WaterReport2004.pdf
12. Convention on Access to Information, PublicParticipation in Decision-Making and Access to Justice in Environmental Matters, (Aarhus, 25 June 1998), 38 ILM 517 (1999).
13. ILO Convention 107 concerning the Protection and Integration of Indigenous and Other Tribal and Semi-Tribal Populations in Independent Countries (26 June 1957).
14. ILO Convention 169 concerning Indigenous and Tribal peoples in Independent Countries (27 June 1989), 28 ILM 1382 (1989).
15. See note 1.

## References

Asian Development Bank, 1995. *Governance: sound development management*. Manila: ADB.

Bluemel, E.B., 2004. The implications of formulating a human right to water. *Ecology Law Quarterly*, 31 (4), 957–1006.

Coalition Eau, 2010. Reconnaissance Internationale du Droit A L'Eau. Available at http://www.coalition-eau.org/spip.php?rubrique1 [accessed 24 March 2010].

Committee on Economic, Social and Cultural Rights, 2002. *General Comment No. 15: the right to water (Articles 11 and 12 of the International Covenant on Economic, Social and Cultural Rights)* (U.N. DOC. E/C.12/2002/11, 26 November 2002).

Cullett, P., and Gowlland-Gaultieri, A., 2005. Local communities and water investments. *In*: E. Brown Weiss, L. Boisson de Chazournes and N. Bernasconi-Osterwalder, eds. *Fresh water and international economic law*. Oxford: Oxford University Press, 303–330.

Gleick, P.H., 1998. The human right to water. *Water Policy*, 1, 487–503.

Global Water Partnership, 2003. Effective water governance: learning from the dialogues. GWP status report. *Third world water forum*, Kyoto, Japan, 16–23 March 2003.

Grimes, H., 2009. Responding to the "water crisis": the complementary roles of water governance and the human right to water. *Water Law*, 20 (2–3), 119–128.

Hardberger, A. 2005. Life, liberty and the pursuit of water: evaluating water as a human right and the duties and obligations it creates. *Northwestern Journal of International Human Rights*, 4 (2), 331–362.

Harlow, C., 2006. Global administrative law: the quest for principles and values. *European Journal of International Law*, 17 (1), 187–214.

ISO, 2007a. ISO 24510:2007 *Activities relating to drinking water and wastewater services – Guidelines for the assessment and for the improvement of the service to users*. Available from: http://www.iso.org/iso/iso_catalogue/catalogue_tc/catalogue_tc_browse.htm?commid=299764published&on&includesc=true

ISO, 2007b. *ISO 24511:2007 Activities relating to drinking water and wastewater services – Guidelines for the management of wastewater utilities and for the assessment of wastewater services*. Available from: http://www.iso.org/iso/iso_catalogue/catalogue_tc/catalogue_tc_browse.htm?commid=299764&published=on&includesc=true

ISO, 2007c. ISO 24512:2007 *Activities relating to drinking water and wastewater services – Guidelines for the management of drinking water utilities and for the assessment of drinking water services*. Available from: http://www.iso.org/iso/iso_catalogue/catalogue_tc/catalogue_tc_browse.htm?commid=299764&published=on&includesc=true

Kingsbury, B., et al., 2005. Global governance as administration: national and transnational approaches to global administrative law. *Law and Contemporary Problems*, 68 (3–4), 1–13.

Kingsbury, B., 2009. The concept of "law" in global administrative law. *European Journal of International Law*, 20 (1), 23–57.

McCaffrey, S., 2005. The human right to water. *In*: E. Brown Weiss, L. Boisson de Chazournes and N. Bernasconi-Osterwalder, eds. *Fresh water and international economic law*. Oxford: Oxford University Press, 93–116.

McIntyre, O. 2012. Environmental protection and the human right to water: complementarity and tension. *In*: L. Westra, C. Soskolne and D. Spady, eds. *Human health and ecological integrity: ethics, law and human rights*. Abingdon: Earthscan/Routledge Taylor & Francis, 225–236.

Morgan, B., 2004. The regulatory face of the human right to water. *Water Law*, 15 (5), 179–186.

Morgan, B., 2006. Turning off the tap: urban water service delivery and the social construction of global administrative law. *European Journal of International Law*, 17 (1), 215.

Murphy, S.D., 2005. Taking multinational corporate codes to the next level. *Columbia Journal of Transnational Law*, 43, 388–433.

Organisation for Economic Co-operation & Development, 1976/2000. *OECD guidelines for multinational enterprises* (revised 2000). Available from: http://www.oecd.org/dataoecd/56/36/1922428.pdf.

Ratner, S.R., 2001. Corporations and human rights: a theory of legal responsibility. *Yale Law* Journal, 111 (3), 443–544.

Rogers, P., and Hall, A., 2003. *Effective water governance.* Technical Committee Background Paper No. 7. Stockholm: Global Water Partnership.

Ruggie, J., 2009. Report of the Special Representative of the Secretary-General on the issue of human rights and transnational corporations and other business enterprises. Business and human rights: towards operationalizing the "protect, respect and remedy" framework. UN Doc. A/HRC/11/13. Available from: http://www2.ohchr.org/english/bodies/hrcouncil/docs/11session/A.HRC.11.13.pdf.

Ruggie, J., 2011. Report of the Special Representative of the Secretary-General on the issue of human rights and transnational corporations and other business enterprises, John Ruggie. Guiding principles on business and human rights: implementing the United Nations "protect, respect and remedy" framework. UN Doc. A/HRC/17/31. Available from: http://www.ohchr.org/documents/issues/business/A.HRC.17.31.pdf.

Schachter, O., 1982. International law in theory and practice. *Recueil des Cours*, 178, 9.

Shelton, D., 2011. Comments on the normative challenge of environmental "soft law". *In*: Y. Kerbrat and S. Maljean-Dubois, eds. *The transformation of international environmental law*. Paris: Pedone & Hart, 61–71.

Smets, H. 2010. The right to sanitation: a new human right in developed countries [online]. *Environmental Policy and Law*, 40 (2–3) 112. Available from: http://iospress.metapress.com/content/v048703626x3/?p=9f80b84eab284cb3a0c4e8eb8173056a&pi=11

Tanzi, A., and Arcari, M, 2001. *The United Nations Convention on the Law of International Watercourses*. The Hague: Kluwer Law International.

Tropp, H., 2007. Water governance: trends and needs for new capacity development. *Water Policy*, 9 (Supp. 2), 19–30.

Tully, S.R., 2004. The contribution of human rights to freshwater resource management. *Yearbook of International Environmental Law*, 14, 101–137.

United Nations Conference on Environment and Development. Agenda 21 (Rio de Janeiro, 13 June 1992). U.N. Doc. A/CONF.151/26 (Vols. I, II & III) (1992).

United Nations Development Programme, 1997. *Governance for sustainable human development.* Geneva: UNDP.

United Nations General Assembly Resolution 217A (III) [Universal declaration of human rights] (New York, 10 December 1948).

United Nations General Assembly Resolution 2200A (XXI) [International covenant on economic, social and cultural rights, international covenant on civil and political rights and optional protocol to the international covenant on civil and political rights] (New York, 16 December 1966).

United Nations General Assembly Resolution A/64/L.63/Rev.1 on The human right to water and sanitation (26 July 2010).

United Nations Human Rights Council Resolution A/HRC/15/L.14 on Human rights and access to safe drinking water and sanitation (24 September 2010).

UN Special Rapporteur on the Human Right to Water and Sanitation. Various. Paris: Office of the United Nations High Commissioner for Human Rights. Available from: http://www.ohchr.org/EN/PublicationsResources/Pages/Publications.aspx.

Williams, M., 2007. Privatization and the human right to water: challenges for the new century. *Michigan Journal of International Law*, 28 (2), 469–506.

World Bank, 1992. *Governance and development*. Washington, DC: World Bank.

# The right to water in a transboundary context: emergence of seminal trends

Christina Leb

The human right to water is receiving increasing recognition in national legislative acts and international instruments. Along with this recognition comes the legal delineation of state obligations with respect to water availability, access and supply, as well as to the protection of these individual water rights. Questions remain, however, as to whether "water-rich" states are under any obligation to provide the resource to "water-poor" states and, if so, to what extent. These questions are explored here with respect to international cooperation obligations under human rights law as well as with respect to the law applicable to transboundary water resources.

## Introduction

Population growth, economic development and climate change are putting increasing stress on global freshwater resources. While the relative impacts of these factors on water demand and availability vary, their cumulative impacts will significantly degrade future water supply conditions (Vörösmarty *et al.* 2000, UNFPA 2001). Water is becoming increasingly scarce in regions that host important and densely populated transboundary river basins and shared aquifer systems. These changes raise the question of how to ensure the satisfaction of basic human water needs in the long term. Water is vital for life and each human being has a fundamental interest in having access to (at least) a minimum amount of safe water. Water scarcity will not only impact on people's livelihoods but may also increase the potential for local and international conflict. International law takes account of the fundamental interests and vital needs of individuals.

The human person was introduced as a focal point of international law through the Charter of the United Nations (UNC), which lists the promotion of respect for human rights and fundamental freedoms as one of the purposes of the UN system (Zemanek 2000). The numerous multilateral human rights instruments adopted after the UNC corroborate this international concern for the well-being of individuals. Human rights apply first and foremost in the relationship between individuals and states. They are formulated as obligations owed by states to individuals in their territory and subject to their jurisdiction as the right-holders, making human rights an issue of primarily domestic implementation.

There is, however, an important international dimension as far as the obligations of states to provide water in sufficient quantity and quality are concerned. States coexist in a condition of hydrologic interdependence. Transboundary water systems (rivers, lakes and aquifers) supply drinking water to people living in different riparian states. The use or abstraction of water by one state in the system may have direct impacts on the quantity and quality of water available to the populations of other riparian states. Wider inter-linkages are created by the global hydrologic cycle that interconnects states beyond individual transboundary water systems. For instance, smoke and dust aerosols originating in West Africa impact cloud formation and rainfall over the Amazon basin, creating a causal link between the cooking habits on one continent and water availability on another (Baars *et al.* 2011). The ability of a state to provide water may thus depend on factors located outside its territory and jurisdiction.

Consequently, the question about the existence of rights and obligations that would enhance the ability of individual states to satisfy their populations' water needs in a state of hydrologic interdependence needs to be analyzed. The question is whether there exist international obligations to satisfy vital water needs (in a transboundary context) which would complement the human rights obligations that apply between states and individuals.

The present article explores this issue, looking at two areas of international law: human rights and the law applicable to the management and use of transboundary water resources. The rules of international humanitarian law, which also address water rights of individuals, are not analyzed here. The term "right to water" is used in a sense that is not limited to the definition of "the human right to water" provided in General Comment No. 15 of the Committee on Economic, Social and Cultural Rights (CESCR 2002), but also includes specific human rights obligations related to water such as those established for women, children and persons with disabilities and rights to water that may have their legal basis outside the body of human rights norms.

The first part of the article outlines the contributions of human rights law to the rights to water of individuals and analyzes the concepts of extraterritorial and international human rights obligations, asking the question about their potential contributions to the satisfaction of basic water needs in a transboundary context. The second part explores the rules applicable to the management and use of transboundary water resources. As has happened in other areas of international law, this body of norms has, over time, turned its attention towards the needs and interests of individuals. Recent developments provide novel approaches of integrating the rights to water of individuals into international water agreements, giving them a transboundary dimension.

## Promoting the satisfaction of vital human water needs through human rights norms

Water needs of individuals have received increased international attention with the mass media's circulation of countless images of people dying of hunger and thirst and of cholera and other waterborne diseases in regions affected by natural catastrophes or other calamities. States have responded both through international development cooperation and through the elaboration of legal standards to mitigate the impacts of such crises. International development efforts to improve basic water supply for people recently marked an important milestone by accomplishing the Millennium Development Goals target of halving the proportion of the population without sustainable access to safe drinking water ahead of the 2015 time limit (UNICEF/WHO 2012).

In international law, particularly in human rights law, there have also been significant improvements with respect to the recognition of rights to water of individuals. Neither the

1948 Universal Declaration on Human Rights nor the two guiding international human rights treaties, the 1966 International Covenant on Civil and Political Rights (ICCPR) and the 1966 International Covenant on Economic, Social and Cultural Rights (ICESCR), mention water.[1] Many years of international debate about the existence of a human right to water, developments in domestic legislation and international law, and case law promoting the protection of water rights of individuals and groups led to the recognition of the human right to water (and sanitation) by the United Nations General Assembly and the Human Rights Council in 2010. The respective resolutions of these two bodies have added an affirmative voice regarding the existence of human rights obligations concerning water.

### *The human right to water: from nowhere to be seen into the spotlight of attention*

As a consequence of the lack of reference to water in the initial human rights treaties, an international debate developed about the existence of human rights obligations with respect to water and on the question of whether the human right to water is a self-standing right. Some authors consider this lack of reference to have been an oversight. According to them, water was, just as air was, too obvious to mention and a human right to water could therefore be assumed (Greatrex 2004, Sacher and Windfuhr 2008). Others are critical of the view that a self-standing human right to water can be derived from other rights; they argue for careful analysis and precise legal definition of the status and nature of human rights obligations with respect to water (Dennis and Stewart 2004, Human Rights Council 2007).

The right to water was explicitly recognized in international declarations such as the 1977 Mar del Plata Action Plan (United Nations 1977) and the 1992 Dublin Statement on Water and Sustainable Development, while multilateral human rights treaties include either specific or implicit human rights obligations with respect to water. Water is for instance mentioned as part of specific obligations in the 1979 Convention on the Elimination of All Forms of Discrimination against Women (Article 14), the 1989 Convention on the Rights of the Child (Article 24) and the 2006 Convention on the Rights of Persons with Disabilities (Article 28). A number of regional human rights instruments equally refer to obligations with respect to access to water; examples are the 1990 African Charter on the Rights and Welfare of the Child (Article 14) and the 2003 Protocol to the African Charter on Human and People's Rights on the Rights of Women in Africa (Article 15). And even though neither the ICCPR nor the ICESCR mentions water in its text, both contain rights that are inherently linked to the protection of basic human water needs. Human rights obligations with respect to water have been derived from the right to life, the right to self-determination, the right to an adequate standard of living and the right to health.

In 2002, the Committee on Economic, Social and Cultural Rights (CESCR) issued its General Comment No. 15, on the right to water. General comments are not legally binding, yet are considered authoritative interpretations of the ICESCR (McCaffrey 2005, Tignino 2011). They guide state parties in the implementation of their human rights and reporting obligations. In General Comment No. 15, the CESCR confirms the existence of human rights obligations with respect to water, deriving them in particular from the rights to an adequate standard of living and to the enjoyment of the highest attainable standard of physical and mental health (Articles 11 and 12 ICESCR). The Committee defines state obligations to "respect, protect and fulfill" the right to water: states have to refrain from interfering with the right; they have to prevent other parties from interfering; and they have to promote, facilitate and provide access to adequate water. This adequacy requirement refers to availability, quality and accessibility (CESCR 2002). States have an immediate obligation to provide a minimum, essential level of safe water and to take meaningful steps towards the realization of the right without discrimination of any kind. In addition, they have a continued duty to achieve the highest attainable standard of realization of the right based on the resources available to them (CESCR 1990, 2002). The language of the document furthermore suggests that the human right to water could be considered a self-standing right. Paragraph 2 of the comment reads: "*The* human right to water entitles everyone to sufficient, safe, acceptable, physically accessible and affordable water for personal and domestic uses" (CESCR 2002, emphasis added). Subsequent paragraphs describing the normative content of the right utilize similarly suggestive formulations.

The 2010 resolutions of the General Assembly and the Human Rights Council offer an additional argument for the proponents of a self-standing human right to water. As with General Comment No. 15, their language suggests the existence of a self-standing, justiciable human right to water. The resolutions are certainly a demonstration of the international commitment to accord a high level of protection to the right to water, yet they do not have legally binding force. A number of states continue to reject the notion of a self-standing right to water (Khalfan 2010). This was also evidenced in the vote in the General Assembly: about one-quarter of the state parties present abstained.

While there is no universal agreement on the existence of a self-standing human right to water, the developments with regard to the right to water have been significant: state obligations towards individuals with regard to water went from not being considered at all in earlier human rights treaties to a clearly stated acknowledgement by the international community that the right to water is indeed fundamental to the fulfilment of all other rights and can be derived from existing human rights.

What does this mean for states that do not have the capacity and resources to achieve the full realization of the right to water? A state that does not take all necessary and feasible steps to meet its obligations violates the human right in question (Dennis and Stewart 2004). This includes the "failure to take into account its international legal obligations regarding the right to water" (CESCR 2002). So, what do these international legal obligations consist of?

### *International protection of the right to water*

While human rights obligations are primarily owed by states to the individuals in their territory and subject to their jurisdiction, they also have an international dimension. This is expressed *inter alia* in the Charter of the United Nations (1945). The UN system was established with an objective to "achieve international cooperation in solving international problems of an economic, social, cultural, or humanitarian character, and in promoting and encouraging respect for human rights" (Article 1 (3) UNC). Member states have assumed a general obligation to cooperate in order to achieve universal respect and observance of human rights and the improvement of living standards (Articles 55, 56). Additionally, states expressed an international legal interest in the implementation of human rights obligations through their ratification of international human rights treaties (Human Rights Committee 2004).

Some states and authors argue that this international legal interest goes as far as to provide an exception to the principle of non-interference in matters of domestic jurisdiction (Article 2 (7) UNC), in cases where massive human rights violations take place, and justifies international humanitarian intervention (Mertus 2000, De Schutter 2009). This interpretation of the UNC is, however, heavily criticized by others, who argue against the legality of such interventions and point to the inherent contradiction in the use of force for humanitarian purposes (Schreuer 1999, Zemanek 2000). A difference has to be noted between the legal interest of a state in human rights compliance by other states and the "injured right-holder" (Zemanek 2000). The right-holders with regard to human rights violations by states are the respective individuals and groups. The general cooperation obligation to encourage respect for human rights under the UNC should be understood as a positive obligation to promote compliance rather than an authorization to collectively punish violation.

International human rights covenants do provide for some extraterritorial and international obligations. This article follows the definitions of Künnemann (2004) according to which "extraterritorial obligations" are state obligations towards individuals and groups outside the respective state's territory, whereas state obligations to cooperate and to provide assistance to each other are referred to as "international obligations". Extraterritorial obligations are owed to individuals; international obligations concern the legal relationship between states. The present analysis is limited to illustrating the core elements of the ICCPR and ICESCR obligations.

*Extraterritorial obligations*

Both the ICCPR and ICESCR include extraterritorial obligations. The human rights and consequently the associated state obligations that are enshrined in these two covenants apply not only to individuals within the territory of a state but also to individuals who are outside a state's territory but under its effective control. This was confirmed by the International Court of Justice (ICJ) in its advisory opinion on the *Legal Consequences of the Construction of a Wall in the Occupied Palestinian Territory* (ICJ 2004). ICCPR Article 2 (1) stipulates that "each state party to the present Covenant undertakes to respect and to ensure to all individuals within its territory and subject to its jurisdiction the rights recognized in the present Covenant. . . ." The Court explained that the criteria of being within a territory of a state and of being under its jurisdiction are not cumulative (ICJ 2004), confirming the argument made by Meron (1995) that the word "and" in ICCPR Article 2 (1) has to be understood as "and/or" in this context. The ICJ further advised that both covenants "are applicable in respect of acts done by a state in the exercise of its jurisdiction outside its own territory" (ICJ 2004). This confirms an international dimension to human rights, which seeks to prevent that a state evades its obligations towards individuals who are effectively under its control but outside its territory. This applies to all human rights obligations encompassed in the rights of the ICCPR and the ICESCR, including those related to water.

Extraterritorial obligations that go beyond jurisdiction and effective control have also been argued for the right to self-determination. This right comprises a right of peoples to dispose of their natural wealth and "in no case may a people be deprived of its own means of subsistence", including water (ICCPR Article 1 and ICESCR Article 1). According to General Comment No. 12 of the Human Rights Committee (1984), all state parties to the international human rights covenants shall take positive measures to facilitate realization and respect for this right "not only in relation to their own peoples but *vis-à-vis* all peoples". This call for positive action is in line with the arguments made for the existing international human rights obligations according to the UNC.

*International obligations*

The existence of international human rights obligations that apply between states has been argued for based on Article 2 (1) ICESCR. This norm, for which no equivalent exists in the ICCPR, requires states "to take steps . . . through international assistance and cooperation" with a view to achieving progressive realization of the covenant rights. It imposes obligations of different natures. On the one hand, it establishes an obligation of the state towards the individuals in its territory and under its jurisdiction to do everything possible, including seeking assistance from other states, to achieve full realization of their rights. On the other hand, it can be argued that this rule imposes a certain duty of assistance. All state parties have agreed to the obligation to seek assistance; consequently they have agreed to an implicit duty to provide assistance. This duty applies between the state parties of the covenant and benefits states that lack the necessary resources to fulfil their human rights obligations (CESCR 1990). The initial state obligation towards individuals extends only as far as "to the maximum of [a state's] available resources" (Article 2 (1)). There is therefore a need to determine what is meant by this availability in order to establish the scope of the implicit international duty to assist. It follows from the text of the covenant that availability includes resources that can be obtained from other states in addition to those available domestically; otherwise, the obligation to seek assistance would be meaningless. At the same time, however, the international duty to provide to another state must be limited; its scope does not extend to the duty to meet all the needs of the state seeking assistance, but to provide only as much as a state is in a position to assist with (CESCR 1990, Craven 1995).

Article 23 ICESCR provides a non-exhaustive description of international actions that can be taken, including "the conclusion of conventions, the adoption of recommendations, the furnishing of technical assistance and the holding of regional meetings". This norm was reiterated by the General Assembly and the Human Rights Council in their 2010 resolutions on the right to water, calling upon "states and international organizations to provide financial resources, capacity building and technology transfer". In General Comment No. 15, the CESCR elaborated that "depending on the availability of resources, states should facilitate realization of the right to water in other countries, for example through provision of water resources, financial and technical assistance, and provide the necessary aid when required" (CESCR 2002). Applied in a transboundary basin context, this would mean that for instance in a basin where significant water imbalances exist, the states which are relatively water-abundant should ensure that the relatively water-scarce state(s) would be able to meet (at least) their core obligations with respect to essential water supply. The duty of assistance with regard to water could also extend beyond transboundary basins and could encompass activities to provide "virtual water" through food aid or trade and to assist in setting up water supply systems or ship bottled or bulk water to states in need, if for example they have been affected by prolonged drought or other natural disasters disrupting water supply. At the same time, this duty can only apply as far as it does not require a relatively water-abundant state to take retrogressive measures with regard to the rights to water of its own population because this in turn would constitute a violation of its own human rights obligations (CESCR 2002).

The wording in General Comment No. 15 – i.e. "should facilitate" rather than "shall provide" – confirms the limited scope of these international duties. Indeed, these international obligations of assistance are controversial; states have not yet invoked a right to assistance as an international human rights obligation owed between states but rather treat them as moral and political aims (Alston and Quinn 1987, von Tigerstrom 2007, Tobin 2012). At the same time, Article 23 makes an important contribution in underlining the potential for mutual support and complementarity of different international law instruments when calling for the conclusion of additional conventions and the adoption of recommendations. The subsequent analysis illustrates that water treaties can be valuable tools for promoting human rights objectives, including the protection of the right to water in a transboundary context.

## Protection of water needs of individuals through international water law

The international law applicable to the management and protection of transboundary water resources comprises a set of principles, rights and obligations that facilitate the promotion of access to basic water supply for all. Human water needs receive protection through the general principles of transboundary water law and through treaties adopted at the regional and at the basin level. Recent water treaties confirm a trend towards greater appreciation of the needs of individuals in international law and, more specifically, towards the protection of vital human water needs through international water law.

### *Customary international law and the consideration of basic water needs*

Codifications such as the 1966 Helsinki Rules (ILA 1967) and the 1997 UN Convention on the Law of the Non-navigational Uses of International Watercourses (UN Watercourses Convention) shed light on the content and scope of the customary principles and rules of international water law, which crystallized over time through consistent state practice on the use of shared watercourses (United Nations General Assembly 1997). While not legally binding – as of July 2012 the UN Watercourse Convention lacks 8 ratifications to enter into force – these documents illustrate the way in which the general principles of transboundary water law take account of the water needs of basin populations. The principle of equitable and reasonable utilization, the obligation not to cause significant harm and the general duty to cooperate all play a role in safeguarding basic water needs of individuals.

#### *General principles of transboundary water law*

The obligation not to cause significant harm demands that basin states exercise due diligence in the utilization of their territory and of the shared water resources that flow within it so as not to cause significant damage in the territory of other basin states (ILC 1994). The threshold set by the term "significant" means that the harm can be measured by objective evidence when it causes a "real impairment of use" (ILC 1994). Thus, a state would have to guard against polluting shared water resources to such an extent that the basin population of another state, which exclusively depends on these water resources, can no longer use these waters to satisfy its basic water needs; where pollution is so heavy that it is no longer safe to drink or to eat fish from these waters, significant harm is caused.

Pollution is a harm that is generally inflicted by upstream on downstream states, given that it takes the course of the water flow. It is important to note, though, that harm does not always flow downstream, as one may think following the laws of physics; it can also flow upstream. Downstream states can affect upstream states over time: the former can foreclose future uses of upstream states by establishing prior water uses and then claiming the rights to such uses (McCaffrey 2007a, Salman 2010). In regions with rapidly growing populations and changing climates, there is a risk that prior uses downstream may affect an upstream state's capacity to satisfy water needs of its population over time. The general principles of transboundary water law apply equally to all basin states. It will therefore have to be assessed whether the foreclosure of future uses actually reaches the threshold of "significant" harm in the given case. In addition, one may have to assess whether the prior uses fall within the ambit of equitable and reasonable use, in particular if one shares the opinion that this latter principle prevails over the obligation not to cause significant harm (McIntyre 2007, Kaya 2008, Salman 2010). If such an interpretation is assumed, prior uses that harm upstream states over time may still be justified as far as they are equitable and reasonable. This issue is also relevant when considering that the various principles of international water law interact with each other in the absence of an explicitly established hierarchy (ILC 1994, Sixth Committee 1996–97, McCaffrey 2007b). In cases where a tension arises over a use that is equitable and reasonable but causes significant harm, the respective states are under an obligation to consult and negotiate with each other based on the general duty to cooperate. Equity and reasonableness apply as guiding criteria in the efforts to achieve a mutually agreed balance between the interests at stake (ILC 1994).

Article 5 of the 1997 UN Watercourses Convention identifies the scope of the principle of equitable and reasonable utilization as an obligation of states to take the interests of other concerned riparian states into account (ILC 1994). It entails a correlative right to an equitable and reasonable share of waters and to the benefits thereof (ILC 1994). The principle addresses the protection of water needs through a variety of factors that have to be weighed against each other. The UN Watercourses Convention (Article 6) and a number of water treaties include illustrative lists of factors that have to be taken into account, including the hydrologic, climatic, economic and social conditions of individual basins. Factors that concern the social and economic needs of all basin states and the water needs of the basin-dependent population have to be weighed against a whole range of other criteria, including prior and future uses, and assessed within the context of the specific basin conditions. Since the principle does not establish a priority between different criteria, it also does not preclude uses that might impact on the capacity to satisfy basic human needs from the water resources that flow in the shared water system.

The question therefore arises whether the legal entitlement to an equitable share of the transboundary waters serves as sufficient guarantee for the satisfaction of basic human needs in a basin state. Experts on this issue come to different conclusions. According to Bourquain (2008), the principle of equitable and reasonable utilization establishes a legal basis from which an obligation to ensure that sufficient quantities of water are left in a transboundary system for the satisfaction of vital human needs, including those of co-riparian populations, can be derived. He argues that "equitable" cannot mean that a threat to human life elsewhere is allowed. The principle therefore contains a prohibition against threatening the life of co-riparian populations through for example the over-abstraction or heavy pollution of shared waters. In a different manner, McCaffrey argues that an obligation to ensure enough water is left in the system for the satisfaction of vital water needs of other co-riparian states does not and cannot exist, because it may be contrary to the hydrologic realities of some basins. In the extreme situation where there is not enough water to cover the needs of the basin population of all riparian countries, a state cannot be obliged to deprive its own people of vital water supplies in order to be able to satisfy the needs of the population of another riparian state (McCaffrey 1992). This would mean that the principle of equitable and reasonable utilization does not endorse, but also does not prohibit, a threat to life elsewhere as an equitable use and that, in the exceptional case in which there is not enough water in the system, states will have to engage in efforts to find a common solution – such as to bring in water from alternative sources – based on the general duty to cooperate.

In conclusion, it can be argued that the general principles do protect vital water needs of basin-dependent populations, even in the absence of a hierarchy between the different factors and criteria that determine equitable and reasonable use. This protection is also taken up in other articles of the UN Watercourses Convention and also in the 2008 Draft Articles on the law of transboundary aquifers, prepared by the International Law Commission (ILC) as a codification of customary rules that apply to the management of groundwater resources (United Nations General Assembly 2008).

*Codification for the protection of vital human needs*

Both the UN Watercourses Convention and the 2008 ILC Draft Articles stipulate that vital human needs have to be given extra attention. The definition of "vital human needs" is less comprehensive than the scope of human rights obligations with respect to food and water; it is limited to sufficient water to "sustain human life, including both drinking water and water required for production of food in order to prevent starvation" (ILC 1994). This definition covers the core obligations of the human right to water, but does not include other elements such as the progressive improvement of conditions, acceptability and cultural values. The approach with regard to the protection of vital human water needs differs somewhat between the convention and the draft articles.

Article 10 of the UN Watercourses Convention stipulates that "in the absence of agreement or custom to the contrary, no use of an international watercourse enjoys inherent priority over other uses". Therefore, in the case of conflict between different uses, the controversy needs to be resolved according to the criteria established by the principle of equitable and reasonable use and the obligation not to cause significant harm, while giving special attention to the basic water needs of individuals. Article 10 re-emphasizes the importance of the factor of "social and economic needs" that is used in determining equitable and reasonable use (ILC 1994). The 2008 ILC Draft Articles do not make this prioritization dependent on a conflict between uses, but give special weight to vital water needs directly during the process of assessing uses. Article 5 (2) reads: "However, in weighing different utilizations of a transboundary aquifer or aquifer system, special regard shall be given to vital human needs." This formulation would arguably provide a somewhat stronger protection than the text of Article 10 of the UN Watercourses Convention; the due consideration of vital human needs early on may already prevent use conflicts. The approach of the draft articles is further strengthened by their Article 17: where an emergency, occurring due to natural causes or human conduct, jeopardizes the satisfaction of vital human needs, states are allowed to take all measures that are necessary to meet these needs for the time of the emergency, notwithstanding the obligations which might arise from the principle of equitable utilization and the obligation not to cause significant harm.

At the same time, the way these parts of the convention and of the draft articles are formulated expresses some shortcomings with respect to the protection of the rights to water of individuals. States have to give "special regard" to vital human needs and "take those needs into account"; the articles do not impose an obligation to provide nor do they impose an obligation to satisfy these needs before others (Bourquain 2008). A stronger legal protection of rights to water is provided through water agreements that have been adopted at the regional and basin level.

***Additional protection through regional and basin-level agreements***

Regional and basin-level water treaties, such as the 2000 SADC Revised Protocol on Shared Water Courses and the 2003 Protocol for Sustainable Development of Lake Victoria Basin, provide some protection of the rights to water of individuals through codification of the general international water law principles. The 1999 London Protocol on Water and Health to the 1992 UNECE Convention on the Protection and Use of Transboundary Watercourses and International Lakes stands out as a regional instrument that promotes the implementation of human rights obligations concerning water without explicitly mentioning them. A close reading of the protocol reveals that it translates the core content of the human right to water into an international water treaty. The protection of human water needs is part and parcel of the protocol's objective to promote human health and well-being.

In the first paragraph of the preamble, the parties to the protocol express their conviction that "water in quantities, and of quality, sufficient to meet basic human needs is a prerequisite both for improved health and for sustainable development". Its articles pick up on the human rights elements of adequacy of supply, non-discrimination and the requirement to adopt national and local actions to achieve access to drinking water and sanitation for all. Equally inherent in the protocol is the idea of international cooperation to pursue the aim to provide safe drinking water. Its rules echo the international cooperation obligations of Article 2 (1) of the ICESCR; however, having its legal roots in international water law, the protocol reaches far beyond the human rights standards for international assistance. It establishes cooperation obligations in support of its transboundary water management objectives and also with regard to the implementation of domestic plans and targets. For instance, state parties sharing transboundary waters have to exchange information and prepare joint water management plans and establish warning systems to respond to water-related diseases. The elaborate international cooperation obligations included in the protocol are the added value that water law brings to the achievement of human rights objectives within the UNECE framework (Tanzi 2010).

Rules establishing a priority – and thus an implicit protection – of water uses for domestic and sanitary purposes are also included in a number of basin-level agreements. In these treaties, state parties take on an international engagement to give precedence to these uses over others. Such engagements form part of the water management principles in these basins; they confirm an agreed priority for human water uses when assessing equitable and reasonable use and they imply certain water quality requirements that have to be observed by the individual state parties. The 1909 Boundary Waters Treaty is interesting in this context, considering that in the early 20th century priority was still given to navigation over other uses in other regions of the world, in particular Europe and parts of Africa. States agreed to prioritize human and domestic uses over economic uses also in the Mekong and the Cuareim river basins. In 1975, the four lower riparian states of the Mekong River – Cambodia, Laos, Thailand and Vietnam – declared their intention to adhere to a priority for domestic and urban water uses in their Joint Declaration of Principles for Utilization of the Waters of the Lower Mekong Basin (Radosevich and Olson 1999). In the 1997 Agreement on the Utilization and Development of the Water Resources of the Cuareim River Basin, Brazil and Uruguay agreed on a clear priority for drinking water supply to riparian population and put state organs under an obligation to augment water supply within their available means.

With regard to drinking water quality standards on transboundary river basins, the most comprehensive rules are probably those adopted in Europe through supra-national legislation. Paragraph 35 of the European Water Framework Directive gives an additional international dimension to certain water quality standards, requiring member states to make efforts for appropriate coordination with non-member states that share the same transboundary basin.

Recent treaties adopted on the African continent are more specific with regard to the human right to water. The 2008 Water Charter of the Niger Basin and the 2002 Senegal Water Charter both give priority to the satisfaction of vital human needs and recognize a fundamental human right to safe drinking water. The Niger Basin states consider the right to water as a principle guiding their cooperation in managing the shared resources (Preamble). The right is defined as the fundamental right to sufficient physically accessible water at affordable cost and of a quality that is acceptable for personal and domestic use by everyone (Article 1). In a similar manner, the Senegal Water Charter recognizes the enjoyment of the right to water as an explicit objective of any repartition of the river's water (Article 4). Both treaties establish a clear priority of human needs for the use of the shared water resources. In the Senegal River case, the obligation to satisfy vital human needs additionally has priority over all technical and economic considerations of water allocation (Articles 6 and 7). In recognizing the fundamental right of each individual to drinking water, these treaties go beyond the protection of vital human needs through general principles and norms of international water law; they establish positive obligations to provide for the satisfaction of vital human needs in a transboundary context and also cover other elements of the human right to water more comprehensively. The respective basin organizations, the Organisation pour la mise en valeur du fleuve Sénégal and the Niger Basin Authority, watch over the implementation of these rights, providing additional institutional protection.

## Conclusion

With growing populations, continued economic development and climate change, international concern about the satisfaction of basic human water needs is on the rise. Two out of three people could be living under conditions of "water stress" in 2025 (FAO 2007). More and more states will no longer be in a condition to satisfy their water needs on their own. Thus, the question about the existence of international obligations that may compensate for the unequal geographic distribution of global water resources gains increasing relevance. The above analysis reveals that human rights law offers only a limited response to this transboundary challenge. While state obligations to respect, protect and fulfil the right to water of individuals are by now well defined, the scope of international obligations of assistance remains vague and controversial. Developments in the area of international water law are gradually filling this gap. The general principles of transboundary water law ensure that the needs of individuals are taken into account by riparian states in the process of developing their shared water resources. These principles and water treaties that prioritize domestic uses and vital human water needs protect the core content of human rights obligations with respect to basic water supply. Still more effective legal protection comes from agreements adopted more recently at the regional and basin levels. The UNECE Protocol on Water and Health and the Niger and Senegal Water Charters stand out as landmark instruments, which combine human rights objectives with regard to water with the more elaborate international cooperation obligations that exist in the area of transboundary water law. These treaties underline the mutually supportive and complementary character of different areas of international law and the intrinsic coherence of the international legal system.

## Acknowledgements

The author would like to thank Ariella D'Andrea and the two anonymous peer reviewers for their constructive comments on an earlier draft of this article. Research for this article was carried out with the support of the Swiss National Science Foundation within the context of a comprehensive analysis of international water agreements and international cooperation obligations.

## Note

1. Details of all statutory references are listed in the Appendix.

## References


Alston, P., and Quinn, G., 1987. The nature and scope of states parties' obligations under the International Covenant on Economic, Social and Cultural Rights. *Human Rights Quarterly*, 9, 156–229.

Baars, H., *et al.*, 2011. Further evidence for significant smoke transport from Africa to Amazonia. *Geophysical Research Letters*, 38, L20802.

Bourquain, K., 2008. *Freshwater access from a human rights perspective*. Leiden: Martinus Nijhoff.

CESCR (Committee on Economic, Social and Cultural Rights), 1990. *The nature of states parties obligations (Art. 2, para. 1): 14/12/90, CESCR General Comment 3. (General Comments)*. UN Doc. E/1991/23.

CESCR (Committee on Economic, Social and Cultural Rights), 2002. *General Comment No. 15. The right to water (arts. 11 and 12 of the International Covenant on Economic, Social and Cultural Rights)*. UN Doc. E/C.12/2002/11.

Craven, M., 1995. *The International Covenant on Economic, Social and Cultural Rights: a perspective on its development*. Oxford: Clarendon Press.

Dennis, M., and Stewart, D.P., 2004. Justiciability of economic, social, and cultural rights: should there be an international complaints mechanism to adjudicate the rights to food, water, housing, and health? *American Journal of International Law*, 98 (3), 462–503.

De Schutter, O., 2009. The status of human rights in international law. *In*: C. Krause and M. Scheinin, eds. *International protection of human rights: a textbook*. Turku/Abo, Finland: Institute for Human Rights, Åbo Akademy University, 39–60.

Greatrex, H., 2004. The human right to water. *Human Rights Research Journal*, 2(1), 1–14.

FAO (Food and Agriculture Organization), 2007. *Coping with water scarcity. Challenge of the twenty-first century*. FAO: Rome. Available from: http://www.fao.org/nr/water/docs/escarcity.pdf [Accessed 2 August 2012].

Human Rights Committee, 1984. *CCPR General Comment No. 12: The right to self-determination of peoples (Art 1)* [online]. 13 March 1984. Available from: http://www.unhcr.org/refworld/docid/453883f822.html [Accessed 20 April 2012].

Human Rights Committee, 2004. *General Comment No. 31: The nature of the general legal obligation imposed on states parties to the covenant*. UN Doc. CCPR/C/21/Rev.1/Add.13.

Human Rights Council, 2007. *Report of the United Nations High Commissioner for Human Rights on the scope and content of the relevant human rights obligations related to equitable access to safe drinking water and sanitation under international human rights instruments*. UN Doc. A/HRC/6/3, 16 August 2007.

Human Rights Council, 2010. *Resolution 15/9 on Human rights and access to safe drinking water and sanitation*. UN Doc. A/HRC/RES/15/9.

International Court of Justice, 2004. *Legal consequences of the construction of a wall in the occupied palestinian territory: advisory opinion of 9 July 2004*. ICJ Reports, New York: United Nations.

ILA (International Law Association), 1967. *Report of the fifty-second conference held at Helsinki, 14–20 August 1966*. London: ILA, 484–533.

ILC (International Law Commission), 1994. Draft articles on the law of the non-navigational uses of international watercourses and commentaries thereto and resolution on transboundary confined groundwater. *Yearbook of the International Law Commission, 1994*, II (Part Two), 89–135.

Kaya, I., 2008. *Equitable utilization: the law of non-navigational uses of international watercourses*. Aldershot: Ashgate.

Khalfan, A., 2010. *UN affirms the right to water and sanitation as legally binding*. [online]. Available from: http://livewire.amnesty.org/2010/10/12/a-historic-step-as-un-affirms-the-right-to-water-and-sanitation-as-legally-binding/ [Accessed 30 July 2012].

Künnemann, R., 2004. Extraterritorial application of the International Covenant on Economic, Social and Cultural Rights. *In*: F. Coomans and M. Kamminga, eds. *Extraterritorial application of human rights treaties*. Antwerp: Intersentia, 201–231.

McCaffrey, S., 1992. A human right to water: domestic and international implications. *Georgetown International Environmental Law Review*, 5 (1), 1–24.

McCaffrey, S., 2005. The human right to water. *In*: E. Brown Weiss, L. Boisson de Chazournes and N. Bernasconi-Osterwalder, eds. *Fresh water and international economic law*. Oxford: Oxford University Press, 93–115.

McCaffrey, S., 2007a. Some developments in the law of international watercourses. *In*: M. Kohen, ed. *Promoting justice, human rights and conflict resolution through international law: liber amicorum Lucius Caflisch*. Leiden: Martinus Nijhoff, 785–791.

McCaffrey, S., 2007b. *The law of international watercourses: non-navigational uses*. 2nd ed. New York: Oxford University Press.

McIntyre, O., 2007. *Environmental protection of international watercourses under international law*. Aldershot: Ashgate.

Meron, T., 1995. Extraterritoriality of human rights treaties. *American Journal of International Law*, 89 (1), 78–82.

Mertus, J., 2000. Reconsidering the legality of humanitarian intervention: lessons from Kosovo. *William and Mary Law Review*, 41 (5), 1743–1787.

Radosevich, G., and Olson, D., 1999. Existing and emerging basin arrangements in Asia: Mekong River Commission case study. *In: Third Workshop on River Basin Institution Development, 24 June 1999, The World Bank, Washington, DC*. Available from: http://siteresources.worldbank.org/INTWRD/918599-1112615943168/20431963/MekgongRiverComCaseStudy.pdf [Accessed 31 July 2012].

Sacher, D., and Windfuhr, M., 2008. The debate on "water as a human right" and its implications for development assistance. *In*: S. Neubert, W. Scheumann and M. Kipping, eds. *Water Politics and Development Cooperation*. Heidelberg: Springer, 147–176.

Salman, S., 2010. Downstream riparians can also harm upstream riparians: the concept of foreclosure of future uses. *Water International*, 34 (4), 350–364.

Schreuer, C., 1999. Is there a legal basis for the NATO intervention in Kosovo? *International Law FORUM du droit international*, 1 (3), 151–155.

Sixth Committee of the General Assembly, 1996–97. *Summary records of meetings Nos. 12 to 25, 48 and 51 to 62, respectively held from 7 to 25 October 1996, on 27 November 1996 and from 24 March to 4 April 1996*. UN Doc. A/C.6/51/SR.12-SR.62

Tanzi, A., 2010. Reducing the gap between international water law and human rights law: the UNECE Protocol on Water and Health. *International Community Law Review*, 12 (3), 267–285.

Tignino, M., 2011. *L'eau et la guerre: éléments pour un régime juridique*. Paris: Éditions Bruylant.

Tobin, J., 2012. *The right to health in international law*. Oxford: Oxford University Press.

UNICEF/WHO, 2010. *Millennium Development Goal drinking water target met*. Joint news release, 6 March [online]. Available from: http://www.who.int/mediacentre/news/releases/2012/drinking_water_20120306/en/ [Accessed 13 March 2012].

United Nations General Assembly, 1997. *Convention on the Law of Non-Navigational Uses of International Watercourses*. UN Doc. A/RES/51/299 (not in force).

United Nations General Assembly, 2008. *Resolution 63/124 on The law of transboundary aquifers*. UN Doc. A/RES/64/124.

United Nations General Assembly, 2010. *Resolution 64/292 on The human right to water and sanitation*. UN Doc A/RES/64/292.

UNFPA (United Nations Population Fund), 2001. *State of world population 2001*. New York: UNFPA.

United Nations, 1977. *Report of the United Nations Water Conference, Mar del Plata 14–25 March, 1977*. New York: United Nations. UN Doc. E/CONF.70/29.

von Tigerstrom, B., 2007. *Human security and international law: prospects and problems*. Oxford: Hart.

Vörösmarty, C., Green, P., and Salisbury, J., 2000. Global water resources: vulnerability from climate change and population growth. *Science*, 289, 284–288.

Zemanek, K., 2000. New trends in the enforcement of *erga omnes* obligations. *In*: J.A. Frowein and R. Wolfrum, eds. *Max Planck yearbook of United Nations law*. Netherlands: Kluwer International, 1–52.

## Appendix: statutory references

African Charter on the Rights and Welfare of the Child (1990) OAU Doc CAB/LEG/24.9/49.

*Ajuste complementario al acuerdo de cooperación entre el gobierno de la Republica oriental del Uruguay y el gobierno de la Republica federativa del Brasil para el aprovechamiento de los recursos naturales y el desarrollo del a cuenca del rio Cuareim* (1997). Available from: http://ocid.nacse.org/tfdd/tfdddocs/575SPA.pdf [Accessed 31 July 2012].

Boundary Waters Treaty - Treaty relating to Boundary Waters, and Questions Arising between the United States and Canada (1909) 79 Legislative Texts 260.

Charte de l'eau du bassin du Niger (2008). Available from: http://www.abn.ne/ [Accessed 30 July 2012].

Charte des eaux du fleuve Sénégal (2002) *In* L. Boisson de Chazournes, R. Desgagné, M. Mbengue and C. Romano. *Protection internationale de l'environnement*, 2nd edition, Paris: Pedone, 301–314.

Charter of the United Nations (1945) 1 UNTS XVI.

Convention on the Rights of Persons with Disabilities (2006) UN Doc. A/61/611.

Convention on the Elimination of All Forms of Discrimination against Women (1979) 1249 UNTS 113.

Convention on the Rights of the Child (1989) 1577 UNTS 3.

Directive 2000/60/EC of the European Parliament and of the Council of 23 October 2000 establishing a framework for Community action in the field of Water Policy (2000) *OJL 327.*

Dublin Statement on Water and Development (1992). Available from: http://www.un-documents.net/h2o-dub.htm [Accessed 31 July 2012].

International Covenant on Economic, Social and Cultural Rights (1966) 993 UNTS 3.

International Covenant on Civil and Political Rights (1966) 999 UNTS 171.

Protocol for Sustainable Development of Lake Victoria Basin (2003) LEX-FAOC041042.

Protocol on Water and Health to the 1992 Convention on the Protection and Use of Transboundary Watercourses and International Lakes, done in London (1999) UN Doc. MP.WAT/2000/1/EUR/ICP/EHCO 020205/8Fin.

Protocol to the African Charter on Human and People's Rights on the Rights of Women in Africa (2003). Available from: http://www.unhcr.org/refworld/docid/3f4b139d4.html [Accessed 31 July 2012].

Revised Protocol on Shared Watercourses in the Southern African Development Community (SADC Revised Protocol) (2000). Available from: http://www.sadc.int/index/browse/page/159 [Accessed 31 July 2012].

Universal Declaration of Human Rights (1948). Available from: http://www.un.org/en/documents/udhr/ [Accessed 31 July 2012].

# Scottish Water: a public-sector success story

Sarah Hendry

**ABSTRACT**
This article explores the governance and regulation of water services in Scotland, including the emergence of a single national provider, Scottish Water; the role of government and its novel Hydro Nation strategy; and the activities of the regulators – for prices, environment, drinking water quality and consumer protection – as Scottish Water moved from amongst the lowest, to amongst the best, of its benchmarked comparators. It examines the recent initiative of a Customer Forum; and draws conclusions on the potential for a public service provider to evidence private-sector efficiency whilst contributing fully to public policy goals.

## Introduction

Scotland is part of the United Kingdom (UK), but is a separate jurisdiction and has always retained a separate legal system; private law, including property law, was protected.[1] Historically, Scotland had a riparian system of water use rights developed from Roman law (Reid & Zimmerman, 2000), and Scots water law has always been distinct from the law in England. As environmental law developed in the nineteenth century, other areas (especially waste management and industrial air pollution) tended to be made UK-wide, and this continued in the twentieth century in implementation of European Union (EU) directives, but this was not the case for water; lack of parliamentary time in the Westminster (UK) Parliament meant that reform of this area, like other areas of purely Scottish law, was often neglected. Since 1999, Scotland has had a devolved Parliament with powers over property law, environmental law and implementation of relevant EU legislation.[2] This gave the new Parliament scope to develop policy and law in relation to different aspects of water, and it has continued to do so, with five primary Acts of Parliament and numerous regulations and policy statements. Some of this work has been especially relevant to water resources, for example the implementation of the EU's Water Framework Directive[3] and Flood Risk Management Directive[4]; but much of it has been concerned with the delivery of urban water services (drinking water and wastewater). The Parliament began this work with an extensive inquiry into the water industry (Scottish Parliament Transport and Environment Committee, 2001) and continues to be keenly involved in this area of reform. Most recently, the Scottish government has developed a Hydro Nation strategy (more on this below) in recognition of the importance of water and water policy to Scotland.

## Water services in Scotland: creating Scottish Water as the public-sector provider

Water services in Scotland are provided by Scottish Water, a public corporation established by statute under the Water Industry (Scotland) Act 2002 (the 2002 Act).[5] It replaced three regional water services authorities, themselves established in 1996 to provide water and sewerage services following abolition of a 'top tier' of regional councils providing strategic services.[6] Thus, Scotland is one of many jurisdictions to have made the move from municipal supply to a separate specialist provider. Scottish Water supplies drinking water to more than 95% of the population, and wastewater services to more than 90%, with the remainder, mostly in remote rural areas, served by small private systems. Scottish Water (like its predecessors) is a vertically integrated supplier, carrying out the whole value chain from abstraction to discharge. This article will analyze the development of Scottish Water and its regulatory and governance framework since its inception.

### *Private-sector participation*

The creation of the three regional authorities in 1996 took place during a time of highly politicized, and much analyzed, reforms in the water services sector in England (Bakker, 2003). In 1989, the 10 regional monopolies in England were divested as part of the Thatcher administration's privatization agenda.[7] This promoted a neoliberal critique of inefficient public monopolies and a global ideological shift towards private-sector delivery of utility services, reflected *inter alia* in the fourth Dublin Principle, that water is an economic good for which full costs should be recovered (Dublin Statement, 1992). The implications of Principle Four are still an important part of the global debate today, and views remain heavily polarized. Although few countries other than England fully divested the asset base of water services, many others have brought in the private sector by other means, including concessions and build-own-operate contracts (Delmon, 2001); these may also cause political tension. The use of the term 'privatization' to mean divestiture as well as any form of private-sector participation is unfortunate and contributes to a lack of clarity in the ongoing debate; this article will use 'private-sector participation' for the latter, and 'divestiture' for the former.

Divestiture of the water industry did not take place in Scotland, for both political and economic reasons (Hendry, 2003). A popular public campaign to 'Save Scotland's Water' and an unofficial referendum organized by Strathclyde Region (the largest of the regional councils then providing the service) demonstrated significant opposition, and in the end the regional boards were seen as less controversial. As these would no longer be part of an elected authority, it was necessary to provide a mechanism to set prices for their services, and this was achieved by the appointment of a water industry commissioner who would advise government on the levels of charges.[8] His appointment marked the beginning of the current regulatory model.

Besides the political motivation, another reason for the divestiture of the English water companies (as with other forms of private-sector participation in other countries) was the need to secure private capital for investment in order to meet the requirements of the EU's environmental directives, including those on Drinking Water Quality[9] and

Bathing Waters,[10] and the Urban Waste Water Treatment Directive.[11] These required significant capital investment, in Scotland as well as England, and a new generation of wastewater treatment plants were built in Scotland under the Private Finance Initiative, a form of public–private partnership. Engineering consortia borrowed private money (at significantly higher rates than public-sector lending) and constructed these plants under build-own-operate-transfer schemes. As with similar schemes in other places, the customer's contact remains with the public provider, and hence there may be little public awareness of the schemes. These did secure capital that was not public debt, and in 2001 the water industry commissioner considered that the schemes offered value for money, but by 2005 Scottish Water's improved performance led to their being seen as overly expensive (Water Industry Commissioner for Scotland, 2005). Nonetheless, some 50% of Scotland's public wastewater plants were built with private capital and are operated by private firms.

## Hydro Nation

In 2010, the Scottish Government launched its Hydro Nation initiative (Scottish Government, 2010). Hydro Nation was conceived as a policy platform to bring together different aspects of the management of Scotland's water resources, in order to maximize the value of that resource base. The first consultation was very much focused on Scottish Water, and addressed, *inter alia*, how Scottish Water could maximize the use of its assets, in order to support various policy objectives. These were seen as including, in the domestic arena, maximizing renewable energy generation; but also, further afield, providing technical advice, and advice on regulation and governance models, to water services in other jurisdictions.

The second consultation was a prospectus for draft legislation (Scottish Government, 2012) resulting in a Water Resources Bill being introduced to Parliament in summer 2012. This included provisions directly relevant to Scottish Water and its activities within the Hydro Nation, and wider measures on the ministers' role, including a duty to ensure the "development of the value" of Scotland's water resources. These proposals were enacted in the Water Resources (Scotland) Act 2013 (the 2013 Act).[12]

### *Scottish Water's role in Hydro Nation*

The Hydro Nation strategy is not exclusively focused on Scottish Water; it is intended and expected to engage with a wide variety of entities, all active in water in Scotland, and to bring diverse benefits. There is recognition of Scottish research expertise across a variety of water-related disciplines, contributing to domestic and international agendas, and of the importance of the domestic supply chain, especially in terms of innovation. Again, innovation is seen as relevant both internally to Scotland and abroad. For Scottish Water, the domestic agenda would involve its contribution to other Scottish government policy goals, including climate change response, biodiversity, flood management and the protection of water-related ecosystem services – as well as economic development and public health – reflected in the specific policy objectives for Scottish Water that frame the strategic review. Some of these objectives would be relevant to Scottish Water's core business of providing water and wastewater services, but Scottish

Water was also given two specific duties under the 2013 Act: to develop the commercial value of its assets,[13] and to promote the use of its assets to generate renewables.[14] These new duties might be effected through one of Scottish Water's subsidiaries, which operate commercially at home (Scottish Water Horizons) and abroad (Scottish Water International). There was also an expanded enabling power to clarify that Scottish Water was able to engage in any non-core activity "that it considers will assist in the development of the value of Scotland's water resources".[15]

'Value' is defined in the 2013 Act as including economic, social, environmental and other benefits, and refers to both monetary and non-monetary worth.[16] The rationale behind the 'value' provisions in the 2013 Act is complex – and tangential to this article (Hendry, 2016a; Martin-Ortega et al., 2013) – but the broad Hydro Nation concept and the central role of Scottish Water are both important. The first indicates clearly the government's understanding of the need to protect and properly manage the water resources for the future, even in a country as blessed with water resources as Scotland. The second demonstrates the extent to which Scottish Water is seen as an active player in the government's policy agendas. All water services providers have a role in enhancing the social and economic development of the citizens they serve; bulk abstractors, and dischargers of wastewater, contribute to resource management and environmental objectives. In England, where the water companies are also vertically integrated and subject to a similar policy and regulatory environment, but where the ownership of the assets is private, the providers also have these roles and they are enforced in similar ways, but there is nothing similar to Hydro Nation; and if there were, it would be difficult to treat private water companies in the same way as Scottish Water. In general, we look to regulation to ensure efficiency and counter monopolistic tendencies in the public and the private sector, but ownership does have some relevance, and in terms of achieving public policy outcomes unhindered by shareholder pressures, this article will argue that a regulated public entity is a valuable contributor.

## Scottish Water's regulatory and governance model

Scottish Water operates in a broad regulatory context, whereby policy objectives and charging principles are set by the government (Scottish Government, 2014a, 2014b). The "principles of charging" are made in a policy statement.[17] The ministerial objectives are made as directions and are binding on Scottish Water and its regulators.[18] The principal (economic) regulator is the Water Industry Commission for Scotland (WICS); the five-person commission replaced the single commissioner under the Water Services (Scotland) Act 2005 (the 2005 Act).[19] Unusually for the public sector, WICS was given powers to set charges, rather than make recommendations to ministers, under the 2005 Act. Scottish Water is also regulated by the Scottish Environment Protection Agency, which authorizes all uses of the water environment, including bulk abstractions, impoundments and river works, as well as wastewater discharges.[20] The current system of water use licensing was introduced in 2005 as part of the reforms to implement the EU's Water Framework Directive, which also mandated a comprehensive system of river basin planning.[21] As a vertically integrated provider making bulk abstractions, Scottish Water is a major stakeholder in this planning process and is responsible for many measures required to improve the ecological status of water bodies under the

Water Framework Directive. There is a separate Drinking Water Quality Regulator, also established under the 2002 Act, and currently, consumer interests are represented by Citizens Advice Scotland. Consumer protection will be addressed further below.

### *The business planning process and a broad approach to economic regulation*

All of the regulators work with government and Scottish Water itself to achieve the requirement of the 2005 Act, which is that services should be provided at the "lowest overall reasonable cost".[22] The system takes a broad approach to economic regulation in that it is accepted that the charges levied, and borrowing raised, must be sufficient not just to cover the costs of service delivery but also to meet all mandatory environmental requirements, as well as the social objectives laid out in the principles of charging.

Scottish Water is regulated using a revenue cap, broadly the 'price cap' system developed in the 1980s in England for the divested utilities, and still used for the English water companies. Price cap regulation operates in defined phases, and currently, a six-year pricing period is used in Scotland. The government operates a 'quality and standards' process, by which all the regulators and the government agree on the technical detail of Scottish Water's investment programme. In essence, Scottish Water develops a draft business plan, containing their proposals for the next strategic review period. WICS assesses this in the light of the ongoing reporting which they (and other regulators) require from Scottish Water, and then issues a draft determination, which is put out for consultation – to the ministers, Scottish Water and the customer representatives, and also published for public comment.[23] The commission must have regard to any directions or other guidance issued by the ministers.[24] Once the determination is finalized, Scottish Water have the right to appeal, to the (UK) Competition and Markets Authority.[25] Otherwise, the determination will apply through that pricing period. There is the possibility of an "interim determination" if circumstances change outside Scottish Water's control (for example, new legislation or some other "external shock"),[26] and WICS may request that Scottish Water so apply.

Scottish Water will finalize its business plan, and then issue a charging scheme annually, within the overall settlement, and WICS will approve that yearly (and may modify it, with reasons).[27] The charging scheme must enable Scottish Water to provide its core services "effectively" and comply with any policy statement.[28] Effective exercise of core functions is then defined as complying with the objectives in the directions, at the "lowest reasonable overall cost".[29]

### *Borrowing, charging and cost recovery*

Currently, Scottish Water raises around £1 billion per year in charges, and a limited amount of government borrowing. In the 2014 principles of charging, the government said it would lend up to £720 million over the six years to Scottish Water – £120 million per year, as advised by WICS – although they subsequently announced there would be no lending in this financial year. Scottish Water pays interest and repayments on these loans; they are not grants, and along with the charges, they are Scottish Water's only income streams to support their core functions. Around 50% of expenditure is capital expenditure. Scottish Water's performance is assessed against a 'basket' of technical measures (the Overall Performance Assessment); the system was expressly designed to be similar to that used in England, so that Scottish Water's performance could be benchmarked against the English companies. In 2002, Scottish Water trailed behind all of its comparators on almost every measure; today, it is in the upper quartile. The improvement has been impressive and will be analyzed further below.

The general principle is that Scottish Water's customers should pay the full costs of the services that are delivered, with broad cost-reflexivity within each of the main customer groups (business and household customers); that is, that each group should broadly meet the full costs of delivery of the service. For household customers, charges are recovered by local authorities along with local council tax; Scottish Water does not bill customers directly. This benefits Scottish Water in a number of ways, especially around costs of collection, and may avoid contentious dealings with customers, but may also tend to reduce visibility and understanding of the system by the public. There is no compulsory metering in Scotland, although households may request a metered supply. Separate water and sewerage charges are banded along with council taxes themselves, so charges are lower for those in properties of lower value. Recognizing diseconomies of scale for rural customers, charges are harmonized across Scotland within the bands, i.e. a Band A household will always pay the same amount, regardless of where the property is. The gives in general some protection to poorer households. For those receiving help with their council tax payment, water and sewerage charges will be reduced, but only by up to 25%. In the past, as water charges are relatively low, affordability has not been considered a particular problem in Scotland. However, as incomes have fallen and benefits cuts work through the system, there is increasing concern over affordability (Citizens' Advice Scotland, 2015; Consumer Futures, 2014). For business customers, water is metered; see further below. There is an exemption scheme for charities, which has just been revised; some support will continue to be available to charities with a turnover of less than £300,000 per year, unless they are operating a shop, cafe or licensed premises.[30]

### *Business customers and licensed providers: the introduction of retail competition*

The 2005 Act introduced some market liberalization for services to businesses. Whilst Scottish Water retains its monopoly over the public networks, with a prohibition on any other person or organization putting water into the system or taking wastewater out, it was decided to open up a market in retail services only, for business customers only – some 150,000 business premises. WICS was tasked with licensing new providers of retail services, with Scottish Water providing wholesale services to the licensees. Scottish Water itself 'hived off' a ring-fenced subsidiary, Scottish Water Business Stream, which is also licensed by WICS. Its competitors are for the most part subsidiaries of the English water companies, and these are now increasing in number, as England is introducing retail competition in 2017. Scottish Water remains a supplier of last resort, and crucially, wholesale prices are published, enabling competitors and customers alike to see clearly what savings (or other benefits) can be achieved by switching.[31] Until very recently, relatively few businesses had transferred – mainly those who had multiple premises and could benefit from more streamlined billing, or those seeking specialist advice on water efficiency (or wastewater management).

Nonetheless, WICS's view is that the presence of competition has forced Scottish Water Business Stream to be more competitive for its remaining customers; WICS (2013a) has estimated the overall savings at £138 million. Recently however, the main public-sector contract under which water is supplied to education, health and local authority premises across Scotland was won by a subsidiary of Anglian Water.

The new contract is worth £80 million per year and will run for four years. The government anticipates £40 million in direct and indirect savings for the public purse overall (Scottish Government, 2015). Yet most of the comment was negative, accusing the government of 'privatizing' Scottish Water (see e.g. McNabb, 2015). In part this reflects the historical campaigns to prevent divestiture noted above, and the continued polarized debate globally around private-sector participation. Some commentary, especially on social media, reflected a lack of understanding of the way the industry is organized in Scotland; but equally, any hint of 'privatization' in any form opens the government to criticism. The opening of the market under the 2005 Act was the initiative of a Labour–Liberal Democrat coalition government in Scotland; once the market was opened it was not possible, under EU competition rules, for the current Scottish National Party government to refuse to offer the service to tender. Governments in Scotland regularly find themselves giving assurances that the service will remain in public hands; governments in many countries that are considering bringing in the private sector face similar issues. As with the public–private partnership schemes for wastewater, here too is some private-sector participation which nonetheless does not affect the fundamentally public nature of Scottish Water and its service.

## Economic efficiency through regulation

Whilst private-sector involvement in service delivery was seen in the 1990s as a means to reduce inefficiency in public monopolies, the difficulties with introducing competition in networked sectors are well understood. Regulation may be seen as a proxy for competition (Department for Business Skills and Innovation, 2011; Ogus, 1994), and arguably it is regulation, rather than ownership, which gives the best chance of improving services and benefiting customers. This has been argued in relation to the divested water companies in England (Renzetti & Dupont, 2006). The challenge in Scotland was different: could a similar regulatory approach, along with benchmarking, be used to drive efficiency whilst still in the public sector?

### *Incentivizing efficiency through reviews of charges, 1999–2014*

When the commissioner began work in 1999, he analyzed the finances of the three regional authorities and the likely levels of price increases for their customers (Water Industry Commissioner for Scotland, 1999). In that first analysis he noted that prices would need to rise substantially for customers of all three regional boards, but especially in the north (where the customer base was smaller and infrastructure more dispersed). This remained the case in his second review, covering 2002–2006 (Water Industry Commissioner for Scotland, 2001). By this time, however, the Scottish government was considering merging the three authorities into one central provider, Scottish Water, following government consultations but also a detailed inquiry by the Scottish

Parliament (Scottish Parliament Transport and Environment Committee, 2001). The commissioner's (influential) view was that to meet investment needs without substantially increasing borrowing, prices would need to rise by 15% on average through the (then four-year) strategic review period, but that this would be reduced to 8% if the bodies were merged. The commissioner also laid out plans for significant efficiency savings, from his benchmarking work against the English water companies. Substantial efficiencies would continue to be required through the establishment of Scottish Water and through the next price review. One important mechanism to incentivize efficiencies was allowing Scottish Water to retain any outperformance within a price review period – just as happened with the English companies.

Another major issue affecting Scottish Water's early years and its relationship with the Commission had to do with data. From 1973 to 1996, water services had been provided by nine top-tier regional councils and three single tier islands councils; and from 1996 by the three regional water authorities. Unsurprisingly, the data available were not always complete or coherent, and the regulator spent much time and effort in the early reviews formalizing reporting through an annual return from Scottish Water, including auditing of the data by a reporter (Black and Veatch).

### *Data, transparency and commercialization*

Data collection for water services is often problematic, and this may be an especial problem in the public sector and where services are delivered by municipalities, with numerous other functions and demands on their resources, often more visible to voters than water services. One response to this is to 'commercialize' or 'corporatize' the service, whereby its revenues and expenditures should be accounted for separately. In some Australian states, for example, larger municipalities are subject to such commercialization.[32] Commercialization of water services is sometimes controversial, as it may be seen as a precursor to privatization (see e.g. McDonald & Ruiters, 2005), and of course this may be the case. However, accounting separation of complex and capital-intensive services is surely useful in itself, enabling better understanding of the costs of delivery over time and aiding transparency and accountability, whether the service remains with a municipality, is transferred to a specific public-sector vehicle, or is to be subject to private-sector participation. It was certainly an essential (and challenging, and time-consuming) part of the process of the transformation of Scottish Water.

### *From confrontation to cooperation: a maturing regulatory relationship*

The 2006–10 strategic review (WICS, 2005) was voluminous, and confrontational. The process saw the resignation of the chairman of Scottish Water, who felt that the organization was unable to make the changes required within the revenue cap applied. By the time of the final determination, the commissioner had been replaced by WICS and the new powers and duties in the 2005 Act were in place, although the retail market was not yet opened. The efficiency gains required by WICS had been extensive, and by now price rises for most customers, business and household, were below the rate of inflation – a reduction in real terms. Throughout this period, Scottish Water continued to be measured against the same performance indicators as the English companies – including drinking water quality, environmental protection and customer satisfaction – and performance was steadily improving. Targets for leakage were introduced during this period.

The 2010–15 review was less confrontational, as the regulatory relationship matured and as Scottish Water made progress. In 2009, in the Draft Determination, WICS (2009) noted that efficiency had improved by 40%. Also in 2009, the Scottish government extended the regulatory period to five years, and in 2013 it was extended to six, so the current period runs from April 2015 to April 2021. A six-year period is synchronous with the timescales for river basin planning and flood risk management planning under the relevant EU directives. It was also felt that a longer period would lead to less 'overhang' in the capital programme. Throughout the whole process since 2002, Scottish Water had occasionally struggled to complete its capital programme by the end of each period. Related to this, the use of fixed pricing periods affects the supply chain, especially in that there is a large amount of activity at the start of each period, and then a long 'tail' towards the end. In Scotland, where works by Scottish Water are a major portion of all construction activity, that is especially problematic, and WICS and the government have been keen to minimize the effect. In the run-up to the 2015 review, a number of new approaches were introduced.

### *The 2015 strategic review*

A number of factors made the 2015 strategic review different, including responses to developments at home and in England. Benchmarking against English service providers had been an important part of the Scottish model, but in England companies were no longer to be assessed against the basket of measures in the Overall Performance Assessment. Instead, the English regulator (the Water Services Regulation Authority, but still known by its previous acronym, OFWAT) was moving to a system much more focused on customer satisfaction. This followed a series of reviews, both of customer representation (Gray, 2011; OFWAT, 2011) and of the regulatory process more generally (Cave, 2009). One change was to establish a body known as the Customer Forum, and this will be considered in the next section; but other changes are also of interest.

Although it was decided to retain the Overall Performance Assessment in Scotland till 2021 in order to measure trends in Scottish Water's own performance, it was also recognized that with attainment in almost all measures in the upper reaches, there was less scope for improvement. Instead, as in England, measures were devised to assess customer satisfaction in both domestic and business customers, and this is an ongoing process in which the Customer Forum has also been involved.

The desirability of 'smoothing' the regulatory process to protect the supply chain has been noted, and more generally it is desirable for water services providers, especially bulk providers, to operate within long-term planning frameworks. Scottish Water has a 25-year Water Resources Plan (Scottish Water, 2009) focusing on availability of resources, but in the run-up to the 2015 review also developed a 25-year Strategic Projection (Scottish Water, 2014a) to address longer-term impacts, such as climate change, which may affect the resilience of the sector.

Another change relevant to 'smoothing' was the introduction of a midway review, especially of the capital programme. Various proposals in Scottish Water's business plan will be revisited in 2018, and either firmed up and implemented, or perhaps repriced or restructured. Again this is seen as a way of creating a more streamlined process, not just for the supply chain but in terms of overall operation. So, instead of disjointed five-year planning, with distinct start and end points, there is now a six-year plan, but within a 25-year projection, and with rolling reviews every three years.

A further initiative by WICS was the introduction of financial "tramlines". These would set upper and lower levels of financial performance in cash and capital value terms (WICS, 2012) within which Scottish Water's performance would be acceptable to its regulators and to the government. The tramlines are specifically mentioned in the current principles of charging, reflecting the government's involvement in, and support for, the process (Scottish Government, 2014b). The government will decide what should happen in the event of any outperformance above the top line and, with WICS, on what remedial steps are necessary if there is a risk of underperformance. Outperformance below the line will (for now) continue to be retained by Scottish Water within a charging period to provide continued incentives for improvement. As with the development of customer satisfaction measures, customer bodies may also have a role to play here.

## Customer representation and the Customer Forum

One innovative initiative in the 2015–21 review was the introduction of a Customer Forum, a body designed to give a better voice to customers, which effectively led to a form of negotiated settlement such as is found in the US (Hendry, 2016a; Littlechild, 2014). In part this was a response to concerns about the burdensome process of the price cap methodology, relevant to a wide range of monopoly services across the UK including the water sector, as well as recognizing that the needs and wishes of customers were not always well met in economic regulation. Information asymmetries in regulating monopolies affect customers most of all; even where voluminous regulatory information is made available by both regulators and service providers, as in Scotland and England, that information is complex, technical and difficult to understand. WICS's view was that a more customer-focused approach would be beneficial, and indeed a similar process took place in England, with the introduction of Customer Challenge Groups (OFWAT, 2011). In both jurisdictions this also reflected what might be termed a 'regulatory plateau', with a relatively stable legislative environment and relatively stable pricing, in the context of very low interest rates and inflation, along with very high levels of attainment under the Overall Performance Assessment.

### *Consumer policy: a moving target*

In Scotland until 2010 there was a designated body representing customers: the Customer Consultation Panels, also known as WaterWatch. These were established under the 2002 Act, replacing earlier Consultative Committees, and their powers were extended under the 2005 Act to include powers of investigation and consumer advocacy, i.e. they had a role in relation to the policy framework. Whereas WICS has a general duty to promote the interests of the customers, the panels had the duty to represent their views and interests. They had a duty to publish reports and powers to make recommendations to the ministers, to all the regulators and to Scottish Water. Scottish Water had a duty to have regard to its recommendations and representations. The convenors of the panels had a duty to investigate complaints, and could make representations, and there was a duty of cooperation between the panels and WICS.

In 2010, following the UK general election, various changes were made. Consumer protection is reserved to the Westminster (UK) government; it will be devolved under the current Scotland Bill[33] (see also Hill, 2014), but the detail of the devolution settlement impacts in multiple ways, especially as competition is – and will remain – a Westminster function. The panels were wound up, but all their functions remained. Their advocacy and policy-related activities were transferred to the National Consumer Council, and their investigative powers for second-tier complaints (those which had been addressed but not resolved by Scottish Water) were transferred to the Scottish Public Services Ombudsman.[34] Subsequently, due to further changes in consumer protection at the UK level, the powers residing with the National Consumer Council in Scotland (renamed, during this period, Consumer Focus and then Consumer Futures) were again transferred, this time to Citizens Advice Scotland.[35] Thus it has been a challenging period of significant change, and it is against this backdrop that the Customer Forum was established, under a cooperation agreement between WICS, Scottish Water and (at that point) the National Consumer Council / Consumer Focus (Customer Forum Cooperation Agreement, 2011). Its costs were met by WICS (itself funded by a levy on Scottish Water), and the cooperation agreement set out its role in price setting and made clear that it would also work with the other regulators. The Scottish government was involved in, and supportive of, the decision to establish the Forum.

### An increased role for the Customer Forum

Initially, the expectation was that the Forum would only be involved in 'discretionary spending', i.e. that portion of Scottish Water's expenditure that did not relate to mandatory requirements under EU or domestic law or ministerial directions. However, it soon became apparent that this would lead to a very narrow role and contribution, especially as the timing of mandatory improvements can make a significant difference to realizing customers' preferences. It was therefore decided that the Forum, and its Engagement (sub) Committee, should be asked to negotiate generally with Scottish Water over the business plan as a whole. WICS took the view that if the Forum and Scottish Water could agree on a business plan within the financial tramlines and other guidance issued by WICS, then WICS would be "minded to accept" that plan (see WICS, 2013b, 2014 for discussion of the process from the regulator's perspective).

The Forum was a non-statutory body, appointed under public-sector rules; as a member of the Forum, I must declare an interest. There were five members representing domestic users, three representing business customers, and our convenor. Business representation came from two of the licensed providers (Business Stream and Anglian Water), as well as a representative of small businesses (initially from the Federation of Small Businesses; latterly from the Scottish Council for Development and Industry). The ordinary members were not intended to represent particular customer groups, but rather to provide a group of laypeople with a variety of relevant skills and experiences, including in consumer protection, marketing, the civil service and the third sector, who would be able to assess the information provided by Scottish Water, and indeed by its regulators, and challenge preconceptions. In particular, the Forum asked (sometimes repeatedly), "Why is [this proposal] in the interests of customers?"

The Forum engaged with Scottish Water, and with the regulators, in various ways; it was consulted over Scottish Water's customer research, and also undertook research of its own. As well as academic analysis (Littlechild, 2014; Hendry 2016b), the Forum produced its own *Legacy Report* (Customer Forum, 2015). This article will not revisit these analyses in detail, but a few observations may serve to indicate the ways in which the Forum's work might be useful to other service providers and in other places.

Support from the regulators, and from the government, was of course essential; the Forum did not seek to supplant in any way the formal statutory responsibilities of the other players. In many ways its non-statutory existence gave flexibility, but everyone involved was mindful of the need for transparency and accountability; activities were minuted, and the convenor gave evidence to the parliamentary committee on our work. WICS provided a series of information notes to support the process, and these notes set out WICS's thinking in terms meant to be understandable to a layperson. Any entity similar to the Forum would require such information and guidance. Scottish Water also provided information in the form of service improvement reports, addressing, *inter alia*, drinking water quality, environmental performance, and other core activities. Unsurprisingly, both WICS and Scottish Water had been more used to addressing their publications to a more technical audience, whether of economists or engineers. The requirement for Scottish Water to set out their thinking in terms the Forum could understand was initially challenging, but as the process developed Scottish Water staff agreed that the production of these reports had also been helpful to them. The service improvement reports were subsequently published as annexes to the Business Plan (Scottish Water, 2014b).

Framing the detail was the desire of all parties to have a less confrontational, more consensual process at an ideal point in a longer-term regulatory cycle, where the adversarial context of a decade earlier seemed to have been replaced by a genuine wish to demonstrate that all parties were seeking the same goal: providing the service at the most reasonable price and balancing the interests of all the parties. At one level the Forum succeeded because it was a better option than failure. Whilst agreement was incentivized and supported – WICS would accept a business plan within the acceptable parameters – failure to agree would be penalized. Both the Forum and Scottish Water would be required to publicly state what each saw as the reasons for disagreement; and at that point, WICS would step in and make a determination under its statutory powers. That was an outcome all parties wished to avoid, and in the end agreement was reached within the parameters and earlier than would have happened under the statutory process. The future of the Forum is not yet clear, and will depend on the outcome of the negotiations over the Scotland

Bill; but the Scottish government is mindful of the success of the Forum and considering whether a similar approach could work for other regulated entities in the future (Hill, 2014). On another level, the work of the Forum and the successful conclusion of its activities can be placed in a supportive political and policy context that recognized the importance of water, and of Scottish Water, to Scotland as a whole. That political context includes the broad regulatory framework and the role of Scottish Water in the Hydro Nation strategy.

## Conclusions

This article has sought to assess the development of Scottish Water as a public provider of water and wastewater services, from its origins in municipal supply, through three regional sectoral authorities, to a single vertically integrated entity serving most homes and businesses in Scotland. The removal of the service from municipal control, and then its merger into a single entity, facilitated the regulator's obtaining improved data on assets, costs and income streams, improving accountability and transparency. These data in turn enabled the significant efficiency gains seen over a decade.

The article has also analyzed the regulatory process over a period of 15 years, in which Scottish Water's performance improved from amongst the worst of its benchmarked comparators to amongst the best. Along with that, its reputation has emerged as a key player in the Scottish government's environmental and social agendas, as reflected in its role in Hydro Nation. Hydro Nation itself indicates the importance of water to Scotland, now and in the future.

The regulatory process involves several regulators, for price setting, environmental protection, drinking water quality and consumer advocacy. Most recently the process has seen the establishment of a Customer Forum charged with achieving a negotiated settlement of the entity's six-year business plan. The success of this initiative demonstrates a mature regulatory environment where all the participants are striving towards the same policy goal, that of providing improved service at the lowest reasonable price, within detailed objectives set by ministers.

There is some private-sector involvement in the service, in the public–private partnership schemes for wastewater and in the licensed competition for retail services for business. Yet Scottish Water has remained in public hands, funded by a limited amount of capital borrowing and by the charges it collects from service users, broadly cost-reflectively but with some cross-subsidy, especially from urban to rural. Overall it has been successfully commercialized, with full cost recovery and transparency.

Often, moves to commercialize a service – for example, to introduce separate accounting procedures where the service is delivered by municipalities – are seen as a precursor to privatization and become part of that polarized debate. A quarter-century after the Dublin Principles, with the need for water services – and investment in those services – as great as ever, it would be beneficial to move away from that polarized debate and accept that some nuance is possible, and desirable. Scottish Water can provide several examples of good practice to others – in the regulatory and governance model, in the engagement of consumers, and in manifesting a mature and cooperative approach to achieving public policy goals within a regulated settlement.

Perhaps most importantly, Scottish Water's story demonstrates that with proper regulation and governance, a public provider can operate as efficiently as the best in the private sector. In a world where private capital is elusive and most water services remain the responsibility of public authorities, that is surely a story worth the telling.

## Notes

1. Treaty and Act of Union with England, 1707.
2. Scotland Act 1998 c.46.
3. Directive on a Framework for Water Policy 2000/60/EC.
4. Directive on Flood Risk Management 2007/60/EC.
5. Water Industry (Scotland) Act 2002 asp.3 (2002 Act).
6. Local Government (Scotland) Act 1994 c.39.
7. Under the Water Act 1989 c.15.
8. Water Industry Act 1999 c.9 Part II.
9. Directives on the Quality of Drinking Water for Human Consumption 1980/778/EEC, 1998/83/EC.
10. Directives on the Quality of Bathing Waters 1976/160/EEC, 2006/7/EC.
11. Directive on Urban Waste Water Treatment 1991/271/EEC.
12. Water Resources (Scotland) Act 2013 asp.5 (2013 Act). The Ministerial duty is in s.1.
13. 2013 Act s.24.
14. 2013 Act s.25.
15. 2013 Act s.23.
16. 2013 Act s.1.
17. 2002 Act s.29D.
18. 2002 Act s.56A.
19. Water Services (Scotland) Act 2005 asp.3 (2005 Act).
20. Water Environment (Controlled Activities) (Scotland) Regulations SSI 2011/209.
21. Water Environment (Scotland) Act 2003 asp.3.
22. 2005 Act s.21.
23. 2002 Act s.29B.
24. 2002 Act s.29C.
25. Water Services etc. (Scotland) Act 2005 (Consequential Provisions and Modifications) Order 2005/3172 (a Westminster order, as it is a reserved function).
26. 2002 Act s.29F.
27. 2002 Act s.29A.
28. 2002 Act s.29C.
29. 2002 Act s.29G.
30. See https://www.business-stream.co.uk/water-rates-charges/water-charge-exemption-scheme.
31. See http://www.scottishwater.co.uk/business/our-services/wholesale-services/wholesale-services-to-licensed-providers.
32. For example in Queensland, under the Local Government (Qld) Act 2009 No.17 and the Local Government Regulation 2012 SL No.236.
33. Scotland Bill 2015 HL Bill 73 2015–16.
34. Public Sector Reform (Scotland) Act 2010 asp.8.
35. See http://www.consumerfutures.org.uk/about-us.

## References


Bakker, K. (2003). *An uncooperative commodity: Privatising water in England and Wales*. Oxford: OUP.

Cave, M. (2009). Independent review of competition and innovation in water markets: Final report. Retrieved August 12, 2015, from https://www.gov.uk/government/uploads/system/uploads/attachment_data/file/69462/cave-review-final-report.pdf

Citizens' Advice Scotland. (2015). Sink or swim: Consumers' experience of water and sewerage debt. Retrieved January 11, 2016, from http://www.cas.org.uk/publications/sink-or-swim-consumers-experiences-water-and-sewerage-debt

Consumer Futures. (2014). Keeping your head above water. Retrieved January 11, 2016, from http://www.consumerfutures.org.uk/files/2014/02/Keeping-your-head-above-water.pdf

Customer Forum. (2015). Legacy report: Lessons learned from customer involvement in the 2015–2021 strategic review of charges. Retrieved August 21, 2015, from http://www.watercommission.co.uk/view_CustomerForum.aspx

Customer Forum Cooperation Agreement. (2011). Retrieved January 11, 2016, from http://www.watercommission.co.uk/view_CustomerForum.aspx

Delmon, J. (2001). *Water projects a commercial and contractual guide*. The Hague: Kluwer Law International.

Department for Business, Skills and Innovation. (2011). Principles of economic regulation. Retrieved January 11, 2016, from https://www.gov.uk/government/uploads/system/uploads/attachment_data/file/31623/11-795-principles-for-economic-regulation.pdf

Dublin Statement. (1992). Dublin statement on water and sustainable development [online]. Retrieved January 11, 2016, from http://www.wmo.ch/pages/prog/hwrp/documents/english/icwedece.html

Gray, D. (2011). Review of Ofwat and consumer representation in the water sector (DEFRA). Retrieved August 12, 2015, from https://www.gov.uk/government/uploads/system/uploads/attachment_data/file/69442/ofwat-review-2011.pdf

Hendry, S. (2003). Scotland's water – Safe clean affordable public. *Natural Resources Journal, 43*, 491–517.

Hendry, S. (2016a). The value of Scotland's water resources legal analysis (CREW Publication). Retrieved March 29, 2016, from http://www.crew.ac.uk/publications/valueofthewaterresources-legalanalysis

Hendry, S. (2016b). The Customer Forum – Putting customers at the centre of regulating water services. *Water Policy*. doi:10.2166/wp.2016.199

Hill, R. (2014). Report of the working group on consumer and competition policy for Scotland. Retrieved December 21, 2015, from http://www.gov.scot/Resource/0048/00488834.pdf

Littlechild, S. (2014). The Customer Forum: Customer engagement in the Scottish Water sector. *Utilities Policy, 31*, 206–218. doi:10.1016/j.jup.2014.10.002

Martin-Ortega, J., Holstead, K., & Kenyon, W. (2013). The value of Scotland's water resources (CREW Publication). Retrieved January 11, 2016, from http://www.hutton.ac.uk/sites/default/files/files/publications/water-resources-bill-leaflet-feb2013.pdf

McDonald, D., & Ruiters, G. (Eds.). (2005). *The age of commodity: Water privatisation in Southern Africa*. London: Earthscan.

McNabb, S. (2015). SNP accused of 'privatising' £360m water contract Scotsman Newspaper 6/10/2015 [online]. Retrieved January 11, 2016, from http://www.scotsman.com/news/politics/snp-accused-of-privatising-360m-water-contract-1-3908667

OFWAT. (2011). Involving customers in price setting – Ofwat's customer engagement policy statement. Retrieved December 21, 2015, from http://www.ofwat.gov.uk/wp-content/uploads/2015/11/pap_pos20110811custengage.pdf

Ogus, A. (1994). *Regulation: Legal form and economic theory*. Oxford: Clarendon Press.

Reid, K., & Zimmerman, R. (Eds.). (2000). *A history of private law in Scotland* (Vol. I). Oxford: OUP.

Renzetti, S., & Dupont, D. (2006). Ownership and performance of water utilities. In J. Chenoweth & J. Bird (Eds.), *The business of water and sustainable development* (Chapter 7). Sheffield: Greenleaf Publishing.

Scottish Government. (2010). Building a Hydro Nation. Retrieved January 11, 2016, from www.scotland.gov.uk/Publications/2010/12/14111932/0

Scottish Government. (2012). Scotland the Hydro Nation. Retrieved January 11, 2016, from www.scotland.gov.uk/Publications/2012/02/9536

Scottish Government. (2014a). *The Scottish water (objectives: 2015 to 2021) directions 2014.* Retrieved 20 July 2016 from http://www.gov.scot/Topics/Business-Industry/waterindustryscot/SWI-1/SWI-P/policy

Scottish Government. (2014b). *Principles of charging statement 2015-21.* Retrieved 20 July 2016 from http://www.gov.scot/Topics/Business-Industry/waterindustryscot/SWI-1/SWI-P/policy

Scottish Government. (2015). Water contract brings £40m in savings. Press Release 6/10/2015. Retrieved January 11, 2016, from http://news.scotland.gov.uk/News/Water-contract-brings-40m-in-savings-1df3.aspx

Scottish Parliament Transport and Environment Committee Report No. 9 of 2001. Report on the inquiry into water and the water industry. Retrieved January 11, 2016, from http://archive.scottish.parliament.uk/business/committees/historic/x-transport/reports-01/trr01-09-vol01-01.htm

Scottish Water. (2009). Water resource plan. Retrieved January 11, 2016, from http://www.scottishwater.co.uk/About-Us/Publications/Key-Publications/25-Year-Water-Resource-Plan

Scottish Water. (2014a). Strategic projection. Retrieved January 11, 2016, from http://www.scottishwater.co.uk/about-us/publications/strategic-projections

Scottish Water. (2014b). Business plan 2015–21 and appendices. Retrieved August 12, 2015, from http://www.scottishwater.co.uk/about-us/publications/strategic-projections

Water Industry Commissioner for Scotland. (1999). *Strategic Review of Charges 2000–2002*, archive document available from WICS on request.

Water Industry Commissioner for Scotland. (2001). Strategic review of charges 2002–2006. Retrieved January 11, 2015, from http://www.watercommission.co.uk/view_Price_Setting_2002-06.aspx

Water Industry Commissioner for Scotland. (2005). *The strategic review of charges 2006–2010 draft determination* (Vol. 5, Chapters 8–11), archive documents available from WICS on request.

WICS. (2005). The strategic review of charges 2006–10: The final determination. Retrieved January 11, 2015, from http://www.watercommission.co.uk/UserFiles/Documents/final%20determination.pdf. Supporting documents are available from WICS on request. P.9

WICS. (2009). The strategic review of charges 2010–14 the draft determination. Retrieved January 11, 2015, from http://www.watercommission.co.uk/UserFiles/Documents/WICS%20DraftDetermination_1.pdf

WICS. (2012). Information note No.7 financial tramlines. Retrieved January 11, 2015, from http://www.watercommission.co.uk/UserFiles/Documents/CustomerForumNote7.pdf

WICS. (2013a). Water and sewerage services in Scotland: An overview of the competitive market. Retrieved January 11, 2015, from http://www.watercommission.co.uk/UserFiles/Documents/WICSHouseofCommonsPaper(C).pdf

WICS. (2013b). Strategic review of charges 2015–21: Innovation and choice. Retrieved August 12, 2015, from http://www.watercommission.co.uk/UserFiles/Documents/scr%20innovation%20and%20change%20proof%204.pdf

WICS. (2014). Strategic review of charges 2015–21: Draft determination. Retrieved August 12, 2015, from http://www.watercommission.co.uk/UserFiles/Documents/scr%20innovation%20and%20change%20proof%204.pdf

# Are we killing the rain? Meditations on the water cycle and, more particularly, on bioprecipitation

Jane Maslow Cohen

This essay reviews the multidisciplinary science of bioprecipitation, using it as a lens through which to envision integrative options for land use and water resource management in a new light. Bioprecipitation is the hypothesis that microbial ice nucleators, including *Pseudomonas syringae*, may be highly adapted causal agents of rain and snow. To the extent that land use policies, including pathogenic eradication campaigns, may inhibit the local production of biotic ice nucleators, they may be responsible for 'killing' a generative source of rain. Such possibilities should invite major interest in this gathering field of research. Assuming that it contributes to a richer comprehension of the hydrological cycle's dependence on circulatory biota, these findings should help to stimulate assimilative, integrated reformulations of land use and water management policies and norms.

A growing body of scientific literature that relates to an hypothesis – actually, a small cluster of hypotheses – known as 'bioprecipitation'[1] offers a path to a new appreciation of human agency as a distinctive cause of local and regional climate change, a shift of focus from change on a global level. It pertains to a subject that is nowhere yet on the public's agenda or that of policy makers, one that could promote an instrumental focus of global significance on the integration of land use and water resource management through a systematic effort to invite the production of rain and the disinvitation – that is, the inhibition – of drought.

The basal understanding at the root of this aptly named idea is that atmospherically transient biological materials, especially microorganisms, are efficient producers of ice crystals in an abundance that fosters their descent to the earth as rain, snow and ice. Within this formative view is a secondary hypothesis. It is that the most efficient contributors to ice nucleation (IN)[2], as this activity is called, may be certain varieties of bacteria that populate many, if not all, regions of the world. These have already been discovered as the nucleus of snow and rain on five continents, with tantalizing frequency in heavily cultivated agricultural regions. Their ancient role in this capacity may soon be demonstrated through the coring of glacial ice (Christner 2012).[3] A third hypothesis that nests inside these two is that a particular bacterium, *Pseudomonas syringae*, is the most proliferative ice nucleator within a range of relatively warm atmospheric temperatures in which other nucleators may lose their effect. Currently, *P. syringae* is ubiquitously known and treated as a plant pathogen, damaging its favoured hosts, certain agricultural crops, by employing its talents for propagating ice.

The practical excitement that lies within the compass of these discoveries resides in the possibility that the most effective bacterial ice-nucleation agents could be cultivated, just as food crops are cultivated, for their value in inducing rain. The dark seed of discovery that may lie further within this domain is that common land use policies may be having just the opposite effect. It is this possibility – that land use policies are inhibiting the proliferation of important ice-nucleating microbiota – that is the inspiration for this essay's title and for its intended work.[4]

In both title and form, the 'Meditations' that constitute a portion of this essay are tempered by the state of the relevant science, which is, in important respects, immature. As one microbiologist has put it, "Recent developments challenge the belief that microorganisms are passive travelers in the atmosphere. Indeed, evidence that microorganisms affect meteorological processes is invigorating research to define the role of biology in atmospheric processes and to understand how those processes might affect the abundance, dispersion, and viability of aerosolized microbes" (Christner 2012) – among these, the kinds of microbes, as he goes on to explain, that have been found to nucleate rain.

The careful use of the term 'might' in this statement sketches the present divide between the known and the unknown. (See further Morris and Sands 2012.) Such a rendering should make it clear just where in the timeline of research any attempt, such as mine, to lasso public land use and water resource policies to the bioprecipitation hypothesis will enter the public arena: earlier than may be due. Still, it is difficult not to become engaged by the excitement to be found within the multidisciplinary science that is at work, as the literature review that follows intends to show.

Most recently, bioprecipitation scientists have also sought to gain popular notice for this research (Fox 2012, Morris and Sands 2012, Sands 2012), a signal that, in their view, its most basic findings are ready to enter the public conversation about climate variables. It may not be too early, after all, for ideas that demonstrate the muscular reformative capacities of the research, especially in relation to land use, to begin to emerge. In fact, they already have (see e.g. Sands 2012).

My goals in writing this essay, then, are two. One is to spread word about bioprecipitation theory into the policy realm because of the paradigm-shifting potential that it reveals. The second is to winkle out some early thoughts about the practical implications of this work.

## Bioprecipitation: a literature review

Conventional explanations of precipitation rely principally on meteorology and cloud physics. They bear major evidence of their dependence on the physical sciences, from which they branch. A standard meteorological definition of "precipitation" is: "Any product of the condensation of atmospheric water vapour that falls under gravity" (American Meteorological Society 2009). The referents within this concise portrayal expand into a congeries of inquiries concentrated, within the atmospheric sciences, on the processes that initiate and direct water vapour saturation and condensation and on the relationship of these processes to such vital influences as temperature, humidity, pollution, and the extent and direction of motion within the cloud-types prevalent at different levels within the atmosphere (see, e.g., Keil *et al.* 2008). Study of these extremely dynamic processes presents significant challenges, not least because cloud systems are too vast to subject to controlled study, as opposed to observation in situ, yet too complex and interdependent to render fully through mathematical models and other bounded means (Stratmann *et al.*, in Heintzenberg and Charlson 2009, p. 150). Recent syntheses of the work being done in cloud physics includes critical analyses of the present state of knowledge in regard to major elements of cloud systems, with explicit attention drawn to knowledge deficits, methodological inconsistencies, and other bases for needed development within the field. Accelerating "perturbations" within the climate system provide a hard spur for improvements to be made (see generally Heintzenberg and Charlson 2009).

Beyond contest or revision lies the fact that water droplets can coalesce on account of collisions with other droplets or on account of the formation of ice crystals within clouds. The crystals can form around nuclei composed of either inorganic matter, such as dust, salt or minerals, or organic matter, such as algae or bacteria. Without the intervention of the biological sciences such as the microbiology, genetics and plant pathology that have spearheaded bioprecipitation research, the physical-science view of ice nucleation seems to stop at this door, behind which the crystallization process is taken to be adventitious, temperature dependence being the major determinant of nuclear type (Fox 2012, p. 42), while efficacy in the nucleation process remains unexplored.

In a curious way, what I might term the 'nuclear indifference' hypothesis has been subject to a kind of crude refutation, although it is not yet complete: Prolonged efforts involving both scientific and practical experiments and applications stretching across decades and continents have, with rare exceptions, failed to yield proof of scientifically-accepted status that the 'seeding' of clouds through the variously-engineered injection into them of specific types of inorganic particles can function to enhance the formation of rain (NRC 2003, Cotton 2009). To be sure, explanations of these failures run in several directions not centred on the organic versus inorganic nature of the seeding agent (Beck 2010; Cotton 2009; Levin 2010). For immediate purposes, it is enough to understand that the widely-noted scientific deficiencies of cloud-seeding experiments involving the introduction of inorganic particles into clouds should cast no aspersions on the development of the field of bioprecipitation which, in its aims, methods and outcomes, is entirely remote from the cloud-seeding oeuvre.

The focus of agronomy on agricultural yield was where the field got its start. Research led to the discovery of diseases that attack specific types of edible plants. Indeed, a single pathogen, the bacterium *Pseudomonas syringae (P. syringae)*, was found responsible, on a non-exclusive but highly persistent basis, for disease vectors that injure (though they typically do not destroy) an impressive variety of plants.[5] The function of much modern agronomic research has been to locate and develop effective agonists to respond to agricultural blights, including those caused by *P. syringae*. This research has led to the discovery and wide dissemination of several effective, though not wholly successful, bactericides (see e.g. Bashan and de-Bashan 2002a, 2002b).

An obvious research question involved the method by which *P. syringae* enters plants with the impressive success that it demonstrates. Two teams working independently in the 1970s discovered that *P. syringae* bases its propagative success on its ability to cause frost damage, primarily to leaves and stems, the torn and split tissue then providing a means for bacterial entry into the host plant, a metabolic activity that a specific range of temperatures was found to promote (see e.g. Arny *et al.* 1976, Vali *et al.* 1976). This, *P. syringae* is able to accomplish through its specialized ability to bind to water, the talent of a rare gene that codes for a crystal-inducing peptide lattice in the outer cell wall, found in just three of the microbe's 600 known strains thus far (Morris *et al.* 2004).

Further investigation included the question of how *P. syringae*, which appears to migrate over vast distances, negotiates its passage through the air. Professor David Sands, who has studied this bacterium for over 40 years, hypothesized that the same property that lends efficacy to its ability to bind water to plant surfaces might also promote *P. syringae*'s ability to mobilize, after dissemination into the atmosphere due to radiative fluxes, and to nucleate ice there. Still, the question remained: How? The concise answer is that *P. syringae* utilizes the same method for catalyzing atmospheric ice crystals as it does for inducing ice nucleation on plants: its highly adapted ability to bind water to the surface of its cell (Christner 2012, Morris and Sands 2012).[6]

It took three decades for bioprecipitation research to reach the discovery just described. The basic hypothesis – that ice nucleation may be spurred by biological agents – was advanced by three different research teams some 30 years ago (Lindow *et al.* 1976, Sands *et al.* 1982; Vali *et al.* 1976). Earlier work had prepared the way (see e.g. Soulage 1957, Schnell and Vali 1972, Maki *et al.* 1974, Vali 1968, 1971). Substantial doubt followed the introduction of this idea. There was widespread scepticism that biotic masses large enough to be of influence could attain sufficient loft to enter, let alone subsist within clouds. Research support dried up. According to Sands, the field "collapsed" (Fox 2012, p. 42). The theory could not be well tested through field study. Its validation through wide replication has required DNA testing not then available. Further evidence, involving satellite-based meteorological technologies, again unavailable earlier, requires funding levels the scientific groups involved in this highly dispersed field have yet to acquire, though a recent, path-breaking study using atmospheric modelling has been performed (Goncalves *et al.* 2011). As the field has grown more ambitious (Christner 2012, Morris and Sands 2012), it has conjoined the biological- and physical-science components of the work. Currently, it involves cross-disciplinary collaborations of geneticists, geophysicists, meteorologists, microbiologists, plant pathologists and statisticians, among others. With the advent of these techniques and the assemblage of these teams has arrived the ability to test appropriately the various embedded claims.

As of this writing, positive findings related to biological ice nucleation have been confirmed by some 40 scientific teams working out of dozens of laboratories in at least six countries. The work has advanced substantially just since 2005. Among others, David Sands and Cindy Morris and their colleagues have published some definitive research (Christner *et al.* 2008a, 2008b). This work has confirmed the general bioprecipitation hypothesis through consistent findings.

Atmospheric scientists have recently added a new and powerful arm to these discoveries. In early research, they have demonstrated the potential of biological agents to act as cloud condensation nuclei (CCN) in addition to their IN activity. Through this added process, biological agents can "contribute to the initial cloud formation stages and the development of precipitation through giant CCN and IN processes" (Möhler *et al.* 2005, 2006). In an early overview paper, the authors have called for a wide-spectrum effort to understand the cloud-formative role of biological agents, through a combination of field study, laboratory experiments and modelling.

Some of the most recent research into biological agent activity has led back to the role of *P. syringae*, which appears to be a rock star (my designation) not only in its microbial class (Christner *et al.* 2008a, 2008b) but among ice nucleators of all types. As a class, biological agents have been found to be the most efficient catalysts for atmospheric ice formation across a range of relatively warm temperatures (warmer than –10 °C) – a discovery of potentially immense import in a warming world. For another, of all the known biological agents, *P. syringae* has been found in the greatest abundance, as well as in the widest distribution in many parts of the world. Since current ability to study the concentration of ice nuclei in the atmosphere is extremely limited, rain and snow samples on land have served as proxies for understanding dispersion and concentration modes and as the basis for predictive modelling. In the case of *P. syringae*, its concentration in particle suspension in rain and snow samples obtained in Antarctica, Louisiana, Montana (Christner *et al.* 2008a, 2008b), the French Alps (Amato *et al.* 2007), the Yukon Territory, and Alberta, Canada (Vali 1971), have dominated all other ice nucleation types, though seasonal differences have obtained. The dominance of *P. syringae* as an ice nucleator has been found despite ecosystem differences, moisture source differences, and differences in air mass trajectories (Christner *et al.* 2008a, 2008b). It has been concluded that biological agents have a "ubiquitous" dispersal (Finlay and Clarke 1999). This is consistent with the highly efficient dispersal properties of *P. syringae*.

Given the routinely inimical treatment within agronomy of *P. syringae* and other biological agents that similarly invade host plants, there are serious choices coming along in the policy realm. As David Sands observes, it behoves us now to understand the relationship between the 'tended' and the unintended consequences of our dealings with earth's interdependent systems.[7] Taking in hand the kinds of questions that a fully systemic approach to the hydrological cycle raises for policy by deploying what is already known about *P. syringae* as an ice-nucleating agent creates a compelling probe. After all, it has been consistently demonstrated that when *P. syringae* is not behaving as a plant pathogen it is – provocatively – leading "other lives" (Morris *et al.* 2007).

Unlike climate change, which is mired in political disadvantage and difficult to test by modest means, the relationship of biological ice nucleators to plant propagation, to bactericidal agents, to wind currents, to tree canopy loss and to soil degradation may lie within relatively easy grasp, though important unanswered questions require the use of sophisticated modelling and other relatively expensive techniques.

What follows in this essay are some of my initial thoughts about what it might mean to take bioprecipitation theory on a tour within the policy realm. Although I assemble all of my ideas except the final one[8] under a meditative canopy, it should be clear which are the more tentative and which are not. Of special interest, even though necessarily tentative, may be my attempt to target improvements in the integration of land use and water policy so that these resource expenditures become sensitive to the incubation, growth, and circumnavigation of the microbiota that may serve a critical function within the earth's hydrological cycle by instigating the production of rain.

## Meditations

(1) The field of bioprecipitation got stopped early in its tracks when funding dried up. It did not gain momentum again for almost 30 years. Today, it faces continuing funding hurdles in the US, in that the multidisciplinary, highly collaborative research that currently drives the field transcends the conventions of traditional, agency-authorized intradisciplinary grant review. For work of this kind to achieve its research targets, some methods of adjustment within the grant review process will need to be found. Without these, the US-based research will need to go beggaring, except in its most modest forms, for investment by foreign governments. But foreign public investment is spread among countries and fields. Forging funding linkages can skew both priorities and temporal attachments among internationally dispersed teams, causing misalignments and delays. This is not the optimal way for the research to proceed.

(2) One area of potential research involves the genetic tagging of individual bacterial cells so that their terrestrial and atmospheric trajectories can be tracked.[9] Given the vastness of distances that could be involved and the plethora of inorganic as well as non-bacterial agents that are known to function as ice nucleators, it is plausible to assume that this line of experimentation could pick up only local and specific regional effects, rather than general ones.[10] But for genetic tracking to make possible more localized understandings of atmospheric patterns that affect when and where biota move within the hydrocyclical stream could be of great importance. Learning the influence and flight paths, for example, of rainforest biota – vast organic empires that seem highly identified with and are potentially responsible for local and regional precipitation systems and events – should carry powerful allure.

As of this time, certain countries within South America are accusing Brazil of land use policies in the Amazonian rain forest – especially deforestation – that they claim engender negative effects on rainfall patterns to the south. Improving the knowledge base at the root of this simmering international crisis could lead to responsibility being rightly assigned and remedial measures being undertaken if necessary to maintain sustainable precipitation yields. This is but one example of the work to be done.

(3) Common land use activities that aim at the eradication of pathogenic organisms such as *P. syringae* may need to be significantly re-examined for their hydrological impacts. Cost-benefit analysis comes to mind, should the field prove up. There follow some illustrative prospects for pursuit.

- Educational outreach should be instigated to instil a willingness on the part of agriculturalists world-wide to suspend attempts to eradicate *P. syringae* and other biological ice nucleators once they become securely identified as such. This would entail a significant change in conventional agricultural norms.[11]
- Experimental plantings of crops that currently attract *P. syringae* should be installed to try to establish peak atmospheric biological agent loads, assuming these can be parameterized. This could be accomplished through programmes that subsidize the development of pathogenic seed coatings, followed by experimental plantings on public lands.[12] Or, large experimental tracts could be planted with 'sacrificial' crops (my term), such as tomato and bean cultivars, to serve as hosts in specially prepared and tended arid regions where commercial and residential agriculture do not compete.[13]
- The regulatory approval of crops genetically engineered for immunity to herbicides and pesticides may require examination through the lens of biological agent-related effects. It is currently predicted that newly engineered crops bearing pesticide immunity will increase the use of such eradicants (Pollack 2012, pp. B1, B5). This could engender severe consequences for local and regional populations of phytopathogenic biological agents, such as *P. syringae*. The regulatory integration of any research that may demonstrate this threat should be pursued.
- Dissuasion and even some levels and amounts of interdiction may need to be undertaken, through educational as well as regulatory means, of biotic habitat destruction through such rife practices as suburban commercial and residential clear-cutting; exurban deforestation; urban and suburban irrigation quotas not appropriately calibrated to tree canopy maintenance, leading

to enhanced fire risk; and wetlands infill. The depredatory treatment of public financing for national and state parks, which can give onto inadequate maintenance, park closings, and eventual canopy loss should be reversed and adequate budgets for these essential public goods restored. It is possible that immense green belts or other natural areas will need to be designed to compensate for the microbiotic losses that may be pulverizing or otherwise radically redistributing atmospheric biological agent activity on account of the land use decisions that are routinely made on the basis of conventional policy modes.

- A decisive suppression should take place of the current recrudescence of the Sagebrush Rebellion – a demand, cast in legislative terms, by a coalition of political groups, (eerily) once a generation, that is designed to force the transfer to the states of the public lands held by the federal government. The movement is gathering force through state legislative activity in the US as of the time of this publication – too soon for a case on behalf of bioprecipitation to weigh in, but not too early for the potential effects of the massive Western land sales contemplated in the legislation to be opposed on account of their potentially devastating effects on drought and desertification in a region increasingly prone to these (Trueblood 1980, H.B. 148 [Utah], S.B. 1332 [Arizona]).
- Serious study should be undertaken of the potential for microbiotic loss through the destruction of viable tree canopy in the public parks on account of park closings, with concomitant losses of habitat maintenance programmes. Associated with this are soil degradation and the destruction of canopic habitats, in the event that public parks were to be sold for private development (deBuys 2011, Farrar 2012, *Trends in State Funding of Parks and Recreation* 2011).

It might be useful to conclude this essay with a more extended note than these tightly bundled 'meditations' are able to provide, one that regards a very long-term matter of acrimonious debate connected to the management of the federal public lands. I offer it as an ongoing reminder that drought and desertification are problems that do not attach, of a sudden, only to the twenty-first century, or just to recent climate change fears. In the case of the public lands, they have been a severely disabling problem for a much more protracted time, yielding the administrative law analogue, it seems, of the mythic Gordian knot. The problem involves chronic over-grazing on these lands. The need for revegetation is acute. This has long been recognized. But it has been and continues to be essentially ignored.

At some point, perhaps the need for a very large habitat for *Pseudomonas syringae* will become the public-policy sword that cuts the Gordian knot by furnishing a game-changing argument about the need to conserve the bioprecipitation potential of undeveloped vegetated land masses – an incalculably valuable asset that far outweighs what any other economic use could provide. The final section, below, outlines this claim.

## Public policy now: an opportunity to refocus the over-grazing debate

There may be a multiplicity of unintended ways that bacterial pathogens are being excluded from their timeless propagation haunts, whether on and within agricultural crops subjected to bactericides; in previously uncultivated, now built-up areas; or in politically endangered tree canopies. As 'ubiquitous' as bacteria are commonly considered, colony size may require a critical mass, and mass may function as a crucial variable in the production of efficacy when biological agents are in the clouds.

Assuming that cumulative eradication may become a concern, the reclamation of very large land tracts to invite bacterial self-propagation would seem a desirable end. Perhaps it may prove to be a necessary one, if the aggregative destruction of ice-nucleating bacteria were demonstrated to be a negative influence on the propagation of rain. In advance of that proof, we may engage in the present full-blown strategy of doing nothing, or we may choose to experiment with solicitations toward the natural order for the production of rain. A bold venture, and the single most efficient change in land use policy, would be to effectuate a change in the use of public lands. Approximately one-third of the land mass of the United States is owned as public land. Of this, the vast majority is found in the western states. Of this western land, most is owned by the federal government for reasons pertaining to grants, cessions and purchases at the founding and during the first half of the nineteenth century (Gates 1968).

Of these federally administered lands, many millions of hectares constitute a part of what is known as the 'public domain', the subset of public lands not committed to a specific purpose (as are, for example, the national parks). These lands are required under federal law to be administered and maintained, primarily by federal agencies, for the public's benefit in ways that are statutorily prescribed, the age of many of the statutes within this skein giving rise to the view that they are the "lords of yesterday" (Wilkerson 1992).

A chronic source of political conflict in regard to this vast swath of the public domain involves over-grazing. There has been a persistent debate about the federal tolerance of over-grazing that has rocked back and forth across its contested factual predicates for well over a century. During that extraordinary span, the relevant analysis has, on one side – the conservationist side – become ever more factually refined. But this has hardly moved the dial in favour of a major public policy change. In fact, it has not moved the dial, in substantive terms, at all. (For a noteworthy example of a private contribution toward policy change involving federal grazing permits, see Carey 2011.) The problem has been described thusly:

> Few natural resources better illustrate the challenges for natural resource law and policy than our nation's public rangelands. It is no coincidence that grazing provided the archetype for Garrett Hardin's tragedy of the commons nor that it has served as the classic example of agency capture. The current debate over rangeland management also provides a fascinating instance of environmental aspirations butting heads with longstanding tradition, culture and reliance interests. (Rasband *et al.* 2004, p. 880)

The story begins not with aspirations or traditions but with law. Due to environmental degradation, the tragedy of the commons in respect to over-grazing became such a full-blown problem by the early twentieth century that a major piece of statutory architecture was put in place to try to control it. The central pillar of the regulatory scheme – still the central pillar today – was the turn to a permitting system administered under federal authority (the 1934 Taylor Grazing Act, hereafter TGA). These standards were echoed in a subsequent statute that now serves as the comprehensive management scheme for federal lands (the 1976 Federal Land Policy and Management Act, hereafter FLPMA).

The federal regulatory regime that controls the destiny of millions of acres of the western public domain was, from the start, the product of a bifurcated scheme of entitlements. On the one hand, the TGA and, later, the FLPMA have insisted that grazing on public lands is not a right but a privilege (TGA sections 315(b), 316) and that the permits and leases under which grazing has statutorily taken place do not confer any property rights to the lands themselves. Nevertheless, the permits and leases have run for generations – not only of cattle but of people who have made multi-generational ranching on these lands into an iconic way of life that is bound up with the history and traditions of the American West.

These permits have functioned, unlike, for example, intellectual property protection, so as to confer their value over an indefinite life, much in the manner of Anglo-American land titles, the traditional paradigm of property rights themselves. When the government buys out permits to transfer or to retire them – the latter, a stratagem that has begun to foster more active conservation of land and water than existed before – the permit-holders are entitled to gain compensation under the federal constitution's takings clause for the value of the benefits that they have lost – a clear example of privilege transmogrified into right.

The TGA and the FLPMA do not mandate that all land within the public domain has to be put to work for grazing or for any other purpose. Rather, the two agencies responsible for managing the lands are required to look to "multiple uses" and "sustained yield" (FLPMA sect. 1732(a)). Because of the entrenched nature of cattle-raising on these lands, however, shifting to other uses has not proven easy to implement or maintain, particularly when competing uses are not compatible with the needs of cattle or the desires of the politically influential ranchers who own them. The authority of the two agencies is derived from the land management right granted to Congress under the property clause of the US Constitution (Art. IV, sect. 3, cl. 2). It includes the right to impose a regulatory carapace over the lands. Its terms include the requirements that "grazing use shall not exceed the livestock carrying capacity of the allotments" and that uses shall not impair watershed function, riparian habitat, water quality or wildlife habitat (FLPMA sections 4130.3-1(a), 4180.1, 4180.2(c)).

Beyond cavil, all of those damaging excesses have become chronic features of the grossly impaired landscape. What is most deeply problematic about cattle-raising, without reference to the "carrying capacity" of land, is that, more than many types of heavy, roaming animals, such as bison, cattle strip away vegetation in a manner that discourages regrowth and compound this problem by compacting the earth. A rival to cattle-raising on federal lands has been the development of recreational activities such as hiking, camping, hunting, fishing and water-craft use. Licenses and admissions charges for these activities have already demonstrated that they are more extensive sources of federal revenue than is cattle-raising. But the development of several of these uses has been impeded by the degraded state of the resources that cattle have polluted or stamped down (Rasband *et al.* 2004). Even with cost-benefit analysis clearly on recreation's side, the lessening of grazing's hold has been hard to effect.

The authors of a major environmental casebook write: "An important environmental concern related to grazing is desertification, [which] refers to prolonged abuse of land that weakens its ability to support plant growth through loss of soil productivity, increased soil deterioration, and loss of biodiversity. Overgrazing . . . is one of the prime contributors to desertification" (Rasband *et al.* 2004, p. 891). The authors of a major report on these conditions have titled their work *Welfare Ranching: The Subsidized Destruction of the American West* (Wuerthner and Matteson 2002). In its dedication, they decry "the use of America's public lands as private feedlots" (frontispiece, unpaginated). The tenor of the book is captured by this fragment from a pungently-worded speech that concludes: "Subsidized Western range beef is a trivial item in the national beef economy. If all of our 31,000 Western public-ranchers quit tomorrow, we'd never even notice" (p. 60). But these ranchers are not quitting en masse. As long as they are determined to remain, the relevant agencies and Congress seem to lack the political will to end public grazing. And the desertification of public lands in the West continues on.

The research question that bioprecipitation brings to bear is whether the process of desertification could be reversed, with or without present grazing patterns in place. There are reasons to suppose that soil degradation and concomitant flora destruction organically sponsor more of the same. Losses of soil and soil nutrients foster a lack of plant growth. By introducing test-site plantings that would be avoided by cattle because of briars, thorns, or taste, it could be cost-efficient to discover whether bacteria would treat these as hosts.[14] Under this hypothesis, the desire, manifested through policy change, to oppose bacterial-pathogenic habitat destruction through over-grazing could become the game-changer within the politics of endless debate over the present and the future of Western public lands. Under newly devised conditions, some grazing could go on; but so could host-plant propagation that could serve as rehabilitated or newly-constructed microbial habitat.

## Conclusion

This essay is funded by the claim that the formation of rain may be linked to the prolificacy of plant pathogens, and that this missing piece of knowledge could be fundamental to water conservation and to land use policy reform. While the theory of bioprecipitation has taken on the contours of a settled account as to its chief finding – that bacteria serve as biological ice nucleators – my attempts to draw significance from this finding have occasioned what are only early thought experiments. Bioprecipitation research has not yet crossed the thresholds of land use or water policy studies, even as a matter of knowledge. I hope to invite vigorous engagement across these frontiers.

## Acknowledgements

I am profoundly indebted to this publication's two anonymous readers and to my *Water International* guide and helpmeet Marcella Nanni, as well as to Professors David Sands, Montana State University; Cindy E. Morris, INRA, Avignon, France; and Gabriel Eckstein, Texas Wesleyan University School of Law; Nico Hauwelt, senior hydrogeologist, City of Austin, Texas; Casey Duncan, Tarleton Law Library, University of Texas School of Law, and David J. Kohtz and Joyce Yinen Young, University of Texas School of Law 2012.

My deep and special appreciation goes to James Nickum, whose interventions re-shaped the essay, and to Stefano Burchi, for the kind of encouragement that only an unusually generous senior statesman could provide.

Responsibility for errors of fact or judgment is my own.

## Notes

1. The term was coined by Professor David Sands in 1978 (Christner 2012).
2. The most common terminological variants are 'ice nucleation active' (INA) and 'biological ice nucleators' (BIN).
3. Telephone interview with Professor Brent C. Christner, August 2012. Notes on file with the author.
4. What would it mean to 'kill' 'the' rain? Could the eradication of a bacterial pathogen from a particular place – or from a multitude of places – have an inhibitory effect on the formation of rain? The answer is, first, not *all* rain: Even a highly proficient variety of ice nucleator is but one among many kinds, as was discussed in the text. Then, too, anything that fails to live, due to its prior inhibition, cannot, of course, be *killed*. But precipitation could be denied its potential for development, at least at some times and in some places, if one of the atmosphere's best-performing ice nucleators, let us say, should get driven from its earthly haunts

in tree canopies, due to clear-cutting and deforestation; among crops, where it gets eradicated or otherwise inhibited; and in and on the ground, where soils get consolidated so forcefully that they can no longer retain moisture or support life. Could the ferocious consolidation of precipitation that has recently been experienced in several regions of the world be causally related to changes in the earthly and sky-bound concentrations and dispersal patterns of CCN- and IN-active biological agents? This idea is already under study, if more tightly conceived than in my general formulation, in relation to desertification and drought. As noted in the text, the study of dynamic cloud systems is extremely complex. Just the relation of cloud chemistry to rising temperatures has been described as 'an enigma' (Gillis 2012, pp. A1, A14). And current climate models are considered to be better at predicting temperature than precipitation (see, e.g., Gertner 2007, p. 74). Gaps such as these in present knowledge leave room for the bioprecipitation hypothesis to roam and for untested further hypotheses concerning distributive gains and losses in precipitation potential linked to microbial massing and other factors to enter.

5. *P. syringae* received sustained research attention, therefore, as an aggressive and successful pathogen responsible for basal kernel blight in respect to barley (Braun-Kiewnick *et al.* 2000); as a principal causal agent of halo blight in many bean cultivars (Bozkurt and Soylu 2011); as a prolific cause of bacterial shoot blight in tea plants, widely infecting them in tea-growing countries such as Japan (Tomihama *et al.* 2009); and as an antagonist of tomato plants, on which it causes a blight called bacterial speck across the world (see e.g. Bashan and de-Bashan 2002a, 2002b).
6. Telephone interview by the author of Professor David Sands, 24 September 2011. Notes on file with the author.
7. Telephone interview by the author of Professor David Sands, 24 September 2011. Notes on file with the author.
8. I break out for extended treatment the perennial public policy debate regarding the over-grazing of the public lands in the final section of the paper.
9. Personal interview of the author with Professor David Sands, 11 January 2012. In Professor Sands's view, genetic tagging is not a distant prospect but a current opportunity awaiting funding for research.
10. Email communication, Professor Cindy E. Morris to Jane Cohen, 19 September 2011. On file with the author.
11. Professor Morris calls on her students to imagine asking a farmer "which situation is preferable: a year with no rain and no plant disease or a year with some rain and some disease. Which of these two options would she or he choose?" (Morris 2011, email communication, see Note 10). According to Professor Sands, crop loss to *P. syringae* amounts to no more than 5% a year. Telephone conversation with the author, 24 September 2011. Notes on file with the author.
12. Professor Sands included this idea in a TED talk given in April, 2012 – one of the three popular introductions to bioprecipitation published or broadcast that month (Sands 2012).
13. In an earlier version of this paper, presented at IWRA World Water Congress XIV, Porto Galinhas, Brazil, September 2011, and published in its proceedings on the World Wide Web, I more fully elaborated the idea of sacrificial cropping. Manuscript on file with the author available via email transmission on request.
14. David Sands has proposed this experiment. Personal interview of the author with Professor David Sands, 12 January 2012. Notes on file with the author.

## References

Amato, P., *et al.*, 2007. An important oceanic source of microorganisms for cloud water at the Puy de Dôme (France). *Atmospheric Environment*, 41 (37), 8253–8263.

American Meteorological Association, 2009. "Precipitation", Glossary of Meteorology, American Meteorological Association.

Arny, D.C., Lindow, S.E., and Upper, C.D., 1976. Frost sensitivity of *Zea mays* increased by application of *Pseudomonas syringae. Nature*, 262, 282–284.

Bashan, Y., and de-Bashan, L.E., 2002a. Reduction of bacterial speck (*Pseudomonas syringae* pv. *tomato*) by combined treatments of plant growth-promoting bacterium, *Azospirillum brasilense*, streptomycin sulfate, and chemo-thermal seed treatment. *European Journal of Plant Pathology*, 108 (9), 821–829.

Bashan, Y., and de-Bashan, L.E., 2002b. Protection of tomato seedlings against infection by *Pseudomonas syringae* pv. tomato by using the plant growth-promoting bacterium *Azospirillum brasilense. Applied and Environmental Microbiology*, 68 (6), 2637–2643.

Beck, R.E., ed., and Kelly, A.K., ed., 2009. Augmenting the available water supply. *In*: R.E. Beck and A.K. Kelly, eds., *Water and Water Rights*. New Providence, NJ: LexisNexis, sec. 3.04, pp. 3–13 to 3–29.

Bozkurt, İ., and Soylu, S., 2011. Determination of responses of different bean cultivars against races of *Pseudomonas syringae* pv *phaseolicola*, causal agent of halo blight of bean. *Euphytica*, 179 (3), 417–425.

Braun-Kiewnick, A., Jacobsen, B., and Sands, D., 2000. Biological control of *Pseudomonas syringae* pv. *syringae*, the causal agent of basal kernel blight of barley, by antagonistic *Pantoea agglomerans. Phytopathology*, 90 (4), 368–375.

Carey, J., 2011. Reducing conflict on public lands [online]. National Wildlife Federation. Available from: http://www.nwf.org/News-and-Magazines/National-Wildlife/Animals/Archives/2011/Reducing-Conflict-on-Public-Lands.aspx [Accessed 25 April 2012].

Christner, B., *et al.*, 2008a. Geographic, seasonal, and precipitation chemistry influence on the abundance and activity of biological ice nucleators in rain and snow. *Proceedings of the National Academy of Sciences*, 105 (48), 18854–18859.

Christner, B., *et al.*, 2008b. Ubiquity of biological ice nucleators in snowfall. *Science*, 319 (5867), 1214.

Christner, B., 2012. Cloudy with a chance of microbes: terrestrial microbes swept into clouds can catalyze the freezing of water and may influence precipitation on a global scale. *Microbe*, 7, 70–76.

Cotton, W., 2009. Weather and Climate Engineering. *In*: J. Heintzenberg and R. Charlson, eds. *Clouds in the perturbed climate system: their relationship to energy balance, atmospheric dynamics, and precipitation*. Cambridge, MA: MIT Press.

deBuys, W., 2011. *A great aridness*. New York: Oxford University Press.

Farrar, J., 2012. Stop diverting park money to unrelated state purposes [online]. *Houston Chronicle*, 24 February. Available from: http://www.chron.com/opinion/outlook/article/Stop-diverting-park-money-to-unrelated-state-3360026.php [Accessed 30 April 2012].

Federal Land Policy and Management Act, 43 U.S.C. §1732 et seq. (1976).

Finlay, B.J., and Clarke, K.J., 1999. Ubiquitous dispersal of microbial species. *Nature*, 400 (6747), 828.

Fox, D., 2012. Life at the edge of space: do high-flying microbes control Earth's weather? *Discover* (4/12), 38–44.

Fox, D., 2012. The clouds are alive. *Discover*, April, p. 38.

Gates, P., 1968. *History of public land law development*. Washington, DC: Zenger.

Gertner, J., 2007. The future is drying up. *New York Times*, 21 October, p. E68.

Gillis, J., 2012. Clouds' effect on climate change is last bastion for dissenters. *New York Times*, 1 May, pp. A-1, A-14.

Goncalves, F., *et al.*, 2011. Effect of bacterial ice nuclei on the frequency and intensity of lightning activity inferred by the BRAMS model. *Atmospheric Chemistry and Physics Discussions*, 11, 26143–26171.

Handwerk, B., 2011. Scientists make dozens of storms in the Abu Dhabi desert? [online]. *National Geographic*. Available from: http://news.nationalgeographic.com/news/2011/01/110118-abu-dhabi-desert-rain-cloud-seeding-controversy/ [Accessed July 2012].

H.B. 148, 2012. General legislative session, State of Utah.

Heintzenberg, J., and Charlson, R., eds., 2009. *Clouds in the perturbed climate system: their relationship to energy balance, atmospheric dynamics, and precipitation*. Cambridge, MA: MIT Press.

Hsu, J., and Bryner, J., 2011. Rainmaking in Middle Eastern desert: success or scam? [online]. *LiveScience*, 4 January. Available from: http://www.livescience.com/10398-rainmaking-middle-eastern-desert-success-scam.html [Accessed 12 July 2012].

Keil, C., *et al.*, 2008. Sensitivity of quantitative precipitation forecast to height dependent changes in humidity. *Geophysical Research Letters*, 35 (9), 1–5.

Keil, C., *et al.*, 2009, American Meteorological Society.

Levin, Z., *et al.*, 2010. Reassessment of rain enhancement experiments and operations in Israel including synoptic considerations. *Atmospheric Research*, 97 (4), 513–525.

Lindow, S.E., *et al.*, 1976. Distribution of ice nucleation-active bacteria on plants in nature. *Applied and Environmental Microbiology*, 36 (6), 831–838.

Maki, L.R., *et al.*, 1974. Ice nucleation induced by pseudomonas syringae. *Applied Microbiology* 28 (3), 456–459.

Möhler, O., *et al.*, 2005. Effect of sulfuric acid coating on heterogeneous ice nucleation by soot aerosol particles. *Journal of Geophysical Research*, 110 (D11), 1–12.

Möhler, O., *et al.*, 2006. Efficiency of the deposition mode ice nucleation on mineral dust particles. *Atmospheric Chemistry and Physics*, 6 (10), 3007–3021.

Morris, C.E., Georgakopoulos, D.G., and Sands, D.C., 2004. Ice nucleation active bacteria and their potential role in precipitation. *Journal de physique IV*, 121, 87–103.

Morris, C.E., and Sands, D.C., 2012. *From grains to rain: the link between landscape, air-borne microorganisms and climate processes* [online]. Available from: from: http://bioice.files.wordpress.com/2012/05/grainsrain_v26apr2012d.pdf [Accessed 30 April 2012].

Morris, C.E., *et al.*, 2007. Surprising niche for the plant pathogen *Pseudomonas syringae. Infection, Genetics and Evolution*, 7 (1), 84–92.

National Conference of State Legislatures, 2011. Trends in state funding of parks and recreation [online]. Available from: http://www.ncsl.org/documents/fiscal/parks2011.pdf [Accessed July 12, 2012].

National Research Council, 2003. *Critical issues in weather modification research*. Board on Atmospheric Sciences and Climate, Division of Earth and Life Studies. Washington, DC: National Academy Press.

Pollack, A., 2012. A battle over an engineered crop: groups oppose a corn impervious to a weed killer. *New York Times*, 26 April, B1.

Rasband, J., Salzman, J., and Squillance, M., 2004. *Natural resources law and policy*. New York: Foundation Press.

Sands, D., 2012. A rainmaker named Sue. *In*: TEDx Bozeman: *Big Sky*, *Big Minds*, March 23, 2012. Bozeman, MT. Available from: http://www.youtube.com/watch?v=_9ZeYxoWsuk [Accessed July 12, 2012]. TEDx Bozeman program available from: http://www.tedxbozeman.com/event/

Sands, D.C., 2012. Speech. *TEDx Bozeman Session 1* [internet broadcast, 23 March]. Available from: http://www.livestream.com/tedx2/video?clipId=pla_9ae4e1d9-6c47-43d6-9a2b-04ee07d4d238 [Accessed 26 April 2012].

Sands, D.C., *et al.*, 1982. The association between bacteria and rain and possible resultant meteorological implications. *Journal of Hungarian Meteorological Service*, 86, 148–152.

S.B. 1332, 2012. Second regular legislative session, State of Arizona.

Schnell, R.C., and Vali, G., 1972. Atmospheric ice nuclei from decomposing vegetation. *Nature*, 236 (5343), 163–165.

Soulage, G., 1957. Les noyaux de congelation de l'atmosphere. *Annales Geophysicae*, 13, 103–134.

Taylor Grazing Act, 1934. 43 U.S.C. sect. 315 et seq.

Tomihama, T., *et al.*, 2009. Environmental control in tea fields to reduce infection by *Pseudomonas syringae* pv. *Theae. Phytopathology*, 99 (2), 209–216.

Trueblood, T., 1980. They're fixing to steal your land. *Field and Stream*, 84 (March) 164–167.

United States Constitution, n.d. Available from: http://www.archives.gov/exhibits/charters/constitution.html

Vali, G., 1968. *Ice nucleation relevant to formation of hail*. Thesis (Ph.D.). McGill University.

Vali, G., 1971. Freezing nucleus content of hail and rain in Alberta. *Journal of Applied Meteorology*, 10 (1), 73–78.

Vali G., *et al.*, 1976. Biogenic ice nuclei. Part II: Bacterial sources. *Journal of Atmospheric Sciences*, 33 (8), 1565–1570.

Wilkerson, C., 1992. *Crossing the next meridian: land, water, and the future of the West*. Washington, DC: Island Press.

Wuerthner, G. and Matteson, M.Y., 2002. *Welfare ranching: the subsidized destruction of the American West*. Washington, DC: Island Press.

# Governing water augmentation under the Watercourse Convention

Rhett B. Larson

**ABSTRACT**

This article evaluates how the UN Convention on the Law of the Non-navigational Uses of International Watercourses can be interpreted and implemented to facilitate responsible and equitable implementation of water augmentation technology. Innovation in technology that has the potential to augment water supplies, like desalination or cloud seeding, may aggravate or mitigate conflict over shared international rivers or lakes. On the one hand, international water disputes may arise due to water scarcity, and augmentation mitigates scarcity. On the other hand, wealthy nations may secure increased water supplies through technological innovations while externalizing the costs of those innovations to poorer nations.

## Introduction

The Convention on the Law of the Non-navigational Uses of International Watercourses (Watercourse Convention) (United Nations, 1997) was adopted by the UN in 1977 for the purpose of protecting and managing freshwater lakes and rivers shared across international borders (Article 1). The convention entered into force 17 August 2014, after Vietnam became the necessary 35th nation to ratify (McCaffrey, 2014). Its entry into force is significant because it arguably now embodies customary international law on apportioning the benefits of internationally shared waters and will serve as a framework for regional transboundary water treaties (Eckstein & Salman, 2015). Nevertheless, important questions remain on how it will be interpreted and implemented. It is unknown, for example, how the convention will be interpreted and its principles implemented with respect to technological innovation impacting the sharing of transboundary waters, including in particular desalination and cloud seeding.

The Watercourse Convention does not explicitly reference technological innovation. Yet new technologies have the potential to either aggravate or mitigate conflict over international watercourses. Desalination is rapidly developing into a cost-effective means of augmenting freshwater supplies and remediating saline contamination in existing supplies (Tsiourtis, 2001). Experts have debated the efficacy of cloud seeding, but increasing streamflow through weather modification continues to be pursued in many parts of the world (Guo, Zheng, & Jin, 2006). As these technologies evolve and expand in use, they will change how nations share water resources, particularly where one riparian state has the resources to avail itself of new technologies that allow it to increase or improve its water supply, while externalizing costs to co-riparian states lacking those same resources (Larson, 2014).

This article evaluates how the Watercourse Convention's principle of reasonable and equitable utilization governs the impact of technological innovations on transboundary water sharing, and proposes an interpretation of those principles to facilitate responsible development and equitable implementation of new or developing technologies in international watercourses. The first section describes how technological innovation can impact the sustainable and equitable allocation of international watercourses among states under the principles of the convention. The second section evaluates how the principles of the convention can be interpreted to avoid conflict and encourage cooperation with respect to water augmentation technology in international basins and even between international basins. The third section proposes a basin-level governance approach under the convention for the management of technological innovation aimed at water supply augmentation.

## The impact of innovation on international watercourses

Technological innovation plays an essential role in water resource development. Innovation allows water to be stored, treated, transported and used in new and more efficient ways. Technological innovation has improved irrigation efficiency, facilitated water recycling, advanced water storage and lowered water consumption in industrial processes and domestic appliances. These technologies facilitate management of existing water sources. Other technologies have the potential to open up new sources of freshwater. Improvements in information and communication technology have the potential to improve water efficiency, more accurately predict droughts and floods, improve cooperation and data sharing between jurisdictions and companies managing water, and facilitate transparency in water governance. It is impossible in one article to address the myriad ways technological innovation impacts water governance. This article focuses on the potential for technological innovation to augment water supplies. This section discusses the international implications of two such augmentation technologies: cloud seeding and desalination.

Desalination already features prominently in both contentious and cooperative efforts to share international watercourses. Desalination is an energy-intensive process for removing salt from water, and generates potentially hazardous brine wastes (Lattemann & Höpner, 2008). While the brine waste disposal, thermal pollution and saline water intake associated with the desalination process raise serious environmental concerns, the major obstacle to implementation of desalination has been energy costs and the greenhouse gas emissions associated with energy consumption (Craig, 2010). The energy costs associated with desalination have historically been so high as to limit its use to extremely water-poor but extremely energy-rich nations (Abderrahman, 2001).

As the state of the art of desalination advances, however, the concerns and costs associated with its energy consumption may become increasingly muted. For example, photovoltaic solar cells and pressure transfer systems (recycling pressure from the waste stream to the production process) conserve energy, reduce greenhouse gas emissions and lower costs (El-Sayed, 2007). When desalination plants are co-located with power plants or wastewater treatment plants, energy costs and consumption are further reduced (Voutchkov, 2004). In combination, these innovations significantly reduce the energy costs and environmental impacts associated with desalination, making it both environmentally and financially feasible for water augmentation (Yamada, Laughlin, & Wood, 1995). For example, the desalination plant on the Llobregat River in Barcelona, Spain, relies on solar power, pressure transfer and co-location, reducing energy consumption from the approximately 25 kwH/m$^3$ typical of desalination plants in the 1980s to approximately 2.5 kwH/m$^3$ today (Cazurra, 2008).

The advancing state of the art in desalination has made it an increasingly viable water supply tool. Nevertheless, its implementation in international basins raises serious and unresolved issues. In 2011, for example, the US federal government explored the possibility of financially supporting two new desalination plants in Mexico to provide water to 300,000 homes on both sides of the US–Mexico border (Spagat, 2011). The proposal sparked controversy due to concerns that the plant developers were attempting to avoid US environmental regulation of the facility (Larson, 2012). The proposal also raised diplomatic complications over the apportionment of the shared Colorado River, because the US government suggested that Mexico forgo some of its rights to that river under the 1944 Rivers Treaty between the two countries in exchange for US support in developing desalination (Spagat, 2011).

The response to this proposal is only one of the most recent examples in a long-standing tension associated with desalination in the Colorado River basin. The salinity in the Colorado River is less than 50 parts per million at its headwaters, yet began to exceed 1200 parts per million at the US/Mexico border in the 1960s due to contamination from agricultural runoff in the US (Lohman, 2003). Mexico argued that salt pollution of the Colorado River violated US obligations under the 1944 Rivers Treaty governing international rights to the river (Judkins & Larson, 2010). After negotiations, the US agreed to maintain salinity in the Colorado River at the Mexican border at just over the salinity behind the US Imperial Dam (Minute 242, International Boundary and Water Commission, 1973). To maintain this salinity, the US federal government authorized the construction (at a cost of $245 million in 1974) and operation of a desalination plant in Yuma, Arizona (Colorado River Basin Salinity Control Act, 1974). However, the operation of the Yuma plant has been sporadic (Judkins & Larson, 2010). When in operation, the plant reduces the salinity of the Colorado River before it reaches Mexico, but it also reduces overall flow to the Colorado River Delta and discharges brine waste to environmentally sensitive wetlands (Tarlock, 2006). When the plant is not in operation, the US attempts to comply with its obligations by diverting agricultural runoff away from the Colorado River to the Cienega de la Santa Clara in Sonora, Mexico, which depends upon this diversion to remain a viable wildlife habitat (Carrillo-Guerrero, Flessa, Hinojosa-Huerta, & López-Hoffman, 2013). The nations of the Colorado River basin now face an intractable dilemma, where desalination and the failure to use desalination both present serious environmental, economic and diplomatic challenges.

Jordan and Israel face similar issues arising from the decision whether to implement desalination in the Jordan River basin (Dreizin, Tenne, & Hoffman, 2008). Under the peace treaty between Israel and Jordan, Israel agreed to desalinate springs in the Jordan River basin to produce potable water for Jordan (Treaty of Peace Between the State of Israel and the Hashemite Kingdom of Jordan, 1995). Until Israel installs desalination facilities for these springs, Israel must supply Jordan with water from the Jordan River. The treaty expressly provides for desalination as a means to increase water supply and apportion rights in the Jordan River. Yet so far Israel has not engaged in desalination on behalf of Jordan, opting instead for continued ad hoc allocation measures on the Jordan River, in part due to environmental concerns as well as energy costs (Mohsen, 2007). The Red Sea–Dead Sea conduit proposed in the 1960s remains a further possibility for providing water to Jordan, Palestine and Israel (Beyth, 2007). This project would also involve large-scale desalination. These proposed desalination projects in the Jordan River basin reflect a common approach to resolving disputes over shared water sources, the theory being that scarcity drives water disputes and that abundance prevents, or at least mitigates, such disputes (Fischhendler, 2008). Nevertheless, water augmentation projects like desalination in the Jordan River still lead to questions about whether and how this augmented water supply would impact apportionment of the benefits of the shared water source.

Cloud seeding presents similar issues to desalination. Like desalination, cloud seeding represents a technological innovation with the potential to augment water supply in an international river basin (Sharon, 1978). Cloud seeding involves dispersing charged particles like silver iodide or frozen carbon dioxide to act as condensation nuclei for the development of rain drops (Rangno & Hobbs, 1995). More freshwater is held in the atmosphere at any given moment than in all rivers combined, and that holding capacity will only increase as global temperatures increase (Karl & Trenberth, 2003). Thus, a technology that makes atmospheric water accessible for potable use is comparable in many ways to technology that makes saltwater accessible for potable use.

Cloud seeding, however, remains a less widely implemented technology than desalination, and its impacts on water supply, environmental quality and climate are less well understood than desalination's (McCaffrey, 2014). Even where increased rainfall has been correlated with cloud-seeding operations, the causal effect of cloud seeding on precipitation is difficult to establish (Levin, 2009). With desalination, there is no question as to the source and cause of the augmented water, and little question as to who is responsible for both positive and negative externalities. But with cloud seeding, it may be difficult, if not impossible, to determine whether seeding operations were the proximal cause of increased precipitation, and even if that could be determined, the extent to which the augmented supply is directly attributable to that particular seeding operation would still be difficult to establish. Despite these uncertainties and limitations, at least 27 nations currently have some cloud-seeding operations, ranging from small-scale commercial operations to large-scale, state-supported water augmentation projects (Simms, 2015). Cloud-seeding operations have been conducted for decades in the states of the Colorado River basin (Simms, 2015). They have also been conducted in the Jordan River basin (Sharon, 1978).

The international implications of cloud seeding have been addressed largely in the context of its role in military operations (Majzoub, Quilleré-Majzoub, Abdel Raouf, &

El-Majzoub, 2009). The US conducted Operation Popeye during the Vietnam conflict, which included cloud seeding to stimulate monsoon events to impede Vietcong transportation (Hauser, 2013). This led to the 1976 Convention on the Prohibition of Military or Any Other Hostile Use of Environmental Modification Techniques (Jensen, 2005). International law has not yet developed to specifically address non-hostile uses of weather modification technology, but the principles of the Watercourse Convention may ultimately have to address the potential impact of cloud seeding on transboundary water apportionment. Does the right to "reasonable and equitable utilization" under the Watercourse Convention change when one state invests in cloud seeding and can prove that such an investment resulted in increased streamflow? While proving such a claim may be difficult today, improved seeding techniques, isotope monitoring and climate models may eliminate or mitigate today's evidentiary obstacles to claiming increased water rights based on weather modification operations.

Unlike desalination, cloud seeding has not yet been directly addressed in bilateral treaties governing the management of these shared watercourses. In the case of both desalination and cloud seeding, one riparian state with greater economic resources can potentially avail itself of new technologies to a degree unavailable to its co-riparian, potentially with positive or negative externalities for the co-riparian. International law must equitably allocate the costs and benefits of these technologies in a way that encourages responsible innovation and implementation of water augmentation technology. Furthermore, the convention contemplates basin-level governance. But impacts of cloud seeding, both positive and negative, may be experienced by non-riparian states who are not parties to any regional water treaty. Given its recent entry into force, and its role as a framework for transboundary river management, the Watercourse Convention provides the best currently available foundation upon which to build international laws governing the impacts of these kinds of technological innovation on transboundary water management.

## Governing innovation under the Watercourse Convention

The silence of the Watercourse Convention on governing technological innovation requires careful interpretation of the convention's text to facilitate responsible implementation of new or developing technologies in international river basins. One of the major purposes of the convention was to establish universal principles to guide states in equitably apportioning the benefits of internationally shared waters, not necessarily the waters themselves (Helal, 2007). The three main principles of the convention are: (1) rights to the equitable and reasonable utilization of the shared watercourse; (2) the obligation to avoid inflicting extraterritorial harms in the use and development of the shared watercourse; and (3) the duty to cooperate and communicate in managing and developing the shared watercourse (Tarlock, 2010). Importantly, the principles of the convention do not apply, regardless of their status as customary international law, where they are inconsistent with the provisions of regional transboundary water treaties (Article 3).

Under the principle of equitable and reasonable use, articulated in Articles 5 and 6, riparian states have the right to make use of an international watercourse, so long as that use is reasonable and equitable. The equity and reasonableness of the use are evaluated based on several factors, including hydrology, population, conservation and the social and economic needs of the riparian state. The convention does not prioritize uses in determining equity and reasonableness; however, "special regard" is to be given to concerns of "vital human needs". Of potential importance to the issue of implementing water augmentation technology like desalination and cloud seeding in international basins, the convention considers the availability of alternatives to using the shared watercourse in evaluating the reasonableness and equity of international water rights.

The convention's second principle, articulated in Article 7, requires riparian states to avoid causing "significant harm" to co-riparian neighbours in their development and use of the shared watercourse. In prohibiting such significant harm, the convention requires that riparian states avoid such harm with "due regard" to rights of equitable and reasonable utilization, and allows for riparian states to compensate or mitigate such significant harm if and when it occurs. This "due regard" language, and the associated option to compensate or mitigate harm to co-riparians *ex post*, has been interpreted to mean that the Watercourse Convention subordinates the duty to avoid significant harm to the right to equitable and reasonable use of the watercourse (Salman, 2007).

The third principle, articulated in Article 8, requires good-faith cooperation between co-riparians in the development of water resources in an international basin. This duty is consistent with the customary international environmental law principle of "good neighbourliness" (Larson, 2012). That duty of communication and cooperation in managing shared watercourses was upheld in the seminal international water law dispute between Hungary and Slovakia over the Gabčikovo-Nagymaros dam project on the Danube (International Court of Justice, 1997).

Nations sharing international watercourses may assert claims against co-riparians for violating principles of the Watercourse Convention in implementing new or developing technologies to the extent these principles bind the co-riparian states (Larson, 2014). To understand how the convention might apply to new or developing technologies, it is helpful first to understand how it could be applied to existing technologies. For example, Benin and Nigeria are bound by the principles of the convention. If Benin, without consulting any of its co-riparians, implemented a run-of-the-river dam on the Niger River that impeded flows to Nigeria, Nigeria could claim that Benin had violated all three principles of the Watercourse Convention by inequitably and unreasonably using the river, causing significant extraterritorial harm, and failing to cooperate in good faith with its co-riparian.

This is a straightforward example of how the convention's principles could be applied to the implementation of technological innovation in an international river basin. However, water augmentation technologies like desalination and cloud seeding may defy such straightforward analysis for several reasons. First, these technologies may effectively import new water to the basin, raising the issue of whether the principles of the convention should apply to the augmented supply at all. Second, negative externalities may be suffered by non-riparian states. For example, cloud seeding may interfere with weather patterns in ways that impact states that do not share a river with the cloud-seeding state (Rangno & Hobbs, 1995). Or desalination may result in pollution of the marine environment or contamination from the disposal of wastes outside a river basin, impacting ecosystems not directly associated with the transboundary water and nations not sharing freshwater sources with the desalinating state (Larson, 2012). The Watercourse Convention contemplates basin-level cooperative management, but water augmentation technology is not inherently limited to intra-basin impacts. Third, water augmentation technologies can create significant harms that might not be the type contemplated by the convention's second principle.

Furthermore, the question remains whether the use of water augmentation technologies by one riparian requires states to forgo appropriations from the shared watercourse in order for utilization to be equitable. As noted above, under Article 6(g), the reasonableness of the use of a shared watercourse is evaluated partially in light of available alternatives. If one state, either because of available resources or because of different geographic conditions, could avail itself of water augmentation like desalination or cloud seeding, must that state forgo use of the shared watercourse, if the technology is of a comparable economic cost? Furthermore, if a state implements water augmentation techniques, does it have an obligation to share the benefits of that augmented source with its co-riparians, or with other states outside of the shared basin who may be impacted by augmentation technology? If the principle of equitable and reasonable utilization under the convention is interpreted to require reapportionment of shared watercourses where one state has technologically augmented its supply, such an interpretation could prove a disincentive to invest in and develop new technologies where they are most needed (Larson, 2014).

For example, the US would arguably be discouraged from investing in research, development and implementation of augmentation technologies if by doing so it would have to forgo rights to shared international watercourses like the Colorado River. On the other hand, if reapportionment is not required to achieve reasonable and equitable utilization, existing inequities will be aggravated where wealthier or geographically advantaged riparian states augment supplies by technological means and less developed co-riparians lack the means to take a similar approach but suffer potential negative impacts from desalination or cloud seeding. For example, Israel could access additional water supplies through desalination or cloud seeding, and partially externalize environmental and energy costs to Jordan through greenhouse gas emissions and marine pollution, and maintain its current allocation on the Jordan River (Larson, 2014). Of course, both the Jordan River basin and the Colorado River basin are governed, at least in part, by existing treaty regimes. Thus, these examples are illustrative only to the extent that those principles are inconsistent with treaty provisions, as these treaties would pre-empt the application of Watercourse Convention principles.

The principle of reasonable and equitable use under the Watercourse Convention is comparable in many respects to common law riparian rights in the eastern United States (Vick, 2012). The adjudication of disputes over riparian rights in US courts, and the interpretation of reasonableness therein, thus may provide some insight and direction on how best to interpret the convention's reasonable and equitable utilization principle in light of water augmentation technologies in international basins. Three cases deciding the reasonableness of a use under common law riparian rights are particularly helpful, given the role of technological innovation in the dispute. These cases illustrate three factors international tribunals and river basin treaty commissions could consider in interpreting the reasonable use principle of the convention in the context of the impact of water augmentation technology on managing shared transboundary waterbodies. These three factors are: (1) the relevance of the technology to the dispute over the shared watercourse; (2) the reasonableness of the technology given the character of the watercourse; and (3) balancing the interest in encouraging innovation and efficiency against the interest in protecting established water rights and avoiding extraterritorial harm (Larson, 2014).

The nineteenth-century case of *Mason v. Hoyle*, in the US state of Connecticut, illustrates the first two factors (14 A. 786 1888). In that case, both parties operated mills on the same stream. The upstream mill owner, Hoyle, implemented a state-of-the-art steam boiler and engine. The new technology allowed Hoyle to close gates on the stream to divert water to fill his reservoir, while still running the mill with the steam engine. When Hoyle closed his gates, streamflow to his downstream neighbour, Mason, was infringed. Mason sued Hoyle, claiming that his riparian water rights were impaired because Hoyle's use was not reasonable. Hoyle relied on an innovative new technology to make more economical use of his water rights, but that technology adversely impacted his downstream co-riparian. The State Supreme Court of Connecticut concluded that the use of the steam engine was relevant to determining water rights, and that Hoyle's use was not reasonable under the circumstances. This illustrates the first factor for adjudicating claims involving technological innovation under the Watercourse Convention: technological innovation that impacts the relative rights to an international watercourse is a relevant consideration in determining reasonable and equitable utilization. Technological innovation is relevant where the technology is a central issue in resolving an international dispute over a transboundary watercourse (Larson, 2014). Thus, even where water supply augmentation technology does not directly impact the shared river itself, and even if it has impacts external to the basin, it is a relevant consideration in determining reasonable and equitable utilization rights in international watercourses if the technology is claimed by either party to potentially aggravate or mitigate disputes over the shared watercourse.

The *Mason* case also illustrates the second factor to consider in interpreting Articles 5 and 6 of the Watercourse Convention in the context of technological innovation: the reasonableness of the use of technology given the characteristics of the watercourse (Larson, 2014). The court in *Mason* drew a connection between reasonableness and fairness. This same connection between reasonableness and fairness is reflected in Article 6 of the convention. That article lays out the factors tribunals and commissions should consider when determining reasonable and equitable utilization (Vick, 2012). These factors include water conservation; the social and economic needs of the riparian states; the geography, climate, ecology and hydrology of the basin; existing and potential uses of the watercourse; and the availability of alternatives to any planned or existing uses. These considerations mirror the language used by the court in *Mason*, which required that water uses, including implementation of new technology, be "adapted to the character and capacity of the stream". Thus, implementation of water augmentation technology in an international basin is reasonable and equitable if that use is effectively adapted to the hydrologic, ecologic, economic and socio-cultural conditions of the basin, and if that technology is implemented with a cost-benefit analysis comparing it to alternative approaches. Evaluating the character and capacity of the stream should also include the need for water augmentation, which would mean an evaluation of population growth, changing consumption and climate patterns, and the extent to which effective conservation or water-recycling alternatives have been implemented that might obviate the need for augmentation.

The third factor to consider in applying the Watercourse Convention to the implementation of water augmentation technology is balancing the competing interests in the watercourse. The competing interests are often promoting innovation and efficiency, weighed against the need to satisfy existing interests in the allocation of benefits, rights and obligations associated with the international watercourse. The case of *Wayman v. Murray City Corp.*, decided by the Supreme Court of Utah in the US, illustrates how this factor can be considered in the context of technological innovation on a shared watercourse (458 P.2d 861 1969). Murray City installed an advanced and powerful pump on its existing well. While the new pump increased capacity, Murray City ultimately did not withdraw any more water than it had in the past. Nevertheless, nearby well owners sued the city, arguing that the more powerful pump reduced the pressure in their wells. Effectively, like Hoyle, Murray City implemented a technological innovation to make better use of its water right, but that innovation impaired the rights of neighbours who shared a water source, like Mason (Larson, 2014).

The court in *Wayman* applied a reasonableness standard, similar to the riparian standard in *Mason*, in holding in favour of Murray City. The court noted that having a right to water does not guarantee that the conditions existing when the right was first asserted will go unchanged. The court in *Wayman* weighed the general interest in promoting innovation and efficiency against the interests of water rights holders in maintaining existing conditions. As in *Wayman*, assessing the reasonableness of implementing water augmentation technology under the Watercourse Convention should require a balancing of interests. Courts should weigh the interests of riparian states adversely impacted by new technologies against the general public interest in encouraging innovation and the specific interests of individual parties to avoid the cost of insuring the technological status quo (Larson, 2014). This should not mean an automatic preference for encouraging investment and innovation in water augmentation. Instead, it should be seen as an attempt to achieve Pareto-optimal outcomes (Larson, 2012). Furthermore, this balancing of interests will not always be similar to that in *Wayman*, where the general interest in innovation outweighed individual interests in predictability and reliability. International tribunals and river basin commissions must also weigh the individual benefits to riparian states from increasing water supplies by technological means against the general interest of clearly assigned and predictable water rights and duties. Preserving the status quo may unfairly privilege existing users in some instances, but it may also serve to create an efficient market by preserving clearly assigned rights. And an effective water market may play an essential role in facilitating the efficient and equitable implementation of water augmentation technology (Chau, 2014). This may mean that some issues associated with water augmentation technology are best resolved through customary international environmental law or through international trade and investment law, when appropriate. These alternative legal approaches may help resolve disputes involving water augmentation technology simply by lowering transaction costs and encouraging nations to facilitate water markets.

The three above-mentioned considerations will assist courts in interpreting the factors comprising reasonable and equitable utilization under Article 6 of the Watercourse Convention in instances when the Watercourse Convention governs water augmentation technology. These three considerations can be summarized into relevance, reasonableness, and equity. First, courts should consider the relevance of the water augmentation technology to the legal conflict (i.e., the extent to which the technology plays a central role in the transboundary water dispute). Second, courts should evaluate the reasonableness of the implementation of the technology (i.e., the extent to which the implementation of the technology is adapted to the character of the stream). Third, courts should balance the interests of the parties and the public in encouraging innovation and efficiency while protecting established rights (i.e., aiming to achieve Pareto-optimal outcomes) (Larson, 2014).

Central to the consideration of each of these factors is the question of whether, or at what point, water generated or made useable by technological means becomes part of the international basin, and thus subject to the Watercourse Convention (Larson, 2014). US water rights principles are illustrative of one interpretive approach to this issue. Many of the United States make a distinction between 'developed water' and 'salvaged water' (Larson, 2012). Developed water is 'new water' imported into, and not previously part of, the basin. Salvaged water, on the other hand, is water that is part of a basin but was unusable because it was inaccessible or polluted. In the case of *Southeastern Colorado Water Conservancy District v. Shelton Farms, Inc.*, the State Supreme Court of Colorado held that increased streamflow achieved by removing invasive species from the river bank was salvaged water (529 P.2d 1321 1975). A party that imports developed water through technological innovation has exclusive right to that developed water. A party that deploys technological innovation to make salvaged water accessible or usable has no such right to that salvaged water, but must use that water within the existing water rights regime. The distinction between developed water and salvaged water, and how rights are apportioned based on that distinction, is a helpful illustration of how water augmentation technology can import 'new' water into a basin, but can also 'liberate' otherwise inaccessible water within a basin. The developed/salvaged distinction is a useful way of thinking about whether and how to apply the principles of the Watercourse Convention to technology that imports new water (developed water) or that liberates inaccessible water (salvaged water).

Desalination could produce developed water in importing desalinated seawater into a basin. It could also produce salvaged water in remediating rivers contaminated by high-salinity agricultural runoff or by treating otherwise unusable brackish groundwater (Larson, 2012). Drawing this distinction in the context of cloud seeding is more complicated. Water vapour moves between basins, and cloud seeding could capture water that came from one basin and would have fallen in another but for technological intervention. The distinction between developed water and salvaged water, while difficult to draw in some cases of water augmentation technology, may still be useful in evaluating how to apply the principles of the convention in that context. For developed water in prior appropriation states, there is an incentive to invest in augmentation, because the investor will reap the benefits by having a superior right. For salvaged water, there is less incentive to remediate contamination or make new sources accessible, because the investor's rights are subordinate to existing rights. The convention should be interpreted and applied in a way that encourages responsible importation of developed water where the character and capacity of the stream require such importation, and facilitate the equitable sharing of that imported water. It should also be interpreted to provide incentives for implementing technology to salvage water.

Where the water generated by technological innovation is developed water, and absent any treaty provision to the contrary, the presumption under the convention should be that the riparian state implementing the technology has exclusive rights to the developed water, with no obligation to forgo rights to the shared watercourse. This presumption should be rebuttable in cases implicating "vital human needs" under Article 10 of the convention. In that case, the state claiming vital human needs must demonstrate an inability to meet certain water security benchmarks (Molle & Mollinga, 2003). What those benchmarks should be, how failure to achieve them should be demonstrated, what reasons for that failure are acceptable, and the amount of water required to meet those benchmarks will be the subject of future articles, as will the question of whether states have an obligation under international law to share the benefits of implementing water augmentation technologies.

Where technological innovation generates salvaged water, the riparian state making use of the technology cannot make a superior claim to any more than the reasonable and equitable use of the shared water source. In this case, the convention's principle of reasonable and equitable utilization should be applied in accordance with the three factors discussed above (relevance, the character of the stream, and efficiency) to determine reasonable and equitable use given the presence of the new technology. However, any salvage efforts to make existing sources accessible or useable in the shared watercourse could be used to create a water quality credits trading market, to encourage investment in water-salvaging technology (Larson, 2013).

The line between salvaged water and developed water can be problematic (particularly for cloud seeding), as can establishing an equitable evaluation of when offsets are required in the augmentation context to meet vital human needs. For cloud seeding, it is impossible currently to determine with any level of precision where a given molecule of water vapour originated, and thus whether seeding operations 'imported' that molecule (making it developed water) or simply returned it to its home basin (making it salvaged water). Because of the limited ability to establish a causal link between seeding operations and increased streamflow, and because of the question of the origins of water induced into precipitation, it is perhaps preferable to treat cloud seeding as salvaged water.

To treat augmented water brought about by cloud seeding as developed water might aggravate water conflict as seeding states attempt to lay claim to the largest amount of water that can possibly be attributed to seeding operations. For now, increased snow pack for water storage and recreation, and weather control to mitigate issues of fog and flood, create sufficient incentives for at least some investment in cloud-seeding research. There is not a compelling need to grant any greater water rights to the seeding state beyond its reasonable and equitable use of the river. As the technology improves, it is possible that the evidentiary issues involved in establishing causation will be overcome, requiring an adaptation of the law to consider whether cloud seeding could be treated in a manner more similar to developed water sources, like desalination. It is therefore essential that the interpretation of the principles of the Watercourse Convention be coupled with improved cooperative transboundary water resource governance, another essential principle of the convention itself.

## Implementing new or developing technologies in international watercourses

The Watercourse Convention not only incorporates principles of reasonable and equitable utilization, but also establishes a framework for cooperative water governance. The principle of cooperation, or 'good neighbourliness', lies at the centre of this governance framework. In the case of water augmentation technology, this principle should be interpreted and applied with three governance elements in mind: (1) adaptive management, (2) collaborative governance, and (3) shared benefits (Larson, 2014). Importantly, the governance contemplated by the Watercourse Convention is generally interpreted as basin-level governance between co-riparians. Both desalination and cloud seeding, however, have impacts beyond the basin. Desalination can impact marine habitat that is not a part of the freshwater basin and generates waste disposal issues that may impact states that do not share the transboundary water; desalination's energy consumption and waste disposal may also have impacts beyond the basin. Cloud seeding may prevent rainfall in another basin, or induce flooding in another basin. To the extent that the convention is applied to water augmentation technology, the principle of reasonable and equitable utilization may require a reinterpretation to a broader governance model to account for the inter-basin impacts of these augmentation techniques.

The first element is adaptive management of water augmentation technology within an international basin. Adaptive management is "a systematic process for continually improving management policies and practices by learning from the outcomes of implemented management strategies" (Paul-Wostl, 2007). The ability to predict the future, including technological innovation, is inevitably limited. Thus, collaborative institutions like international river basin commissions should periodically re-evaluate their policy and management decisions to adapt to changing conditions, including the advancing state of the art of water augmentation technology. Adaptive management is an iterative approach to sustainable development that extends from daily management decisions to overall economic, legal and political strategies (Larson, 2012). The Watercourse Convention implicitly calls for adaptive management, because determinations of reasonableness and equity under Article 6 require consideration of dynamic systems, including population, climate, and potential future uses of the watercourse, that necessitate adaptation. Adaptive management functions well only when governments ensure that the institution implementing this management strategy is accountable to the public and well-funded – governance institutions are likely to be slow to adapt without expertise and resources (Larson, 2014).

The second element, therefore, involves developing effective collaborative governance. Just as with adaptive management, collaboration requires accountability and adequate funding for transparent reporting and expertise. Absent transparency, accountability and a largely disinterested, scientifically driven approach to management, nations may be reluctant to cooperate in a multilateral governance regime. The greater challenge to management of water augmentation technology, as noted above, is that it may not lend itself well to basin-level governance. Costs of transparency and accountability are greater with increasing transaction costs, and those costs increase as more states have a vested interest in water management. Based on the "internalization prescription for government jurisdiction", governing authority over natural resources should be assigned "to the smallest unit of government that internalizes the effects of its exercise", a principle often referred to as subsidiarity (Cooter, 2000). Spillover goods include water resources that cross jurisdictional boundaries. Jurisdictional boundaries should correspond to the geographic contours of such spillover goods (Larson, 2015). As these goods move between jurisdictions, assigning the appropriate level of governance is difficult, particularly as governments can externalize costs, like pollution originating in one jurisdiction and flowing downstream to another, or externalize water scarcity by damming rivers and cutting off flow to downstream riparians.

To achieve internalization of costs, when effects of resource management spill over jurisdictions, a cooperative intergovernmental commission should regulate such externalities (Cooter, 2000). Its boundaries should correspond to the basin or sub-basin, the smallest unit of governance that internalizes the effects of water management. The model for this type of approach is the multi-state river basin commission, which is a common feature of inter-jurisdictional water resource management under the Watercourse Convention framework (Hall, 2010). The convention explicitly references such commissions, stating that riparian states "may consider the establishment of joint mechanisms or commissions ... to facilitate cooperation" (Article 8(2)). The difficulty with water augmentation technology, however, is that it often implicates interests beyond those of the basin – including marine ecology and climate systems that cross watersheds. Thus, in some instances of water augmentation, broader governance structures may be necessary. One possible approach is to retain a presumption in favour of basin- or sub-basin-level governance, but to allow a process in which other states intervene and seek stakeholder input in management decisions for any governance decision implicating inter-basin impacts.

The third element of effective governance of augmentation technology in international basins is shared benefits. The idea is derived from welfare economics, and considers water a valuable commodity with many competing possible uses, and the benefits of those uses, rather than simply the water itself, shared across all stakeholders with interest in the shared resource (Tarlock & Wouters, 2007). The principle of equitable and reasonable utilization under the Watercourse Convention expressly contemplates sharing the benefits of developing a shared watercourse, and is interpreted to encourage each riparian state to engage in development where they have a comparative geographic or hydrologic advantage. For example, water can generate hydroelectric power in upstream areas and irrigate crops in downstream areas. Rather than each state attempting to fully develop the resource for energy and food, co-riparian states should engage in water resource development in accordance with their comparative hydrologic advantages, and then share benefits across the basin. The shared-benefits approach ensures that uses occur in the most appropriate geographic, cultural and economic scope, and then equitably distributes benefits derived from the shared resource across all member states and stakeholders (Larson, 2012). A governance regime that aims for shared benefits is more likely to be perceived as legitimate by member states and other stakeholders, and that legitimacy is important to secure adequate funding, and for adaptive management, and for institutional buy-in for cooperative management (Larson, 2014). The convention implicitly encourages implementation of shared benefits, as it addresses the reasonableness and equity of not only water distribution but also the benefits of the development of the shared watercourse (Larson, 2014).

Governance regimes operating under the principles of the convention should encourage implementation of water augmentation technology in a way that facilitates sharing benefits. This third element already forms part of the existing approach under the convention, but is emphasized here to note that individual riparian states will have unique geographic, hydrologic and socio-economic conditions that may weigh in favour of different approaches to augmentation. Each should implement the approach best suited to those conditions. Upland jurisdictions have a comparative advantage in cloud seeding (given the potential for mountain snow pack and thus greater water storage), and delta jurisdictions have greater potential for sea desalination. International river basin commissions operating under the principles of the convention should encourage sharing benefits from water augmentation implanted in the appropriate geographic setting. A focus on shared benefits in evaluating the advisability of water augmentation projects will hopefully avoid unsustainable augmentation projects implemented despite geographic and ecologic obstacles. Such projects might include large-scale inter-basin bulk water transfers or large-scale brackish groundwater desalination.

One way to integrate these considerations into the governance of water augmentation technologies like cloud seeding and desalination is to authorize these governance institutions with an oversight role in permitting the implementation of water augmentation technologies when they have meaningful extra-territorial implications (Larson, 2012). The commission should be regarded as holding a fiduciary obligation to manage the shared resource for the benefit of all member states and stakeholders, with a duty comparable to that owed by a trustee to multiple beneficiaries of the trust corpus (Larson, 2015). This fiduciary duty should help address concerns associated with sacrificing sovereignty to the commission. So long as the commission regulates technology in accordance with its fiduciary duty, the permitting regime for water augmentation technology will be overseen and enforceable by the commission itself. Also, commission permits for water augmentation will require regular disclosures by the augmenting parties to facilitate transparency, with regular review periods for permit renewals, allowing stakeholder input. Permitting requirements can establish, wherever appropriate, offsets from the shared watercourse to account for augmented supplies, or to require transboundary benefit sharing from water resource development. Such a permitting approach would ideally facilitate responsible development and implementation of water augmentation technologies at the international level (Larson, 2014).

## Conclusion

Water resource management requires adaptation, as it involves governing a dynamic system that integrates virtually all sectors of the economy, as well as significant ecological and cultural resources. Water resource management must also adapt to the dynamic technological environment, where innovation will allow previously inaccessible water resources to be developed. Such innovation has the potential to generate, aggravate, or mitigate international water resource conflict, which could grow as water variability is impacted by global climate change. Climate change, along with population growth and economic development, may strain water resources in some regions to the point that conservation alone will not be an adequate management strategy. Augmentation of water supply by technological means may be necessary in some settings. In those cases, the Watercourse Convention, as the framework for international water rights, should be interpreted to account for the relevancy of the technology to transboundary water disputes, the reasonableness of the technology given the character of the watercourse, and the efficiencies gained from implementing the technology. This interpretive approach should be taken under a cooperative and adaptive basin-level governance structure that is appropriately empowered to encourage sustainable and equitable implementation of new and developing technologies in international river basins.

## References

Abderrahman, W. (2001). Energy and water in arid developing countries: Saudi Arabia, a case-study. *International Journal of Water Resources Development, 17*, 247–255. doi:10.1080/07900620120031306

Beyth, M. (2007). The red sea and the Mediterranean-dead sea canal project. *Desalination, 214*, 365–371. doi:10.1016/j.desal.2007.06.002

Carrillo-Guerrero, Y., Flessa, K., Hinojosa-Huerta, O., & López-Hoffman, L. (2013). From accident to management: The Cienega de Santa Clara Ecosystem. *Ecological Engineering, 59*, 84–92. doi:10.1016/j.ecoleng.2013.03.003

Cazurra, T. (2008). Water reuse of South Barcelona's wastewater reclamation plant. *Desalination, 218*, 43–51. doi:10.1016/j.desal.2006.12.019

Chau, J. (2014). Water markets and the U.N. watercourses convention. *Georgetown International Environmental Law Review, 27*, 179–198.

Colorado River Basin Salinity Control Act. (1974). Pub. L. No. 93-320, 88 Stat. 266.

Cooter, R. (2000). *The strategic constitution*. Princeton: Princeton University Press.

Craig, R. K. (2010). Water supply, desalination, climate change, and energy policy. *Pacific McGeorge Global Business & Development Law Journal, 22*, 225–255.

Dreizin, Y., Tenne, A., & Hoffman, D. (2008). Integrating large scale seawater desalination plants within Israel's water supply system. *Desalination, 220*, 132–149. doi:10.1016/j.desal.2007.01.028

Eckstein, G., & Salman, S. (2015). Concluding thoughts on the implications of the entry into force of the United Nations watercourses convention. *Water Policy, 17*, 181–186. doi:10.2166/wp.2014.009

El-Sayed, Y. M. (2007). The rising potential of competitive solar desalination. *Desalination, 216*, 314–324. doi:10.1016/j.desal.2007.01.009

Fischhendler, I. (2008). Ambiguity in transboundary environmental dispute resolution: The Israeli-Jordanian water agreement. *Journal of Peace Research, 45*, 91–109. doi:10.1177/0022343307084925

Guo, X., Zheng, G., & Jin, D. (2006). A numerical study of cloud seeding by silver iodide and liquid carbon dioxide. Atmospheric Research, 79, 183–226. doi: 10.1016/j.atmosres.2005.04.005

Hall, N. (2010). Interstate water compacts and climate change adaptation. *Environmental & Energy Law & Policy, 5*, 237–323.

Hauser, R. (2013). Using twentieth-century U.S. weather modification policy to gain insight into global climate remediation governance issues. *Weather, Climate, and Society, 5*, 180–193. doi:10.1175/WCAS-D-11-00011.1

Helal, M. S. (2007). Sharing blue gold: The 1997 UN convention on the law of non-navigational uses of international watercourses ten years on. *Colorado Journal of International Environmental Law & Policy, 18*, 337–378.

International Boundary and Water Commission, United States and Mexico. (1973). Minute 242: Permanent and definitive solution to the international problem of salinity of the Colorado River.

International Court of Justice. (1997). Gabčikovo-Nagymaros Project, Hungary v Slovakia, Judgment, Merits, ICJ GL No 92, [1997] ICJ Rep 7, [1997] ICJ Rep 88, (1998) 37 ILM 162, ICGJ 66 (ICJ 1997), 25th September 1997, International Court of Justice [ICJ].

Jensen, E. T. (2005). The international law of environmental warfare: Active and passive damage during armed conflict. *Vanderbilt Journal of Transnational Law, 38*, 145–185.

Judkins, G. L., & Larson, K. (2010). The Yuma desalting plant and Cienega de Santa Clara Dispute: A case study review of a workgroup process. *Water Policy, 12*, 401–415. doi:10.2166/wp.2009.084

Karl, T. R., & Trenberth, K. E. (2003). Modern global climate change. *Science, 302*, 1719–1723. doi:10.1126/science.1090228

Larson, R. (2012). Innovation and international commons: The case of desalination under international law. *Utah Law Review, 2012*, 759–813.

Larson, R. (2013). Orphaned pollution. *Arizona State Law Journal, 45*, 991–1032.

Larson, R. (2014). The U.N. Watercourse convention and technological innovation. *London Journal of International Law, 1*(221). Retrieved from http://ssrn.com/abstract=2279582

Larson, R. (2015). Interstitial federalism. *UCLA Law Review, 62*, 908–968.

Lattemann, S., & Höpner, T. (2008). Environmental impact and impact assessment of seawater desalination. *Desalination, 220*, 1–15. doi:10.1016/j.desal.2007.03.009

Levin, Z. (2009). Report on the state of cloudseeding for rain enhancement. *The Cypress Institute.* Retrieved from http://www.cyi.ac.cy/system/files/Report_EEWRC_omformat.pdf

Lohman, E. (2003). Yuma desalting plant. *Southwest Hydrology, 2003*, 20–23.

Majzoub, T., Quilleré-Majzoub, F., Abdel Raouf, M., & El-Majzoub, M. (2009). "Cloud Busters": reflections on the right to water in clouds and a search for international law rules. *Colorado Journal of International Environmental Law & Policy, 20*, 321–354.

*Mason v. Hoyle.* (1888). *Connecticut Supreme Court Reporter, 56*, 255.

McCaffrey, S. (2014). The entry into force of the 1997 watercourses convention. *Water Policy, 16*, 1202–1203. doi:10.2166/wp.2014.008

Mohsen, M. (2007). Water strategies and potential of desalination in Jordan. *Desalination, 203*, 27–46. doi:10.1016/j.desal.2006.03.524

Molle, F., & Mollinga, P. (2003). Water poverty indicators: Conceptual problems and policy issues. *Water Policy, 5*, 529–544.

Paul-Wostl, C. (2007). Transition towards adaptive management of water facing climate and global change. *Water Resource Management, 21*(49–62). doi:10.1007/s11269-006-9040-4

Rangno, A. L., & Hobbs, P. V. (1995). A new look at the Israeli cloud seeding experiments. *Journal of Applied Meteorology, 34*(1169–1193), 1995. doi:10.1175/1520-0450(1995)034<1169:ANLATI>2.0.CO;2

Salman, S. (2007). The Helsinki rules, the UN watercourses convention and the Berlin rules: Perspectives on international water law. *International Journal of Water Resources Development, 23*, 625–640. doi:10.1080/07900620701488562

Sharon, D. (1978). Rainfall fields in Israel and Jordan and the effect of cloud seeding on them. *Journal of Applied Meteorology, 17*, 40–48. doi:10.1175/1520-0450(1978)017<0040:RFIIAJ>2.0.CO;2

Simms, V. (2015). Making the rain: Cloud seeding, the imminent freshwater crisis, and international law. *The International Lawyer, 44*, 915–937.

*Southeastern Colorado Water Conservation District v. Shelton Farms, Inc.* (1975). *Colorado Supreme Court Reporter, 187*, 181.

Spagat, E. (2011, October 15). Mexico's newest export to US may be water. *Associated Press.* Retrieved from http://archive.boston.com/news/nation/articles/2011/10/15/mexicos_newest_export_to_us_may_be_water/

Tarlock, A. D. (2006). Possible lessons from a comparison of the restoration of the Danube and Colorado deltas. *Pacific McGeorge Global Business & Development Law Journal, 19*, 61–78.

Tarlock, A. D. (2010). Four challenges for international water law. *Tulane Environmental Law Journal, 23*, 369–408.

Tarlock, A. D., & Wouters, P. (2007). Are shared benefits of international waters an equitable apportionment? *Colorado Journal of International Environmental Law & Policy, 18*, 523–536.

Treaty of Peace between the State of Israel and the Hashemite Kingdom of Jordan. (1995). *International Legal Material, 34*, 43.

Tsiourtis, N. X. (2001). Desalination and the environment. *Desalination, 141*, 223–236. doi:10.1016/S0011-9164(01)85001-3

United Nations. (1997). Convention on the Law of Non-Navigational Uses of International Watercourses, 1997. UNGA Res. 51/229, Supp. No. 49.

Vick, M. J. (2012). The law of international waters: Reasonable utilization. *Chicago-Kent Journal of International & Comparative Law, 12*, 141–178.

Voutchkov, N. (2004). Seawater desalination costs cut through power plant co-location. *Filtration & Separation, 41*, 24–26. doi:10.1016/S0015-1882(04)00317-9

*Wayman v. Murray City.* (1969). *Pacific Reporter* 2d 458, 861.

Yamada, R.R., Laughlin, J.K., & Wood, D.K. (1995). Co-located seawater desalination/power facilities: Practical and institutional issues. *Desalination, 102*, 279–286. doi: 10.1016/0011-9164(95)00064-9

# Joint infrastructure and the sharing of benefits in the Senegal and Niger watersheds

Mara Tignino

**ABSTRACT**

Riparian states of the Senegal and Niger watersheds have developed specific techniques for the management of water infrastructure. A common feature in both watersheds is the sharing of benefits from water facilities. Niger River basin states are still at the beginning of a shared vision process for jointly managed infrastructure and equitable benefit sharing, while Senegal River basin states have led the way in innovative forms of shared ownership and governance. Environmental protection and public participation are increasingly included in the development of joint infrastructure, but more could be done to strengthen these aspects of river governance.

## Introduction

The Niger and Senegal Rivers are the first- and second-most important watercourses in West Africa. Spanning over 4200 kilometres, the Niger River is the third-longest river in Africa, after the Nile and the Congo. The Niger River basin overlies the territories of Benin, Burkina Faso, Cameroon, Chad, Côte d'Ivoire, Guinea, Mali, Niger and Nigeria – nine states in total. More than 100 million people live in the basin area. All nine of the river's riparian countries have been included in its legal framework since development of the river began in 1963 (Salman, 2007). The Senegal River is over 1800 kilometres long and covers a drainage area of 300,000 square kilometres. Its basin is a fifth of the size of the Niger River basin, stretching over the territories of four states: Guinea-Conakry (or Guinea), Mali, Mauritania and Senegal.

The riparian states of the Senegal River signed various international agreements to establish transboundary cooperation in the basin in the period following their independence in the 1960s.[1] In 1963, all four countries signed the Convention Relating to the Development of the Senegal River, establishing an inter-state committee for development of the Senegal River basin. In 1968, the four heads of state decided to replace this inter-state committee with the Organization of Riparian States of the Senegal River (OERS).[2] The operation of this new organization was stymied by ideological divergences between Guinea and Senegal. In 1972, the OERS was replaced by the Senegal River Development Organization (OMVS).[3] The OMVS did not include Guinea as a member until Guinea finally acceded to the OMVS in 2006.

The General Act of the Berlin Conference on West Africa of 1885 opened the Niger River (along with the Congo River) for navigation to all basin states under the respective control of France, the United Kingdom and Germany. In contrast to the Senegal River watershed, where riparian states share a cultural and linguistic heritage, wide differences in culture, language and administration exist today in the Niger River basin. Notwithstanding these differences, the Niger basin states adopted several international treaties shortly after their independence from European colonial powers. The first such treaty was the 1963 Act Regarding Navigation and Economic Co-operation between the States of the Niger Basin, in which riparian states undertook to establish an intergovernmental organization "entrusted with the task of encouraging, promoting and co-ordinating the studies and programmes concerning the exploitation of the resources of the Niger River basin" (Article 5). Such an organization came into being a year later through the adoption of the Agreement Concerning the Niger River Commission and the Navigation and Transport on the River Niger. In 1980, the riparian states of the Niger River concluded the Convention Creating the Niger Basin Authority (NBA). As a result of this convention, the NBA replaced the Niger River Commission in 1980. The NBA's initial remit included the development of water resources in areas such as energy, fisheries and industry. The NBA's role in ensuring coherent development of national water resource policies, protection of the environment, fulfilment of the rights of local populations and reinvigoration of cooperation on basin-related issues has been further strengthened through agreement on the Niger Basin Water Charter, signed in 2008.

A specialized regime for the conception, construction and operation of river infrastructure thus exists for both the Senegal and Niger Rivers, in the form of the OMVS and NBA, respectively. Infrastructure on either international watercourse may serve multiple purposes, including producing hydroelectric energy, storing water for irrigation, supplying drinking water to riparian populations, facilitating navigation and preventing flooding. The two basins have chosen different methods to manage water infrastructure. On the Senegal River, OMVS members have developed jointly managed facilities such as the Diama and Manantali Dams. Dam projects also exist on the Niger River, but they are largely national infrastructure projects. After describing the main elements of the legal regimes that govern facilities on the two rivers, this article will compare the relationship between river management models and the distribution of benefits in each context. It will then examine the mechanisms adopted to ensure environmental protection and public participation, both of which are integral to the implementation of integrated water resources management.

## The legal and institutional framework governing joint watercourse infrastructure

The management of the Senegal and Niger Rivers is based on the formalization of riparian state relations through legal instruments and institutions. Since the beginning of the 2000s, the Senegal and Niger basin states have pursued an approach based on a 'shared vision'. This approach draws on numerous principles of integrated water resources management, notably a commitment to public participation in decision making to allow basin issues to be considered contextually as part of the broader development goals of all the riparian states (GWP & INBO 2009; Mbengue, 2014; Sangbana, 2016). In the Senegal River watershed, cooperation is driven by a desire to balance careful stewardship and efficient use of transboundary water resources (Godana, 1985). To achieve these goals, riparian states regard themselves as part of a community of interests and rights around shared resources and therefore embrace a communal, integrated approach to managing them.

In both the Niger and Senegal River watersheds, jointly managed facilities and sharing of benefits are means to put this community of interests into concrete practice. The seeds of this community of interest were planted in the early 1970s. The initial traces of a specialized regime for hydraulic facilities on the Senegal River can be discerned in Article 3 of the Agreement on the Establishment of an Inter-state Committee for Development of the Senegal River Basin. Although this agreement was adopted in 1964, it was not until 1972 that the Senegal River's riparian states adopted the two cardinal instruments of the river's governance regime: the Convention Concerning the Creation of the Senegal River Development Organization and the Convention Concerning the Status of the Senegal River. In the Senegal River context, not only did the normative and institutional framework which emerged over time make the Senegal watershed a forerunner in international water governance, but the heft and substance of its governance regime also made it a pacesetter (Boisson de Chazournes, 2013).

A noteworthy feature of the governance regime of the Senegal River is its provisions for the creation of the status of 'common works'. The management of common works became the subject of two international conventions, the Convention Concerning the Legal Status of Common Works of 1978 (hereinafter Common Works Convention) and the Convention Regarding Financing Arrangements for Common Works of 1982. Under the auspices of these instruments, the OMVS members put in place a unique system of interstate agencies for the management and operation of two dams on the Senegal River: the Manantali and Diama Dams. The 1978 and 1982 conventions were supplemented by two additional agreements in 1997. The 1997 agreements established separate, autonomous agencies responsible for the management and operation of these two facilities: the Convention Creating the Diama Management and Operation Agency (hereinafter SOGED Convention) and the Convention Creating the Manantali Energy Management Agency (hereinafter SOGEM Convention).

In contrast, in the Niger River basin, there is no international instrument specifically dedicated to shared infrastructure. Provisions in existing instruments provide signs of a regime still under construction. For example, the 1964 Agreement Concerning the Niger River Commission and the Navigation and Transport on the River Niger enjoined the Niger River Commission to "collect, evaluate and disseminate basic data on the whole of the basin, to examine the projects prepared by the riparian states, and to recommend to the Governments of the riparian states plans for common studies and works for the judicious utilization and development of the resources of the basin" (Article 2(c)). The Niger River's governance regime has been strongly shaped by the shared vision process initiated in 2002 (Sangbana, 2011). The shared vision laid out the strategic framework for the sustainable development of the Niger Basin. This regime comprises both "common works" and "works of common interest" which will permit benefit sharing among the Niger basin states.

Among the most important results of the shared vision process are the Niger Basin Water Charter and the Sustainable Development Action Plan (SDAP) and its associated investment programme (2008–2017). These instruments set out a general approach to basin development, an approach negotiated and accepted not only by all member states but also by other actors who are users of basin resources. The aim of these instruments is to develop the potential of the basin as a whole and generate the greatest possible advantages for the Niger's riparian states and the populations that depend on its resources.

## Infrastructure in the Senegal and Niger basins: collective governance in practice

### *The Senegal basin*

The Senegal River basin is the site of a number of common works which have been the object of cooperative management by the OMVS. These projects bear witness to the efficacy of collective river governance. The OMVS developed the Diama and Manantali Dams, as well as the port facilities between Saint-Louis and Kayes. Infrastructure projects began in the 1970s in the hopes of overcoming annual irregularities in river levels, reducing flooding, and addressing the significant rainfall deficit.

The Diama Dam is located near the delta of the Senegal River, between Senegal and Mauritania. This project was designed for flood control, irrigation and drinking water provision and to prevent the intrusion of saltwater into the lower valley. The Diama Dam permits the irrigation of 120,000 hectares and satisfies the water needs of several urban and rural populations. This dam is the main source for water withdrawals in Senegal and Mauritania, not only for farming (in particular cereals and traditional crops) but also to supply water to many urban and rural centres (OMVS SOGED, 2006).

Alongside the Diama Dam project, the OMVS also developed the Manantali Dam project. This included construction of a dam on the Bafing River (a tributary of the Senegal River in Mali) and the construction of 1300 km of transmission lines to Bamako, Mali. The Manantali Dam irrigates an area of over 3750 km$^2$ and allows navigation between the cities of Saint-Louis and Kayes. In addition to its irrigation and navigation functions, the Manantali Dam provides about 800 million kilowatt hours per year to the cities of Bamako, Dakar and Nouakchott. The Manantali power station went unfinished until 2002. The delay in construction was largely attributable to the border conflict between Mauritania and Senegal in 1989. The killing of Senegalese farmers by Mauritanians triggered a conflict that cost the lives of hundreds of people. The two states engaged in occasional armed skirmishes and nearly entered full-scale armed conflict.

Senegal, Mauritania and Mali draw major benefits from the Diama and Manantali Dams. Beyond the Diama and Manantali Dams, common infrastructure on the Senegal River includes the sea ports at Saint-Louis and Kayes in Senegal. Since its founding in 2011, the Agency for the Management and Exploitation of Navigation on the Senegal River (SOGENAV) has sought to improve navigation between the fluvial ports in

Senegal and Mauritania. Its tasks include the management and administration of activities related to navigation and transportation on the river.

As none of the riparian states was able to finance these projects on its own, the creation of the OMVS provided a means of mobilizing financial assistance from the international donor community. Twelve donors contributed to a loan totalling USD 620 million (Yu, 2011). The three states negotiated a cost-sharing schedule based on projected benefits and tailored to their differing needs. This *clé de répartition* was adopted by the Council of Ministers of the OMVS in 1985 (Hensengerth, Dombrowsky, & Scheumann, 2012). The cost schedule should be considered a success to the extent that the countries agreed on not only how to share expected benefits from cooperation, but also how to allocate costs in accordance with expected benefits.

### *The legal regime governing the Diama and Manantali Dams*

The joint ownership of the dams through the OMVS makes the Senegal River a unique case of substantive transboundary cooperation on the management of water infrastructure. The Diama and Manantali Dams on the Senegal River are governed and managed according to rules laid out in the Common Works Convention and the Convention Regarding Financing Arrangements for Common Works.

According to the Common Works Convention, a facility is common works whenever it is the object of a legal instrument declaring it to be "common and indivisible property" of the member states. The term "legal instrument" comprises any "convention or resolution adopted by a deliberative body of the OMVS" (Articles 1 and 2). Thus, whether a facility constitutes a common work cannot be deduced from anything about the facility itself; it does not depend, for example, on the facility's territorial location. The convention does set out a non-exhaustive list of works designated as common: the Manantali Dam, the Diama Dam, the Saint-Louis sea-river port, the Kayes river port, and any ancillary facilities or annexes to these works (Article 3).

The Common Works Convention stipulates that any infrastructure designated as common becomes and remains during its lifetime the common and indivisible property of the OMVS member states (Articles 2 and 3). Where a common project is constructed (or partially constructed) on the territory of one state, that state must not create any obstacle for other states with a co-ownership interest to exercise their use, enjoyment and administration rights (Article 6). Furthermore, common works are exempt from requisition, confiscation, expropriation or any other form of seizure or forced takeover by the legislative, executive or judicial powers in any of the co-owner states (Article 5). On the financial level, OMVS member states act as co-guarantors for the repayment of any loans extended to the organization for the construction or operation of jointly managed facilities (Mbengue, 2004). Each co-guarantor state is obligated to contribute to the organization's financial resources. In particular, each member must take all appropriate measures to contribute the necessary funds to the organization, within the appropriate period, to prevent a default with creditors. A state breaching this obligation will bear the entirety of financial charges arising from the delay, including expenses related to any work and engineering contracts on joint facilities. Another aspect of the financial component is the capacity of the members of the organization to borrow individually or jointly for the construction or operation of common works.

Member states pay back the loans to the organization, which controls and manages joint infrastructure on behalf of the OMVS's member states (Convention Regarding Financing Arrangements for Common Works, 1982).

The operation and maintenance of common works may be delegated to autonomous agencies overseen by the OMVS (Article 15). The establishment of these agencies demonstrates the uniqueness of the system created by the Common Works Convention. These agencies are constituted as public, inter-state bodies and tasked with the sharing of economic benefits among OMVS members.

The jointly managed facilities in the Senegal basin could serve as a model of cooperation for other international watercourses in Africa. A tripartite project modelled on the Senegal River example could have been beneficial for the Grand Ethiopian Renaissance Dam project. The concerns of Egypt, Sudan and Ethiopia could have been addressed through joint ownership, management and funding of the project. Although Ethiopia proposed a joint ownership regime, Egypt and Sudan rejected this option (Salman, 2016).

### *The Niger basin*

The development of the Niger River provides a further illustration of the benefits of multi-state collective cooperation and management of river basins. Crossing some of the least developed African countries and four landlocked countries (Burkina Faso, Chad, Mali and Niger), the Niger River offers significant development opportunities for watercourse states. Some of these opportunities are directly linked to the river, like production of energy, irrigation and food production, whereas others are 'beyond the river', like the promotion of economic integration and the fight against poverty (Andersen, Golitzen, Ousmane, Jarosewich-Holder, & Olivry, 2005).

The most significant outputs from the shared vision process led by the NBA are the SDAP and the Niger Basin Water Charter. The NBA spearheaded a strategic framework for the sustainable development of the Niger River basin: that is the SDAP. One of the pillars of the SDAP concerns the development of socio-economic infrastructure (Niger Basin Authority (NBA), 2007).[4] In this regard, the NBA has three priorities before 2025: (1) infrastructure of common interest and/or with transboundary benefits; (2) infrastructure favouring local socio-economic development; and (3) infrastructure ensuring access to basic services. The considerations at play in choosing these priority areas were: (1) maintaining sufficient flow rates during low-water periods; (2) minimizing the impact on wetland zones, particularly the Inland Delta; (3) developing agriculture via irrigation; and (4) generating hydroelectric power. On the basis of these criteria, the parties adopted a development plan for the 2025 horizon that included both the rehabilitation of existing dams (the Kainji and Jebba in Nigeria and the Lagdo in Cameroon) and the development of three new installations (Fomi, Taoussa and Kandadji) (Brachet & Dessouassi, 2008). This particular development plan was considered not only most beneficial for the basin as a whole but also best for the Inland Delta, aggregate electric production and respect for minimum flow rates (Niger Basin Authority (NBA), 2007).

In contrast to the collective water facilities on the Senegal River jointly managed by OMVS member states, the planned infrastructure in the Niger basin is for the most part national works. The Niger Basin Water Charter includes two types of joint infrastructure: common works and works of common interest. Common works, which are to be found on the Niger River, are "a facility which the member states of the NBA have decided through a legal act to make their common and indivisible property" (Article 1.21). Works of common interest are "a facility being of interest to two or more member states of the NBA, for which they have mutually agreed on coordinated management" (Article 1.22). The designation of infrastructure as either common works or works of common interest is dependent on the adoption of legislation by the parties. Only common works carry with them the idea of indivisible, communal property of all the member states. Works of common interest are restricted to a limited number of NBA member states.

The SDAP includes proposals for 26 new dam projects on the Niger River. Several of these projects warrant special attention for their transboundary impacts, like the proposed Fomi Dam in Guinea, the Taoussa Dam in Mali and the Kandjadji Dam in Niger. Funding for these three dams has primarily been provided by the World Bank and the African Development Bank.

The Fomi Dam project is a crucial test of the implementation of the SDAP, due to its potential for both transboundary benefits and potential harms to nearby communities. This multipurpose dam, located in Guinea, would allow Mali to double the area of the arable land and regulate river flow for downstream countries (NEPAD 2014). As conceived in 2015, the Fomi Dam would provide 6.1 billion cubic metres of water storage in Guinea, generate 90 MW of renewable energy and maintain dry-season environmental flows. The Fomi Dam could contribute to food security, production of electricity, and water storage during dry seasons. The implementation of this project is supposed to include a decision-making process involving potentially affected communities. The project seems likely to displace many people in Guinea and could have negative environmental impacts in the inner Niger Delta in Mali, an area that is a source of livelihood for around one million people (World Bank, 2014). Because of these risks posed by the construction of the dam, an environmental and social impact assessment and a mitigation action plan are necessary to advance the project.

This dam is a joint project between Guinea (upstream) and Mali (downstream). Guinea and Mali would have to mitigate the impacts, including on livelihoods, the resettlement of communities in Guinea and the environmental and social impacts in the Niger Inner Delta in Mali (World Bank, 2015). The two states established an Inter-Ministerial Committee in 2014 to oversee the joint implementation of the project, with the NBA serving as an observer.

The SDAP also includes a component focusing on facilities that are in urgent need of rehabilitation, including the Kainji and Jebba Dams in Nigeria and the Lagdo in Cameroon. There is also mention of the finalization of studies on planned facilities in the basin (i.e. the Taoussa, Kandadji and Fomi Dams) to ensure that they comply with environmental and social norms and thus facilitate consensus among financial partners on their development.

As yet, the Niger basin lacks any infrastructure designated as common works in the sense envisioned by the Niger Basin Water Charter. This suggests that Niger River basin states are reluctant to participate in full-blooded joint cooperation regarding river infrastructure. Instead, the Fomi, Taoussa and Kandadji Dams will be works of common interest (Brachet & Dessouassi, 2008). Given that the plan that included these three national dams of common interest was preferred over alternatives in economic, social and environmental dimensions, it seems that the configuration of the basin – in which nine states share more or less direct access to the river's principal course – may never make the establishment of common works a meaningful possibility. The idea that sub-basin commissions are necessary for the success of the NBA's mandate – an idea that finds expression in the Niger Basin Water Charter – weighs heavily in favour of a sub-regional approach to basin management (Article 1.4).

## The sharing of benefits in the Senegal and Niger basins

The distinctions between the governance systems for the Senegal and Niger Rivers continue in the area of benefit sharing. The principle of benefit sharing consists in ensuring that the direct benefits of an installation are distributed in such a way as to balance the share of investments contributed by each state (Brachet & Dessouassi, 2008). Equitable division varies depending on whether the installation in question is designated as common works or works of common interest. With regard to common works, the fundamental principle associated with the status of common and indivisible property entails each state's individual right to an indivisible share, and a collective right to the use, enjoyment and administration of the shared facility. Each state is furthermore required to guarantee the enjoyment of these rights by all the others. The sharing of benefits and costs is dealt with differently in the Senegal and Niger basins depending on whether an installation is designated as common work or work of common interest. As further elaborated below, the distinctions between these legal categories come to the fore at a temporal level, as questions arise about how new users should bear the costs already incurred to build and maintain jointly managed installations.

### *The Senegal basin*

The division of benefits is organized on an equitable basis in the Senegal River basin, under the auspices of the OMVS. In the OMVS, the division of benefits is tightly linked to the status of common works, designated as indivisible property of OMVS members. The operative principle is set out in the Common Works Convention. Article 12.1 states that "investment costs and operational fees are to be allocated between the co-owner states on the basis of the benefits that each co-owner state will have from the operation of common works". Each co-owner, moreover, guarantees repayment of the loans extended to the OMVS by the international donor community for the construction of the relevant infrastructure.

The central importance of sharing benefits, costs and fees can be illustrated by the ongoing issues raised by Guinea's late accession to the Senegal River Water Charter. Guinea, a riparian state of the Senegal River, acceded to the charter in 2006. According to the accession treaty, Guinea also became a party to the Common Works Convention. In the accession treaty, however, no mention is made of the Convention Regarding Financing Arrangements for Common Works nor of the 1997 instruments concerning the management and operation of the Diama and Manantali Dams. As a result, the

Conference of Heads of State and Heads of Government is charged with ruling on Guinea's future share of costs, benefits and expenses (Guinea Accession Treaty, Article 6). Further, the conditions of construction and operation for future common works are to be decided on a case-by-case basis with respect to Guinea (Article 5). Doubt surrounds the degree to which Guinea may benefit from common works constructed prior to its accession to the treaty. The four riparian states are currently negotiating a schedule of costs and benefits that includes Guinea. The inclusion of Guinea in the OMVS requires a revision of the 1985 schedule key.

The principle of shared management of common works finds further institutional expression in the creation of public, inter-state organizations to operate and manage these works, as in the case of SOGED and SOGEM and their responsibilities regarding the Diama and Manantali Dams on the Senegal River. The differences between the agencies provide an informative case study about the practical implementation of benefit sharing. SOGED draws its financing primarily from its exclusive right to sell water of the Senegal River to all users other than those related to the production of electricity (SOGED Convention, Articles 12 and 13). OMVS member states, in turn, must ensure that their own users pay for water provided by SOGED (Article 13). The OMVS maintains an oversight role to ensure that SOGED is fully paid. For SOGEM, in contrast, the principal source of revenue is its exclusive right to produce, transmit and sell the electricity generated by the jointly owned facility. SOGEM determines electricity prices in agreement with the Council of Ministers. SOGEM is responsible for managing these resources and for ensuring the provision of services through the Manantali facilities. SOGEM can also lease all or part of these installations or transfer operation of them to a third party in return for a fee (SOGEM Convention, Article 12.1). As in the case of the Diama Dam, the OMVS must assure itself that the national electricity agencies pay the full cost for energy provided through SOGEM. The most salient difference between the SOGED and the SOGEM lies in the specific range of activities of the two agencies. The two institutions are complementary in supplying water and energy to the population of the Senegal watershed. Benefit distribution is ensured through the sale of water and energy to local users.

### *The Niger basin*

The idea of shared benefits also informs the Niger River basin regime. In Senegal, the principle of shared benefits takes shape around the pursuit of an equitable division that aligns investment costs and direct benefits drawn from particular installations. The principle of shared benefits manifests quite differently when it comes to works of common interest, as aptly illustrated by the management of the Niger River basin. In this case, the division of benefits is restricted to a subset of the member states. This status is to be attributed only to infrastructure shared among a collection of states smaller than the entire membership of the NBA.

However, a closer examination of the practical implementation of the benefit-sharing regime indicates a different outcome, one that distributes benefits more widely to all of the Niger basin's nine riparian states. The computation of earnings for all member states takes into account the economic returns not just from a single facility but from the "entire collection of works of common interest" as foreseen in the chosen basin development scenario, which includes the Fomi, Taoussa and Kandadji Dam projects. The construction of these three projects will make it possible to satisfy the minimum flow objectives for the Niger as a whole and upstream hydro-agricultural needs (Niger Basin Authority (NBA), 2007). It was only by comparing the potential benefits of each dam and their combined contribution to the basin as a whole that the experts concluded that the construction of these three dams would be the most beneficial for the social and economic development of basin's states.

In the case of the Niger basin, benefit sharing among the nine states consists in part in ensuring that the direct benefits generated by projects are distributed in such a way as to strike a balance between the share of benefits received and the share of investments made by each state. The NBA is the body overseeing the implementation of the three projects, and its aim is to generate as many benefits as possible for all member states. In the next years, the NBA could play a role in the adoption of an instrument to share costs and benefits between all or a subset of the riparian states.

## The inclusion of environmental and local community considerations in benefit sharing

There have been efforts in both the Niger and Senegal River contexts to ensure environmental protection, distribution of benefits to local populations, and local participation in decision making.

Dams can contribute significantly and positively to human development. However, they can also have negative impacts on the environment and the living conditions of local populations, or even necessitate forced resettlement. Negatively impacted populations too rarely see a net benefit from the resulting economic and social development. Instead, the main recipients of a dam's benefits often live in cities and urban areas far away from the dam site. Proactive measures are needed to include local populations in benefit-sharing mechanisms and to distribute the economic advantages of dam projects to the local communities. In this sense, benefit sharing comprises not only inter-state relations but also interaction between states and local populations. This latter aspect of benefit sharing can strengthen public participation in the decision-making process for the construction of dams in the two basins.

Civil society must be a part of the decision process in relation to the construction of dams on both the Senegal and the Niger rivers. This inclusion is reflected in the institutionalization of public participation via the creation of consultation fora, such as the local coordination committees in the context of the OMVS and national user organizations in the context of the NBA (Dessouassi, 2015; Sangbana, 2015).

For a long time, neither the protection of the environment nor public participation played much of a role in either the Senegal or the Niger basin's legal regime. Water management projects were considered solely from the perspective of the development needs of individual states (Sangbana, 2015). In response to the negative impacts these projects had on the environment and on local populations, the members of the OMVS and of the NBA were led to include these aspects more directly in their governance instruments (Boinet, 2013; Ndiaye, 2003).

The 2002 Senegal River Water Charter dedicated a chapter to the environment and another to institutions charged with management of water and the environment, as well as making reference to the polluter-pays principle (Article 18), the principle of prevention (Article 16), and the determination of shared objectives regarding water quality (Article 16). Similar intentions were affirmed in the Niger Basin Water Charter, which includes principles of international environmental law ranging from the precautionary principle (Article 6) and the prevention principle (Article 7) to the polluter-pays principle (Article 8).

Both the Senegal and Niger basin projects attempted to distribute the projects' benefits to local populations. Following the construction of infrastructure for the exploitation of the Senegal River (Diama and Manantali Dams), the OMVS invested in socio-economic development and environmental protection of the Senegal River basin. In addition to these macroeconomic programs, OMVS promoted actions to improve the standard of living, incomes and productivity of local people. Thus, the people already benefit from the potential for water supply and energy generated by the dams. With regard to the Niger River, the shared vision framework emphasizes solidarity and community of interest among states sharing this river basin. Member states recognize that each one of them must enjoy a reasonable and equitable share of water resources in order to contribute to the reduction of poverty, to food security and to the protection of the environment (Principle 3 of the Declaration of Paris).

Institutional cooperation mechanisms like the OMVS (in Senegal) and the NBA (in Niger) provide a framework for coordination among riparian states. This not only reduces the numerous risks raised by dam construction but also facilitates a redistribution of benefits among all the states sharing an international watercourse. Furthermore, these mechanisms promote a distribution of 'immaterial' benefits related to the rights of local populations potentially affected by dam construction (Geneva Water Hub/ Platform for International Water Law 2015). As examples, consider the OMVS's investment in environmental protection measures following the construction of the Manantali and Diama Dams, and the NBA's efforts to improve standards of living through the construction of hydroelectric projects such as the Kandadji Dam in Niger (Skinner, Niasse, & Haas, 2009).

With regard to environmental monitoring, members of the Senegal River system have committed themselves to "monitor all activity that could noticeably affect the characteristics of the river regime, the health of the waters, the biological characteristics of its flora and fauna, its bodies of water or its environment in general" (Senegal River Water Charter, Article 16.2). As for the Niger River, the objective of the Niger Basin Water Charter is to promote "a sustainable, equitable and coordinated utilisation of water resources in the Niger river watershed" (Article 2.1). Furthermore, the state parties must "manage the Niger Basin's water resources so as to maintain the quantity and quality of these resources at the highest possible level" (Article 10).

Since the beginning of the investment programmes, the states along the Senegal River have paid attention to the environmental impacts of dam construction. A case in point is the Programme for the Mitigation and Monitoring of Environmental Impacts (PASIE) related to the construction and operation of the Manantali Dam. PASIE was developed in light of the domestic legislation of the OMVS member states and the standards of the World Bank, the African Development Bank and other donors (Uhlir, 2000). The objective was to set out norms to prevent negative impacts on the environment from the Manantali Dam. In the Niger basin, the World Bank is playing a central role in bringing the planned projects into compliance with international social and environmental standards. The bank requires studies of environmental impacts, technical feasibility and resettlement planning. Through the 2002 adoption of the Senegal River Water Charter, the parties redoubled their commitments to sustainable development and public participation. The charter put in place mechanisms to ensure improved monitoring of the aquatic ecosystems on the one hand and, on the other, to bring together a diversity of relevant actors – local populations, policy makers and technical experts – in river management (Article 13). Despite the nature of the Senegal River regime as a pioneer of joint management, much more could be done to ensure environmental protection and promote public participation (Sangbana, 2015; Vick, 2006).

These environmental obligations are accompanied at the institutional level by the establishment, at the OMVS in 2000 and the NBA in 2006, of environmental observatories, whose objective in each context is to monitor environmental developments in the watershed. The observatories provide state parties the information necessary to measure the environmental impacts of dams and other hydraulic infrastructure and to undertake actions to attenuate negative environmental effects. In the case of the Senegal River, the observatory publishes an annual report for policy makers on the state of the environment. The Niger-based observatory ensures the distribution of hydrologic, meteorological, demographic, social and environmental information. Its principal aim is to improve coordination at the basin level through the systematic production of forms of environmental information useful in the facilitation of regional cooperation. The Niger Basin Water Charter aims at facilitating the inclusion of environmental considerations and local community concerns in the sharing of benefits. To this end, each of the nine riparian states has a National Focal Point charged with ensuring the coordination of NBA activities at the state level and with engaging local civil society actors (Niger Basin Water Charter, Article 2). The parties have put in place national user organizations aimed at facilitating effective involvement by non-state actors in the shared vision process. The nine national user organizations and the Regional User Organization facilitate relations among users and other stakeholders in environmental protection and the development of the Niger basin (Dessouassi, 2015).

## Conclusion: the broader impact of joint management of watersheds

The formalization of riparian state relations through legal instruments and institutions is the principal mechanism for the management of the Senegal and Niger Rivers. Riparian states of the Senegal and Niger watersheds have developed innovative techniques for the management of their collective water infrastructure. Both the Senegal and Niger watersheds are managed according to the principle of benefit sharing, but this is realized in different ways in the two river basins. In the Senegal River basin, infrastructure is often co-owned and co-managed as 'common works', whereas in the Niger River basin, infrastructure tends to be national works or 'works of common interest'. Although Niger River basin states appear reluctant to participate in full-blooded joint cooperation regarding river infrastructure, the governance regimes nevertheless display some common attributes. The sharing of benefits and costs depends on whether an installation is designated as common work or work of common interest, a distinction that comes to the fore most visibly with regard to the rights and responsibilities of treaty parties who acceded after infrastructure projects were developed. In both the Niger and Senegal River basin joint management schemes, attention has been paid to environmental impacts and the importance of public participation. However, much more could be done on these important aspects of shared governance of water infrastructure.

West Africa is a vulnerable region. The dams planned for the Senegal and Niger Rivers could either trigger tensions in West Africa or be instruments of peace and stability between riparian countries. Since 2006, Mali has been under the threat of simmering armed conflict in its northern region. In Nigeria, Boko Haram is responsible for several attacks against the population. The conflict in Libya is also a threat to West African countries, especially those in the Sahel region. As a forum of dialogue between countries, mechanisms of transboundary watercourses such as the OMVS and NBA may contribute to the promotion of peace and stability in the region. Benefit sharing laws and institutions strengthen inter-state cooperation and promote the inclusion of local populations in the benefits of socio-economic development. These mechanisms are not only indispensable in the planning and implementation of dams but may also undergird regional conflict prevention.

## Notes

1. Details of all statutory references are listed in the Appendix.
2. Organisation des états riverains du bassin du fleuve Sénégal (OERS).
3. Organisation pour la mise en valeur du fleuve Sénégal (OMVS).
4. The other two aspects are (1) protection of resources and ecosystems, and (2) capacity building of actors and involvement of those actors in integrated water resources management.

## Acknowledgements

The author wishes to thank the team of the Platform for International Water Law and the Geneva Water Hub of the University of Geneva for their insights into the legal and institutional frameworks governing the Senegal and Niger basins. This article benefited from the results of a roundtable expert meeting on Cooperation and Benefit-Sharing in the Senegal and Niger River Basins held at the World Meteorological Organization in September 2015.

## References

Andersen, I., Ousmane, D., Jarosewich-Holder, M., Olivry, J.-C., & Golitzen, K. G. (2005). *The Niger River Basin: A vision for sustainable management.* Washington DC: World Bank. Retrieved April 24, 2016, from, http://www-wds.worldbank.org/external/default/WDSContentServer/WDSP/IB/2005/12/08/000090341_20051208084115/Rendered/PDF/345180PAPER0NR1Basin01OFFICIAL0USE1.pdf

Boinet, E. (2013). *Hydropolitique du fleuve Sénégal. Limites et perspectives d'un modèle de coopération* [Hydropolitics of the Senegal River. Limitations and perspectives of a cooperation model]. Paris: L'Harmattan.

Boisson de Chazournes, L. (2013). *Fresh water in international law.* Oxford: Oxford University Press.

Brachet, C., & Dessouassi, R. (2008). *A shared vision of the Niger Basin.* Paris: French Agency for Development & Niger Basin Authority.

Dessouassi, R. Y. (2015). État du processus de gestion intégrée des ressources en eau dans le bassin du Niger: Expériences de l'implication des acteurs non-étatiques et de la prévention des conflits entre usagers de ressources [The status of the integrated water resources management process in the Niger Basin: Experiences in the integration of non-state actors in the prevention of conflicts between the resource users]. In M. Tignino & K. Sangbana (Eds.), *Public participation and water resources management: Where do we stand in international law?* 69–76. Paris: UNESCO-IHP. Retrieved December 22, 2015, from http://www.gwp.org/Global/ToolBox/References/Public%20Participation%20and%20Water%20Resources%20Management%20(UNESCO,%202015).pdf

Geneva Water Hub & Platform for International Water Law. (2015). *Synthesis report of the Round Table on Cooperation and Benefit Sharing in the Senegal and Niger River Basins.* Retreived December 22, 2015, from http://www.unige.ch/droit/eau/actus/2015/synthese/Rapport_synthese-en_240915.pdf

Global Water Partnership (GWP) & International Network of Basin Organizations (INBO). (2009). Handbook for integrated water resources management in Basins. Retrieved December 22, 2015, from http://www.inbo-news.org/inbo/publications-and-documents/handbook-for-integrated-water-resources

Godana, B. A. (1985). *Africa's shared water resources: Legal and institutional aspects of the Nile, Niger and Senegal River systems.* London: F. Pinter.

Hensengerth, O., Dombrowsky, I., & Scheumann, W. (2012). *Benefit-sharing in dam projects on shared rivers* (Discussion paper 6/2012). Bonn: German Development Institute.

Mbengue, M. M. (2004). Le statut du fleuve Sénégal: Visages actuels [The status of the Senegal River: Current aspects]. In L. Boisson de Chazournes & M. A. Salman (Eds.), *Water resources and international law* (pp. 473–532). Leiden: Martinus Nijhoff.

Mbengue, M. M. (2014). A model for African shared water resources: The Senegal River legal system. *Review of European, Comparative & International Environmental Law, 23*(1), 59–66. doi:10.1111/reel.2014.23.issue-1

Ndiaye, E.-H. M. (2003). Le fleuve Sénégal et les barrages de l'OMVS: Quels enseignements pour la mise en œuvre du NEPAD? [The Senegal River and the dams of the OMVS: What lessons learnt for the implementation of NEPAD?]. *VertigO-La revue en sciences de l'environnement, 4* (3). Retrieved December 22, 2015, from, http://vertigo.revues.org/3883

New Partnership for the Development of Africa (NEPAD). (2014). *Programme for infrastructure in Africa: 2014. PIDA-PAP implementation progress report.* Retrieved March 15, 2016, from: http://www.nepad.org/sites/default/files/PIDA%20PAP%20Progress%20Report%202014_Final_26012014.pdf

Niger Basin Authority (NBA). (2007). Plan d'Action de Développement Durable (PADD) du Bassin du Niger [Niger Basin sustainable development action plan]. Synthesis report. Retrieved December 22, 2015, from http://www.abn.ne/attachments/article/72/Synthese%20du%20PADD.pdf

Organisation for the Senegal River Development (OMVS). (2006). SOGED. Retrieved March 15, 2016, from http://www.portail-omvs.org/sites/default/files/fichierspdf/c2-barrage_diama_-_role_et_enjeux_-juin_2006_-_vf_1.pdf

Salman, M. A. (2007). Niger river. In International Water Resources Association (Ed.), *Max planck encyclopedia of public international law.* Oxford: Oxford University Press. Retrieved April 24, 2016, from http://www.salmanmasalman.org/wp-content/uploads/2013/01/Niger-River-Article-MEPL.pdf

Salman, M. A. (2016). The grand Ethiopian Renaissance dam: The road to the declaration of principles and the Khartoum document. *Water International, 41*(4). Retrieved April 24, 2016, from, http://www.tandfonline.com/doi/abs/10.1080/02508060.2016.1170374

Sangbana, K. (2011). La gestion intégrée des ressources en eau partagées et les organismes de bassin en Afrique: Le cas de l'Autorité du bassin du Niger et de l'Autorité du bassin de la Volta [Integrated management of shared water resources and the african basin organisations: The case of the Niger Basin Authority and the Volta Basin Authority]. In French Society of International Law (Ed.), *Actes du colloque de la Société française pour le droit international* [Proceedings of the Conference of the French Society of International Law] (pp. 235–244). Paris: Pedone.

Sangbana, K. (2015). La participation du public dans le cadre de l'Organisation pour la mise en valeur du fleuve Sénégal (OMVS) [The participation of the public within the Organisation for the Development of the Senegal River Basin (OMVS)]. In M. Tignino & K. Sangbana (Eds.), *Public participation and water resources management: Where do we stand in international law?* 77–83. Paris: UNESCO-IHP. Retrieved December 22, 2015, from http://www.gwp.org/Global/ToolBox/References/Public%20Participation%20and%20Water%20Resources%20Management%20(UNESCO,%202015).pdf

Sangbana, K. (2016). *La protection des eaux douces transfrontières contre la pollution: dimensions normatives et institutionelles* [The Protection of transboundary fresh water against pollution: normative and institutional dimensions]. (Unpublished doctoral thesis). University of Geneva, Geneva.

Skinner, J., Niasse, M., & Haas, L. (Eds.). (2009). *Sharing the benefits of large dams in West Africa.* London: International Institute for Environment and Development. Retrieved December 22, 2015, from http://pubs.iied.org/pdfs/12555IIED.pdf

Uhlir, F. (Ed.) (2000). *Scientific data for decision making towards sustainable development. Senegal River Basin case study. Summary of a workshop.* Washington DC: National Academies Press.

Vick, M. J. (2006). The Senegal River Basin: A retrospective look and prospective look at the legal regime. *Natural Resources Journal, 46*(1), 211–243.

World Bank. 2014. Niger River Basin management project to support institutional strengthening of the Niger Basin Authority and enhance benefit sharing around the planned Fomi dam. Retrieved March 22, 2016, from http://www.worldbank.org/en/news/feature/2014/07/08/niger-river-basin-management-project-to-support-institutional-strengthening-of-the-niger-basin-authority-and-enhance-benefit-sharing-around-the-planned-fomi-dam

World Bank. 2015. Fact sheet. The World Bank's engagement on the Fomi Dam. Retrieved March 22, 2016, from http://www.worldbank.org/content/dam/Worldbank/document/Africa/Guinea/fomi-fact-sheet-2015.01.pdf

Yu, W. H. (2011). *Benefit sharing in international rivers: Findings from the Senegal River, the Columbia River Basin, and the Lesotho Highlands Water Project.* Washington DC: World Bank.

## Appendix: Statutory references

General Act of the Berlin Conference on West Africa. 1885. Retrieved from http://africanhistory.about.com/od/eracolonialism/l/bl-BerlinAct1885.htm [accessed 15 March 2016].

Revision of General Act of Berlin and General Act and Declaration of Brussels. Saint-Germain-en-Laye. 1919. Retrieved from https://www.loc.gov/law/help/us-treaties/bevans/m-ust000002-0261.pdf [accessed 15 March 2016].

Act Regarding the Navigation and Economic Cooperation between the States of the Niger Basin. 1963. 587 UNTS 9. Retrieved from https://treaties.un.org/doc/Publication/UNTS/Volume%201346/volume-1346-I-22674-English.pdf [accessed 22 December 2015].

Convention Relating to the Development of the Senegal River [Convention sur le développement du fleuve Sénégal]. 1963. Retrieved from http://iea.uoregon.edu/pages/view_treaty.php?t=1963-BamakoSenegalRiverBasin.FR.txt&par=view_treaty_html [accessed 8 January 2016].

Agreement Concerning the River Niger Commission and the Navigation and Transport on the River Niger 1964. 587 UNTS 19. Retrieved from http://www.fao.org/docrep/w7414b/w7414b06.htm [accessed 22 December 2015].

Convention relative au statut du fleuve Sénégal [Convention concerning the Status of the Senegal River]. 1972. Retrieved from: http://www.fao.org/docrep/w7414b/w7414b07.htm [accessed 22 December 2015].

Convention portant création de l'Organisation pour la mise en valeur du fleuve Sénégal [Convention concerning the Creation of the Senegal River Development Organisation (OMVS)]. 1972. Retrieved from http://www.fao.org/docrep/w7414b/w7414b08.htm#TopOfPage [accessed 22 December 2015].

Convention relative au statut juridique des ouvrages communs [Convention on the Legal Status of Common Works]. 1978. Retrieved from http://iea.uoregon.edu/pages/view_treaty.php?t=1978-StatutOuvragesCommuns.FR.txt&par=view_treaty_html [accessed 22 December 2015].

Convention Creating the Niger Basin Authority. 1980. Retrieved from http://gis.nacse.org/tfdd/tfdddocs/418ENG.pdf [accessed 15 March 2016].

Convention Establishing the Niger Basin Authority. 1980. 1346 UNTS 207. Retrieved from http://iea.uoregon.edu/pages/view_treaty.php?t=1980-NigerBasinAuthority.EN.txt&par=view_treaty_html [accessed 22 December 2015].

Convention relative aux modalités de financement des ouvrages communs [Convention regarding Financing Arrangements for Common Works]. 1982. [on file with the author].

Convention révisée portant création de l'autorité du Bassin du Niger [Revised Convention Establishing the Niger Basin Authority]. 1987. Retrieved from http://www.fao.org/docrep/w7414b/w7414b0e.htm [accessed 22 December 2015].

Convention portant création de l'agence de gestion et d'exploitation de Diama [Convention creating the Diama Management and Operation Agency]. 1997. Retrieved from http://iea.uoregon.edu/pages/view_treaty.php?t=1997-DiamaRiverDevelopment.FR.txt&par=view_treaty_html [accessed 22 December 2015].

Convention portant création de l'Agence de gestion de l'énergie de Manantali [Convention creating the Manantali Energy Management Agency]. 1997. Retrieved from http://iea.uoregon.edu/pages/view_treaty.php?t=1997-ManantaliRiverDevelopment.FR.txt&par=view_treaty_html [accessed 22 December 2015].

Charte des eaux du fleuve Sénégal [Senegal River Water Charter]. 2002. Retrieved from http://www.portail-omvs.org/sites/default/files/fichierspdf/charte_des_eaux_du_fleuve_senegal.pdf [accessed 22 December 2015].

Paris Declaration, Management and Governance Principles for the Sustainable Development of the River Niger Basin, 26–17 April 2004. Retrieved from http://www.abn.ne/images/documents/textes/paris_declaration.pdf [accessed 22 March 2016].

Charte de l'eau du bassin du Niger [Niger Basin Water Charter]. 2008. Retrieved from http://www.abn.ne/attachments/article/39/Charte%20du%20Bassin%20du%20Niger%20version%20finale%20francais_30-04-2008.pdf [accessed 22 December 2015].

# Ad hoc legal mechanisms governing transboundary aquifers: current status and future prospects

Laura Movilla Pateiro

**ABSTRACT**

In recent years, different international institutions have repeatedly called for states to enter into agreements on the transboundary aquifers they share. Nevertheless, very few agreements have been established. This article examines the few ad hoc legal mechanisms that are in existence, and identifies some possible reasons for states' reluctance. Finally, this article suggests that there is a need for the international community to stimulate a more cooperative approach to the management of this natural resource based on the preventive and precautionary principles.

## Introduction

Approximately 600 transboundary aquifers and groundwater bodies have been identified (IGRAC, 2015) since their first inventory was published in 2009 (Puri & Aureli, 2009). As awareness of their importance has evolved, these resources have received increasing attention from global and regional organizations, legal scholars and academic institutions (Movilla Pateiro, 2014). In this context, and given the scarcity of bilateral and multilateral legal mechanisms on groundwater, international organizations have encouraged states to negotiate and conclude agreements addressing their transboundary aquifers. As a result, on a few occasions states have entered into arrangements that can be viewed as ad hoc in that they are concluded by the affected parties of a particular aquifer. However, since the 1982 agreement creating a joint authority on the Nubian Sandstone Aquifer System, very few ad hoc legal mechanisms have been established. Moreover, the conclusion of new agreements came to a momentary halt after the negotiation of the treaty on the Guarani Aquifer in 2010. Nevertheless, the 2014 Memorandum of Understanding (MoU) on the Iullemeden-Taoudéni/Tanezrouft Aquifer System and the 2015 agreement on the Al-Sag/Al-Disi Aquifer may have slowly reactivated the process.

Against this background, the goal of this article is threefold: (1) to analyze existing legal mechanisms on transboundary aquifers; (2) to identify why there are so few agreements on this resource; and (3) to reflect on the prospects for future legal mechanisms in this field. To achieve these goals, this article first examines the invitation to states from international organizations to engage in development of transboundary aquifer agreements. Secondly, it examines and compares the existing bilateral and multilateral agreements on specific aquifers. The next section analyzes why states may still be reluctant to conclude such legal mechanisms and offers some suggestions regarding the main causes. Finally, this article recommends a cooperative approach based on the preventive and precautionary principles and offers some guidelines on the way forward.

## The international call for states to conclude agreements on specific transboundary aquifers

In the context of increasing attention and knowledge on transboundary aquifers, there have been calls in recent years from different international fora encouraging states to conclude agreements and arrangements for this resource. Issued in 1989, the Bellagio Draft Treaty, a model treaty crafted by scholars, aimed to create a model legal instrument for transboundary aquifers in critical areas "in view of this lack of legal and institutional arrangements in most cases, and the weakness of the few existing institutions dealing with international aquifers" (Hayton & Utton, 1989, p. 664). More recently, there are several examples of this same call upon states within the framework of the UN system.

For example, the 1997 UN Watercourses Convention establishes that states may enter into one or more agreements "which apply and adjust its provisions to the characteristics and uses of a particular international watercourse *or a part thereof*" (Art. 3.3, emphasis added). The term "watercourse" includes the groundwater connected to the surface water (Art. 2.a). Therefore, Article 3.3 may be interpreted as also referring to an agreement covering solely groundwater – "a part thereof" – connected to the surface water. For its part, the Resolution on Confined Transboundary Groundwater, adopted by the UN International Law Commission (UNILC) in 1994, encouraged states, among other things, to conclude agreements with other states in which confined – meaning not connected to surface water – transboundary groundwater is located (Par. 2).

The subsequent 2008 UNILC Draft Articles on the Law of Transboundary Aquifers (Draft Aquifer Articles) stated that "for the purpose of managing a particular transboundary aquifer or aquifer system, aquifer states are encouraged to enter into bilateral or regional agreements or arrangements among themselves. Such agreements or arrangements may be entered into with respect to an entire aquifer or aquifer system or any part thereof or a particular project, programme or utilization" (Art. 9). The Draft Aquifer Articles were submitted to the UN General Assembly (UNGA), which adopted a resolution that included this project in its annex. The resolution also encouraged the states "to make appropriate bilateral or regional arrangements for the proper management of their transboundary aquifers, taking into account the provisions of these draft articles" (A/RES/63/124). UNGA repeated this call in 2011 (A/RES/66/104) and 2013 (A/RES/68/118), with the latter resolution recommending that states use the Draft Aquifer Articles "as guidance". In addition, during the debates on the future of this text most states (both those that endorsed the idea of a global convention on the basis of the Draft Aquifer Articles and especially those that did not) expressed their support of the conclusion of ad hoc agreements on transboundary aquifers.[1]

For its part, the 1992 Convention on the Protection and Use of Transboundary Watercourses and International Lakes of the United Nations Economic Commission for Europe (UNECE Water Convention) also sets forth that the riparian parties "shall enter into bilateral or multilateral agreements or other arrangements, where these do not yet exist, or adapt existing ones, where necessary to eliminate the contradictions with the basic principles of this Convention, in order to define their mutual relations and conduct regarding the prevention, control and reduction of transboundary impact" (Art. 9.1). They "shall specify the catchment area, or part(s) thereof, subject to cooperation", which may also include groundwater, as covered in the scope of the convention (Art 1.1). More specifically, the 2012 UNECE Model Provisions on Transboundary Groundwaters state that "the Parties shall establish arrangements for the exchange of information and available data on the condition of transboundary groundwaters, including available data on the parameters prescribed in Provision 3, as well as information on the status of use of transboundary groundwaters" (Provision 6).

Therefore, there is a clear call to states requesting that they enter into legal arrangements or agreements on the transboundary aquifers they share. Global and regional legal instruments may be adapted or used as guidance. However, the response of states to this call has not thus far been very enthusiastic.

## Existing ad hoc legal mechanisms on transboundary aquifers

Only at the end of the nineteenth century do the first (and subsidiary) references to wells and springs appear in some international treaties. Afterwards, specific references to groundwater and its relation to surface water can be found since the beginning of the twentieth century, although only as a minor concern (Caponera & Alhéritière, 1978). However, references to groundwater in agreements and scholarly articles have increased over time and gained prominence (Burchi & Mechlem, 2005; Matsumoto, 2010). In this sense, groundwater has been mentioned in 14% of the transboundary freshwater treaties concluded since 1950, with increasing consideration in most recent years (Giordano et al., 2014). However, in the majority of cases, surface water is still the main subject of regulation. Thus, although groundwater is sometimes included in the scope of some agreements, provisions on its specific governance are for the most part missing.

Only five particular transboundary agreements count on a legal mechanism concerted among their aquifer states: the Nubian Sandstone Aquifer System and the North-Western Sahara Aquifer System in North Africa; the Iullemeden-Taoudéni/Tanezrouft Aquifer System in West Africa; the Guarani Aquifer System in South America; and the Al-Sag/Al-Disi Aquifer in the Middle East. A few other agreements, such as the one on the Franco-Genevese Aquifer in Europe, have been concluded by subnational units of states.

The Nubian Sandstone Aquifer System is the largest fossil aquifer in the world. It covers an area of approximately 2.6 million km$^2$ underneath Chad, Egypt, Libya and Sudan. In July 1992, Egypt and Libya signed an agreement creating a joint authority for its study and development, and Sudan and Chad joined in the agreement in 1996 and 1999, respectively.[2] Subsequently, between 1998 and 2002, the four states were engaged in the project Regional Strategy for the Utilization of the Nubian Sandstone Aquifer

System, developed with the support of the Centre for Environment and Development for the Arab Region and Europe and funded by the International Fund for Agricultural Development. Within this framework, the four countries signed two informal agreements for the monitoring and exchange of data and information on the aquifer on 5 October 2002.[3] Other international projects have since been developed.[4]

The North-Western Sahara Aquifer System, which is also a nonrecharging fossil aquifer, underlies a total area of over one million km$^2$ in Algeria, Tunisia and Libya. In 1999, the three states decided to engage in a project together with the Observatoire du Sahara et du Sahel (OSS) to improve knowledge about the aquifer and provide for its joint management. During development of the agreement, representatives of the three states agreed to the establishment of a consultation mechanism in a meeting at the headquarters of the FAO in Rome in December 2000.[5] This agreement was followed in 2006 by a declaration establishing a "permanent consultation mechanism" for the aquifer system, which became operational in 2008 and is hosted by OSS. The project has had three phases so far and has led to operational recommendations for its sustainable management.[6]

Also in the African continent, the Iullemeden-Taoudéni/Tanezrouft Aquifer System (ITAS) includes the Iullemeden Aquifer System and the Taoudeni/Tanezrouft Aquifer System, underlying Algeria, Benin, Burkina Faso, Mali, Mauritania, Niger and Nigeria. It is a renewable transboundary aquifer system with an extension (2.5 million km$^2$) similar to that of the Nubian. Between 2004 and 2009, the three main aquifer states of the Iullemeden (Mali, Niger and Nigeria) developed a project financed by the Global Environmental Facility (GEF) and implemented by OSS. The project enhanced the knowledge of the aquifer system and identified its transboundary risks.[7] After its conclusion, the three states published a declaration and an extensive MoU regarding setting up a consultative mechanism for the management of the Iullemeden, and a road map also was agreed.[8] Following the confirmation of a hydrological connection between the Iullemeden and the Taoudeni/Tanezrouft, a new project on the whole aquifer system and the River Niger was launched in 2010.[9] In this context, a new MoU comprising the seven aquifer states was designed during the meeting of the second Council of Ministers of the GICRESAIT Project in Abuja, Nigeria, 28 March 2014.[10] Notably, this document is very similar to the previous MoU on the Iullemeden.

For its part, the Guarani Aquifer System represents another vast underground reserve of freshwater. It covers an area of more than 1 million km$^2$ beneath the surface of Argentina, Brazil, Paraguay and Uruguay and holds approximately 30,000 km$^3$ of water. On 2 August 2010, the four states signed a treaty on this aquifer during a meeting of the heads of state of MERCOSUR in the city of San Juan, Argentina.[11] However, this treaty is not yet in force since it has only been ratified by Argentina and Uruguay. Previously, a project for the environmental protection and sustainable development of the Guarani Aquifer was developed by the four countries, implemented by the Organization of American States and supported by GEF.[12]

On 30 April 2015, Jordan and Saudi Arabia entered into an agreement for the management and utilization of the groundwater in the Al-Sag/Al-Disi Layer,[13] a transboundary fossil groundwater body that is part of the broader Saq-Ram Aquifer System. Both the Al-Sag/Al-Disi Layer and the Saq-Ram Aquifer System are on the border between the two countries. This brief agreement contains rules about management of the Al-Sag/Al-Disi Layer, including temporary restrictions on abstractions in some of its areas. Documents produced by the UN Economic and Social Commission for Western Asia (ESCWA) include references to an MoU signed in 2007 between the Jordanian Ministry of Water and Irrigation and the Saudi Arabian Ministry of Electricity and Water on this same aquifer for technical purposes (ESCWA, 2012, p. 59; ESCWA and BGR, 2013, p. 312).

As stated earlier, an agreement was concluded by the subnational units of two neighbouring states and concerns the Genevese Aquifer, an aquifer of an approximate extension of 19 $km^2$ and an area of about 30 $km^2$ shared by the French department of Haute-Savoie and the Swiss Canton of Geneva. As a consequence of its overexploitation and depletion, both subnational entities decided to sign an agreement for the establishment of an artificial recharge mechanism in 1978.[14] A new convention replaced this first agreement on 18 December 2007, extending for another 30 years with almost the same terms that entered into force on 1 January 2008. In this second instance the signatories were the Canton of Geneva as Swiss representative, and the community of communes of the region of Annemasse and the community of municipalities of Genevois and Viry representing the French partner.[15]

Subnational transboundary arrangements between local authorities seem to be frequent in Europe (Eckstein, 2013). Another example would be the 1967 agreement between the regions of Flanders (France) and Wallonia (Belgium), which proposed to reduce the groundwater abstraction from the carboniferous limestone aquifer they share.[16] Subnational arrangements also exist in North America along the United States–Mexico border, such as between the El Paso Water Utilities Public Service Board and the Junta Municipal del Agua y Saneamiento de Juárez,[17] and between the state of Washington and the province of British Columbia.[18]

## Some remarks on the few existing ad hoc legal mechanisms on transboundary aquifers

After a close examination and comparison of the number, context, content and form of these existing ad hoc legal mechanisms on transboundary aquifers, the following observations can be made.

First, to date very few of these legal mechanisms have been concluded with legally binding force, especially when compared with the hundreds of existing treaties on surface water (Giordano et al., 2014). The small number is particularly surprising if we take into account the large number of transboundary aquifers and groundwater bodies that have already been catalogued, the "silent revolution" of increasing groundwater utilization in the last decades (Llamas & Martínez-Santos, 2005), and the growth of knowledge regarding this resource.

Second, a geographical pattern is hard to determine. The current agreements belong to different regions, and the aquifers have different hydrogeologic features. While a serious water crisis could be in general ruled out in Europe, the agreement on the Franco-Genevese Aquifer arose in the context of the aquifer's endangerment. For its part, the South American region to which the Guarani Aquifer belongs is especially rich in both surface and groundwater despite some localized problems. In contrast, the other aquifers are located in arid and semi-arid areas of the planet. In addition, with the exception of the Al-Sag/Al-Disi Aquifer, the ad hoc legal mechanisms agreed between states have not been concluded in a context of serious depletion or pollution of the aquifer. Instead, they reflect a preventive approach to its governance. However, those concluded at the local level usually constitute reactive responses to actual problems on the aquifer.

Third, a number of the aquifers that are the subject of ratified legal mechanisms (Nubian, North-Western Sahara, Al-Sag/Al-Disi) are mainly nonrenewable aquifers, even though this kind of aquifer represents only a small proportion of the total number of the world's aquifers that are already identified. At the same time, most of the aquifers with negotiated agreements (Nubian, North-Western Sahara, ITAS, Guarani) are some of the largest aquifer systems in the world (WHYMAP and Margat, 2008). Therefore, a particular focus on nonrenewable aquifers and those that are the largest (either renewable or nonrenewable) can be noted.

Fourth, in most cases these agreements resulted from projects with funding and assistance from international organizations that wanted to generate more knowledge about the aquifer and surface–groundwater exchange.

Fifth, the legal nature and status of the ad hoc legal mechanisms have varied. Most of them are not treaties, in the strict sense of the 1969 Vienna Convention on the Law of Treaties (Art. 2.1.a). On the contrary, they constitute what may be called informal agreements, non-normative agreements, gentleman's agreements, or MoUs. That is why the UNGA and the UNILC refer to "agreements or arrangements" in their recommendations to states regarding transboundary aquifers. These 'soft' legal mechanisms have an uncertain and ambiguous legal nature, and they are negatively defined by the fact that they are not treaties. The main consequence is that they are not governed by international law. Thus, they do not invoke international obligations, and therefore, neither do they invoke international responsibility in the event of noncompliance. Their success depends on the continued mutual interest and good faith of the parties (Remiro Brotóns, 1990). In this sense, the 1982 agreement creating the joint authority on the Nubian Sandstone Aquifer System, the 2010 treaty on the Guarani aquifer, and the 2015 agreement on the Al-Sag/Al-Disi Layer can hardly be considered real treaties in this field. Besides, in some cases the title of the agreement or its form (both evidence of the legal nature of an agreement) may be confusing. This is the case for the 2014 MoU on the ITAS. On the one hand, it is called a memorandum of understanding. On the other hand, its provisions are similar to those of a treaty, in that it contains articles on amendments and revisions, termination, entry into force, authentic texts and depository government. Clearly, those agreements concluded by sub-state authorities cannot be considered treaties under the 1969 Vienna Convention.

Finally, the scope and content of these agreements vary considerably. Some of them are mostly technical (Genevese, Al-Sag/Al-Disi), some are fairly simple (Nubian, North-Western Sahara) and others contain very detailed provisions (ITAS, Guarani). However, the lowest common denominator of all the ad hoc agreements is the establishment of some kind of obligation or mechanism for the exchange of data and information, or its generation. In addition, most of them (Nubian, North-Western Sahara, ITAS, Guarani, Genevese) establish an institutional mechanism, one of the closest forms of cooperation. This may be explained by the technical nature of the cooperation in this field, including the above-mentioned generation and exchange of data and information. These activities are facilitated by the existence of an institutional mechanism. The institutional mechanism may also be assigned more complex powers, such as the design of an annual aquifer utilization programme (Genevese) or the settlement of disputes (ITAS, Guarani).

## Why are states reluctant to govern transboundary aquifers through ad hoc agreements?

Despite the increasing attention groundwater has received in recent years, the previous sections have shown states' limited response to further the conclusion of agreements on specific transboundary aquifers. The goal of this section is to identify the main cause of this reluctance on the part of states. Evidently, cooperation on transboundary aquifers is affected by the same difficulties that arise for surface transboundary water resources, such as problematic political relations, conflicting interests, or the increasing inclusion of water in states' national and international security agendas. However, formal transboundary cooperation also faces additional challenges arising from the particular features of a hidden resource such as groundwater. These challenges include the lack of information and awareness of groundwater, the strong connection of this resource with state sovereignty, and the lack of a clear global legal framework.

In the first instance, the remaining knowledge deficit on transboundary aquifers is obvious and critical. Appropriate legal frameworks cannot be designed if their object is not well known or understood, and uncertainty hinders the application of any existing norms (Milman & Ray, 2011). It is true that more information about groundwater is available today than only a few years ago, thanks in particular to several international organizations and programmes that have enhanced its research, such as the UNESCO International Hydrological Programme, its International Groundwater Assessment Centre and the Internationally Shared Aquifer Resources Management Programme. Nonetheless, there is still a significant lack of knowledge, largely due to the cost and complexity of monitoring large aquifer systems (Richey et al., 2015). Additionally, in the case of transboundary aquifers, regular monitoring and exchange of data and information are needed from both sides of the border. This introduces additional difficulties, and the consequence of uncertainty can be compounded when there is no consensus on data or even methods to acquire the data, which may reduce the value of cooperation and increase unilateral decisions (Eckstein, 2013; Milman & Ray, 2011).

Second, sometimes transboundary aquifers are simply not a priority for the states because there are no urgent problems relating to their exploitation, or there is no contamination, or there is plenty of surface water. Then there is no compelling need to resort to the expense of monitoring or exploiting groundwater. Worst, most states probably do not have information regarding their groundwater problems, or are completely indifferent to the transboundary consequences of its national exploitation.

Third, the special connection of this resource with the principle of state sovereignty may favour unilateral actions. The "territorial attraction" that states usually exercise on water resources (Ruiz-Fabri, 1990) is even higher in the case of aquifers. This is mainly because aquifers are made up of two different parts: the geological formation and the groundwater within the geological formation. Sovereign rights are more defensible for the geological formation than for the groundwater, since groundwater is fluid and flows beneath different states' territories. This may have two main consequences: the lack of interest of states in establishing cooperative agreements; and the restatement of the sovereignty principle in the agreements they conclude. Examples of support for the sovereignty principle can be found in the 2010 treaty on the Guarani Aquifer and, previously, in the controversial Article 3 of the 2008 Draft Aquifer Articles, which provides that "each aquifer State has sovereignty over the portion of a transboundary aquifer or aquifer system located within its territory. It shall exercise its sovereignty in accordance with international law and the present draft articles." The states' sovereignty is limited by the Draft Aquifer Articles, which is why some consider it a balanced provision (Yamada, 2011). However, the affirmation of sovereignty over the water in aquifers has caused a major controversy among scholars. Many of them considered it at odds with the evolution of international water law, which supports both limited sovereignty over shared natural resources and the principle of equitable and reasonable utilization (McCaffrey, 2011; McIntyre, 2011). A declaration of sovereignty over this natural resource discourages cooperation and fosters unilateral action instead of joint initiatives (McCaffrey & Neville, 2010). On the contrary, for some authors, the mentioning of sovereignty is either confusing and unnecessary (Dellapenna, 2011; Mechlem, 2009) or simply fails to add anything new (Del Castillo-Laborde, 2010). However, provisions restating states' sovereignty over the part of an aquifer located in their territory may also make the agreement more attractive to some states (Maxwell, 2011; ORSAM, 2012).

Lastly, there is still an unclear global legal framework to take as a reference in groundwater governance. Although significant advances have been made in the last few years, the international law governing transboundary aquifers is still nascent and consists of several international legal instruments of different nature, scope and content at the global and regional levels. Briefly, the 1997 UN Watercourses Convention applies only to groundwater connected to surface water, has no substantive rules taking into account the special features of transboundary aquifers and is far from being a universal convention. The first two remarks also hold true for the 2000 SADC Revised Protocol on Shared Watercourses. For its part, until recently the 1992 UNECE Water Convention had only a regional, pan-European scope of application. Other texts are not legally binding, such as the 1994 UNILC Resolution on Confined Transboundary Groundwater, the 2012 UNECE Model Provisions on Transboundary Groundwater, the 2008 Draft Aquifer Articles, and texts developed by academic institutions, like the 1966 Helsinki Rules, 1986 Seoul Rules and 2004 Berlin Rules of the International Law Association (ILA) or the previously mentioned 1989 Bellagio Draft Treaty. While the differences among all these legal instruments do not make them incompatible *a priori*, they are likely to cause confusion and even rejection (see e.g. the inadequacies between the UN Watercourses Convention and the Draft Aquifer Articles – McIntyre, 2011; Traversi, 2011). On the contrary, the non-binding nature of most of these instruments may give them broader acceptability. In any case, the unique characteristics of transboundary aquifers will require adaptation of the provisions of the above-mentioned documents.

Consequently, it may be argued that all the mentioned factors have contributed to states' reluctance (or at least precaution) to accept governance of transboundary aquifers through ad hoc legal mechanisms.

## Towards a more cooperative approach: some reflections on the way forward

Despite the existing situation, described above, several factors recommend the strengthening of cooperation and the enhancement of the legal framework on transboundary aquifers through treaties and informal arrangements (Mechlem, 2011). The hundreds of identified aquifers, the qualitative importance of this vital resource, its presence in the subsoil of almost all non-island states (Puri & Aureli, 2009), its increasing utilization in the last decades (Llamas & Martínez-Santos, 2005), its use as a source of drinking water by nearly 50% of the world's population (IGRAC, 2015) and the overexploitation of one-third of the largest transboundary aquifers (Richey et al., 2015) all testify in favour of a preventive and precautionary cooperative approach. However, given the lack of concrete knowledge of this resource, execution of agreements for its governance may not be considered an immediate priority by some states. In an ideal scenario, a step-by-step approach should be adopted, whereby knowledge is first acquired, followed by a focus on regulation. The requisite knowledge may be obtained through an agreement for that specific purpose, or through another form of cooperation, such as an international project.

International treaties have historically been the primary source of international water law, and this is still true today. This is mainly due to the particular characteristics of international watercourses, since each of them has its own specific geographic, hydrographic, hydrological, hydrogeological, climatic, ecological, political, social and cultural features (Schwebel, 1979). Hence, regulation of international watercourses requires ad hoc solutions, which has led to a "systematic bilateralization" of the relations between the states concerned (Ruiz-Fabri, 1990, p. 818). The particular characteristics of each international watercourse are also one of the key reasons for the traditional difficulties in identifying general customary rules and achieving a successful global convention in this field. Transboundary aquifers seem particularly prone to one-off solutions, because of the different composition and structure that respective subsoils may have, such as sand and gravel, sandstone, karst, volcanic or basement aquifers (Margat & Van der Gun, 2013) and the variety of possible transboundary implications (Eckstein & Eckstein, 2003). Therefore, tailored legal mechanisms are even more necessary for each aquifer.

Obviously, the conclusion of an ad hoc legal mechanism does not automatically imply meaningful cooperation. The effectiveness of cooperation will depend ultimately on the will of the parties to fulfil and implement the agreement. Nevertheless, the existence of an agreement may be, at least, an indicator that some steps have been taken to establish joint governance over a transboundary aquifer. Furthermore, the potential benefits of bilateral or multilateral treaties on specific aquifers are numerous. They enhance cooperation; they promote and reflect awareness of the particular circumstances of a given aquifer; they provide certainty; they may provide an institutional structure, as well as dispute-settlement mechanisms; and they usually reflect at least some form of preventive or precautionary approach to aquifer problems. At the same time, informal agreements may be preferred by states if they want to blur the recognition of the international legal personality of one of the parties, if they want to increase the margin of discretion in fulfilling commitments, or if they want to avoid the legal consequences of noncompliance or the requirements of its domestic legal systems (Remiro Brotóns, 1990). Besides, informal agreements are simple, flexible, and rapid and have a low public profile, so they may be more attractive to states than a binding treaty. Either way, both kinds of legal instruments, treaties and informal agreements, may arguably be specially recommended in certain situation, such as when groundwater bodies are above a certain size, have a particular problematic nature, or are separated from the surface water body or from other groundwater bodies, especially fossil or nonrenewable groundwater. Including procedural obligations and especially the regular exchange of data and information and monitoring are of particular importance for the sake of improving the knowledge on a particular aquifer.

Cooperation through international projects has gradually increased in the last years (Stephan, 2010). In fact, as stated by Mechlem (2011, p. 217), most of the existing agreements on transboundary aquifers "were preceded by externally funded projects which had a large component of knowledge generation and study of the shared resource and which aimed to collect and exchange data and information.... Such a phase also seems an essential prerequisite to establishing a basis for further cooperation and to building trust before moving towards a more formal agreement." Thus, projects on single aquifers are currently being developed, such as the Dinaric Karst Transboundary Aquifer System in the Balkans, the GGRETA Project on the Trifinio Aquifer in Central America, the Stampriet-Kalahari/Karoo Aquifer in South Africa and the Pretashkent Aquifer in Central Asia. However, they have not yet led to any legal mechanisms for transboundary governance. Other projects have had more general objectives, such as the Groundwater Governance—A Global Framework for Action project (2011–2015), supported by GEF and implemented by FAO, jointly with the UNESCO International Hydrological Programme, the International Association of Hydrogeologists and the World Bank. Its aim was to improve groundwater governance at local, national and transboundary levels by designing a global framework for action that includes responsive laws and regulations.[19] In any case, to be successful, these international projects need to seek engagement and representation from local, national and regional stakeholders and decision makers in order to continue beyond the formal projects' completion (Sindico & Hawkins, 2015).

Furthermore, as stated earlier, the international encouragement to enter into agreements on particular aquifers has been addressed to the *states* sharing them. However, as has been pointed out, some subnational entities, e.g., in the case of the Franco-Genevese Aquifer, have also signed agreements that have been proved very effective in practice. Moreover, there are probably numerous examples of local transboundary governance of groundwater all over the world that may not be in the form of agreements or arrangements. Thus, a local 'bottom-up' approach or 'locally specific arrangements' should not be ruled out in some circumstances, especially if the communities involved are mainly local, such in the case of the US–Mexico border (Eckstein, 2013), or if the respective states do not have a particular interest in formal regulation. In these cases, the recommendation to states should be that of facilitating them, or at least not hindering them. Therefore, the principle of subsidiarity, in which management decisions and regulation should take place at the lowest effective level of governance (as is frequently recommended in the case of water management), may also apply in the governance of some transboundary aquifers.

In sum, a wider cooperative approach to the management of transboundary aquifers should be encouraged by the international community, beyond the simple incentive to conclude international agreements and arrangements. The preventive and precautionary principles should be the theoretical foundation. As is well known, the preventive principle, which is deeply related to the obligation of not causing significant damage to other states or the environment, constitutes a core part of international environmental law itself. In turn, the precautionary principle constitutes a notable step in the evolution of environmental concerns that does not require (as the preventive principle does) solid scientific evidence for measures to be taken. The precautionary principle appeared in the international arena in the 1980s and was formulated in the 1992 Rio Declaration on Environment and Development to mean that "where there are threats of serious or irreversible damage, lack of full scientific certainty shall not be used as a reason for postponing cost-effective measures to prevent environmental degradation" (Principle 15). Since then, this principle has been invoked in a larger number of international declarations and treaties. Its legal status as a principle of international customary law is not entirely clear, although it is deeply rooted in the European Union's environmental law. In any case, despite the *a priori* clear difference between both principles at the theoretical level, sometimes it is difficult to distinguish them in practice (Fitzmaurice, 2009).

Nevertheless, the precautionary principle in particular is called to play an important role in the specific regulation of transboundary aquifers as a result of the uncertainty in many cases of the features of this resource and the consequences of its ongoing exploitation or related activities. Climate change only increases the uncertainty about the nature and quality of aquifers. The precautionary principle may therefore be useful not only in the traditional context of prevention, reduction and control of pollution, but also as an approach to general management of transboundary aquifers. For example, Article 2 of the Draft Aquifer Articles says that "aquifer states shall take a precautionary approach in view of uncertainty about the nature and extent of a transboundary aquifer or aquifer system and of its vulnerability to pollution". Similar statements can be found in Provision 5 of the 2012 UNECE Model Provisions, which also include examples of measures that can help to meet that goal, as well as Article 12 of the 2008 ILA Berlin Rules. Further, Article 38 of the 2008 ILA Berlin Rules refers to the precautionary principle as a general management paradigm: "States, in accordance with the precautionary approach, shall take early action and develop long-term plans to ensure the sustainable use of groundwater and of the aquifers in which the groundwater is contained." Another example of the precautionary approach to the management of this resource can be found in the agreement on the Guarani Aquifer, which Villar and Costa Ribeiro (2011, p. 656) define as a "new paradigm for transboundary groundwater management" because it "represents the intention of parties to construct a new paradigm in the sense that it takes place in the absence of groundwater conflict. It is the first time that transboundary groundwater has been considered in a preventive/precautionary context."

The challenge in the governance of transboundary aquifers will be finding a way to foster this preventive and precautionary cooperative approach. Among other things, this strategy will necessarily require enhanced knowledge about the resource, an increase in public awareness of the benefits of cooperation, an understanding of the consequences of not taking preventive and precautionary actions, an exchange of good practices, international funding, and, in the interest of efficiency, more coordination to avoid overlaps between donors and implementing agencies.

## Conclusion

Despite the call on states to conclude agreements and arrangements on particular transboundary aquifers, ad hoc legal mechanisms have been established for less than 1% of them. These few agreements have been created both by states and by local entities, and they have a heterogeneous legal nature, status, scope, content, motivation and geographic location. However, they have in common the establishment of obligations or mechanisms for the generation and/or exchange of data and information, and most of them provide an institutional framework. Moreover, a majority of the agreements focus on aquifers that are very large and contain nonrenewable groundwater.

This apparent reluctance of states to govern transboundary aquifers through specific legal mechanisms derives from the particular difficulties faced in cooperation over transboundary water resources in relation to groundwater. These include, among other things, the lack of information and awareness on this resource, the strong connection of aquifers with the principle of state sovereignty, and an unclear global legal framework to use as a reference.

The future challenge will be to stimulate a preventive and precautionary cooperative approach to the governance of transboundary aquifers that leads to new and more robust agreements and arrangements. In this regard, the international encouragement to conclude agreements should also consider local border communities. This encouragement should also be accompanied by agreed international projects (including governments, population and other stakeholders) that aim to enhance the knowledge on this resource and develop international legal mechanisms for their sound management and protection.

## Notes

1. *The Law of Transboundary Aquifers: Report of the Secretary-General*, 29 June 2011, A/66/116 and A/66/166/Add.1; *The Law of Transboundary Aquifers: Report of the Secretary-General*, 22 July 2013, A/68/172.
2. An English translation is available in Annex 7 to the GEF project document, 'Formulation of an Action Programme for the Integrated Management of the Shared Nubian Aquifer', available at http://iwlearn.net/iw-projects/2020/project_doc/nsas-project-document/view.
3. *Programme for the Development of a Regional Strategy for the Utilisation of the Nubian Sandstone Aquifer System (NSAS) – Terms of Reference for the Monitoring and Exchange of Groundwater Information of the Nubian Sandstone Aquifer System Agreement No 1 – Terms of Reference for the Monitoring and Exchange of Groundwater Information of the Nubian Sandstone Aquifer System Agreement No 2 – Terms of Reference for Monitoring and Data Sharing, Tripoli, 5 October 2000.* Text available in Burchi and Mechlem (2005), pp. 4–6.
4. Between 2005 and 2011 the four states developed a second project: 'Formulation of an Action Programme for the Integrated Management of the Shared Nubian Aquifer', funded by GEF and executed by the International Atomic Energy Agency, with support from UNESCO; see http://iwlearn.net/iw-projects/2020. More recently, the ministers in charge of water resources in the four states and the chairperson of the joint authority

signed a Regional Strategic Action Plan on 18 September 2013, available at http://iwlearn.net/iw-projects/2020/reports/regional-strategic-action-programme-for-nubian-aquifer-system.

5. 'Establishment of a Consultation Mechanism for the Northwestern Sahara Aquifer System (SASS)', 2002. It was endorsed by Algeria on 6 January 2003, Tunisia on 15 February 2003 and Libya on 23 February 2003. Excerpts available in Burchi and Mechlem (2005), pp. 6–8.
6. http://sass.oss-online.org/en.
7. http://www.oss-online.org/fr/projet-sai.
8. Bamako Declaration of the Ministers in Charge of Water Resources of the Countries Sharing the Iullemeden Aquifer System (Mali, Niger, Nigeria), Bamako, Mali, 20 June, 2009; Memorandum of Understanding to the Setting Up of a Consultative Mechanism for the Management of the Iullemeden Aquifer System, Bamako, 20 June 2009.
9. Joint and Integrated Water Resources Management of the Iullemeden -Taoudeni /Tanezrouft Aquifer Systems and the Niger River (GICRESAIT Project); see http://www.oss-online.org/fr/projet-gicresait.
10. 'Memorandum of Understanding for the Establishment of a Consultation Mechanism for the Integrated Management of the Water Resources of the Iullemeden, Taoudeni/Tanezrouft Aquifer Systems (ITAS), (Algeria, Benin, Burkina Faso, Mali, Mauritania, Niger, Nigeria)', 2nd Council of Ministers GICRESAIT Project, Abuja, Nigeria, 28 March 2014, available at http://www.internationalwaterlaw.org/documents/regionaldocs/Iullemeden_MOU-2014.pdf.
11. 'Acuerdo sobre el Acuífero Guaraní, hecho en San Juan, República Argentina, el 2 de agosto de 2010'; English translation available at http://www.internationalwaterlaw.org/documents/regionaldocs/Guarani_Aquifer_Agreement-English.pdf.
12. See 'Strategic Action Program (SAP) for Guarani Aquifer', which summarizes the main results and recommendations of the project: http://iwlearn.net/iw-projects/974/reports/strategic-action-program/view.
13. 'Agreement between the Government of the Hashemite Kingdom of Jordan and the Government of the Kingdom of Saudi Arabia for the Management and Utilization of the Ground Waters in the Al-Sag/Al-Disi Layer', Riyadh, 30 April 2015. Unofficial English translation at http://www.internationalwaterlaw.org/documents/regionaldocs/Disi_Aquifer_Agreement-English2015.pdf.
14. 'Arrangement relatif a la Protection, a l'Utilization et a la Realimentation de la Nappe souterraine franco-suisse du Genevois, septembre 1977'. Unofficial English translation at http://www.internationalwaterlaw.org/documents/regionaldocs/franko-swiss-aquifer.html.
15. 'Convention relative a la protection, a l'utilisation, a la realimentation et au suivi de la Nappe Souterraine Franco-Suisse du Genevois, fait à Genève le 18 décembre 2007'. Unofficial English translation at http://www.internationalwaterlaw.org/documents/regionaldocs/2008Franko-Swiss-Aquifer-English.pdf.
16. Agreement referred to by Margat and Van Der Gun (2013), p. 233; Académie de l'Eau, BRGM, OIEAU, and UNESCO (2011), p. 87.
17. 'Memorandum of Understanding between City of Juárez, Mexico Utilities and the El Paso Water Utilities Public Services Board of the City of El Paso, Texas', 6 December 1999, http://internationalwaterlaw.org/documents/regionaldocs/Local-GWAgreements/El_Paso-Juarez_MoU.pdf.
18. 'Memorandum of Agreement Related to Referral of Water Right Applications Related to the Transboundary Abbotsford-Sumas Aquifer between the State of Washington as Represented by the Department of Ecology and the Province of British Columbia as Represented by the Minister of Environment, Lands and Parks', 10 October 1996, http://internationalwaterlaw.org/documents/regionaldocs/Local-GW-Agreements/1996-BC-WAWater-Right-Referral-Agreement.pdf.
19. The resulting reports are available at http://www.groundwatergovernance.org/.

## Acknowledgements

This article benefited from research carried out alongside Francesco Sindico, Renee Martin-Nagle and Stephanie Hawkins at the Strathclyde Centre for Environmental Law and Governance, Glasgow, UK. The author's research visit at SCELG was possible thanks to a Royal Society Edinburgh/Caledonian Research Fund European Visiting Research Fellowship and a Xunta de Galicia Postdoctoral Fellowship.

## References

Académie de l'Eau, BRGM, OIEAU, and UNESCO. (2011). Vers la gestion concertée des systèmes aquifères transfrontaliers. Partie I: Constat préliminaire- Analyse générale. Retrieved from http://www.afd.fr/jahia/webdav/site/afd/shared/PUBLICATIONS/RECHERCHE/Scientifiques/A-savoir/03-Doc-pr%C3%A9pa-1-A-Savoir.pdf

Burchi, S., & Mechlem, K. (2005). *Groundwater in international law. Compilation of treaties and other legal instruments.* Rome: DL Service, UNESCO/Food and Agriculture Organization.

Caponera, D. A., & Alhéritière, D. (1978). Principles for international groundwater law. *Natural Resources Journal, 18*, 589–619.

Del Castillo-Laborde, L. (2010). The law of transboundary aquifers and the Berlin Rules on water resources (ILA): Interpretative complementarity. In *UNESCO-IAH-UNEP. Pre-proceedings of ISARM international conference, "Transboundary aquifers: Challenges and new directions" [CD-ROM]*, 6–8 December 2010. Paris: UNESCO.

Dellapenna, J. W. (2011). The customary law applicable to internationally shared groundwater. *Water International, 36*(5), 584–594. doi:10.1080/02508060.2011.599025

Eckstein, G. (2013). Rethinking transboundary ground water resources management: A local approach along the México-U.S. border. *Georgetown International Environmental Law Review, 25*(1), 95–128.

Eckstein, G., & Eckstein, Y. (2003). A hydrogeological approach to transboundary ground water resources and international law. *American University International Law Review, 19*, 201–258.

ESCWA. (2012). *ESCWA water development report 4: National capacities for the management of shared water resources in ESCWA member countries.* New York, NY: United Nations.

ESCWA & BGR. (2013). *Inventory of shared water resources in Western Asia.* Beirut: United Nations.

Fitzmaurice, M. (2009). *Contemporary issues in international environmental law.* Cheltenham: Edward Elgar.

Giordano, M., Drieschova, A., Duncan, J., Sayama, Y., De Stefano, L., & Wolf, A. (2014). A review of the evolution and state of transboundary freshwater treaties. *International Environmental Agreements: Politics, Law and Economics, 14*(3), 245–264. doi:10.1007/s10784-013-9211-8

Hayton, R., & Utton, A. (1989). Transboundary groundwaters: The bellagio draft treaty. *Natural Resources Journal, 29*, 668–722.

IGRAC. (2015). *Transboundary aquifers of the world - The special edition for the 7 World Water Forum 2015.*

Llamas, M., & Martínez-Santos, P. (2005). Intensive groundwater use: A silent revolution that cannot be ignored. *Water Science and Technology Series, 51*(8), 167–174.

Margat, J., & Van der Gun, J. (2013). *Groundwater around the World: A Geographic Synopsis.* Leiden: CRC Press.

Matsumoto, K. (2010). Treaties with groundwater provisions. In J. Delli Priscoli & A. T. Wolf (Eds.), *Managing and transforming water conflicts* (pp. 278–307). Cambridge: Cambridge University Press.

Maxwell, N. (2011). The nubian sandstone aquifer system: Thoughts on a multilateral treaty in light of the 2008 UN resolution on the law of transboundary aquifers. *Texas International Law Journal, 46*, 397–409.

McCaffrey, S. (2011). The international law commission's flawed draft articles on the law of transboundary aquifers: The way forward. *Water International*, *36*(5), 566–572. doi:10.1080/02508060.2011.597094

McCaffrey, S. C., & Neville, K. J. (2010). The politics of sharing water: International Law, sovereignty, and transboundary rivers and aquifers. In K. Wegerich & J. Warner (Eds.), *Politics of water: A survey* (pp. 18–44). London: Routledge.

McIntyre, O. (2011). International water resources law and the international law commission draft articles on transboundary aquifers: A missed opportunity for cross-fertilisation? *International Community Law Review*, *13*(3), 237–254. doi:10.1163/187197311X582386

Mechlem, K. (2009). Moving ahead in protecting freshwater resources: The international law commission's draft articles on transboundary aquifers. *Leiden Journal of International Law*, *22* (4), 801–821. doi:10.1017/S0922156509990239

Mechlem, K. (2011). Past, present and future of the international law of transboundary aquifers. *International Community Law Review*, *13*(3), 209–222. doi:10.1163/187197311X582278

Milman, A., & Ray, I. (2011). Interpreting the unknown: Uncertainty and the management of transboundary groundwater. *Water International*, *36*(5), 631–645. doi:10.1080/02508060.2011.599782

Movilla Pateiro, L. (2014). *El Derecho Internacional del Agua: los Acuíferos Transfronterizos.* Barcelona: J. M. Bosch Editor.

ORSAM. (2012). *An evaluation on the draft articles on the law of transboundary aquifers (Turkish- English)* (ORSAM Report No: 104 - ORSAM Water Research Programme Report No 12, February 2012). Retrieved from http://www.orsam.org.tr/en/enUploads/Article/Files/201229_1orsamreport104o.pdf

Puri, S., & Aureli, A. (2009). *Atlas of transboundary aquifers.* Paris: UN Educational, Scientific and Cultural Organization (UNESCO).

Remiro Brotóns, A. (1990). De los tratados a los acuerdos no normativos. In Ministerio de Asuntos Exteriores (Ed.), *La celebración de tratados internacionales por España. Problemas actuales* (pp. 23–48). Madrid: Ministerio de Asuntos Exteriores.

Richey, A. S., Thomas, B. F., Lo, M.-H., Famiglietti, J. S., Swenson, S., & Rodell, M. (2015). Uncertainty in global groundwater storage estimates in a total groundwater stress framework. *Water Resources Research*, *51*(7), 5198–5216. doi:10.1002/2015WR017351

Ruiz-Fabri, H. (1990). Règles coutumières générales et droit international fluvial. *Annuaire français de droit international*, *36*(1), 818–842. doi:10.3406/afdi.1990.2994

Schwebel, S. (1979). *First report on the law of the non-navigational uses of international water-courses, by Mr. Stephen Schwebel, Special Rapporteur.* Document A/CN.4/320 and Corr.1. Extract from the Yearbook of the International Law Commission 1979, vol. II(1). Retrieved from http://legal.un.org/docs/?path=./ilc/documentation/english/a_cn4_320.pdf&lang=EFS

Sindico, F., & Hawkins, S. (2015). The guarani aquifer agreement and transboundary aquifer law in the SADC: Comparing Apples and Oranges? *Review of European, Comparative & International Environmental Law*, *24*(3), 318–329. doi:10.1111/reel.12139

Stephan, R. M. (2010). La coopération transfrontalière sur les eaux souterraines: un processus en évolution. *Dynamiques-internationales 2.* Retrieved from http://dynamiques-internationales.com/wp-content/uploads/2010/01/DI2-R-Stephan-011011.pdf

Traversi, C. (2011). The Inadequacies of the 1997 convention on international water courses and the 2008 draft articles on the law of transboundary aquifers. *Houston Journal of International Law*, *33*(2), 453–488.

Villar, P. C., & Ribeiro, W. C. (2011). The agreement on the guarani aquifer: A new paradigm for transboundary groundwater management? *Water International*, *36*(5), 646–660. doi:10.1080/02508060.2011.603671

WHYMAP & Margat. (2008). *Groundwater resources of the world.* Large aquifer systems. Available from: http://www.whymap.org/whymap/EN/Downloads/Global_maps/whymap_largeaquifers_pdf.pdf?__blob=publicationFile&v=3

Yamada, C. (2011). Codification of the law of transboundary aquifers (Groundwaters) by the United Nations. *Water International*, *36*(5), 557–565. doi:10.1080/02508060.2011.597345

# Adaptation in the Tisza: innovation and tribulation at the sub-basin level

Shanna N. McClain, Carl Bruch and Silvia Secchi

**ABSTRACT**

This article explores the elements limiting adaptive governance in the Tisza sub-basin, considers policy options available to the sub-basin, and concludes that more attention must be paid to frameworks governing adaptation in transboundary sub-basins where resources are limited.

The Tisza is the largest sub-basin in the Danube River basin, and faces increasing water management pressures exacerbated by climate change. The Tisza countries have experienced challenges with managing climate change adaptation in a nested, consistent and effective manner pursuant to the EU Water Framework Directive. This is due to inefficiencies in climate change adaptation, such as weakened vertical coordination.

## Introduction

The Tisza sub-basin is larger than most European river basins; however, its management is underdeveloped, especially compared with overall Danube basin management. Flooding and surface water pollution present the greatest challenges to the surface waters of the Tisza, and concern regarding climate change is driving interest in managing these pressures in a consistent and effective manner. Drawing from interviews conducted in the Danube basin and the Tisza sub-basin, and from literature on adaptive governance, this article explores whether, in the context of adaptation, bottom-up governance can be successful in the Tisza, and discusses the broader implications for climate change adaptation at the sub-basin level.

Industrial and agricultural production significantly decreased in the Tisza as a result of political and economic change in the last two decades; however, many industrial sites lack fully implemented municipal sewage treatment or were left abandoned and continue to present serious pollution and accidental risk (Becker, 2005). In 2000, a series of significant flood events in Baia Mare, Romania, led to a breach in a mine tailings dam, resulting in the release of over 100,000 $m^3$ of cyanide, copper and other heavy metals into the Tisza, eventually travelling to the Danube and through to the Black Sea (Csagoly, 2000).

The process of adaptation enables a system to better cope with, manage or adjust to a changing condition, hazard, risk, or opportunity, such as those present in the Tisza sub-basin. Adaptive strategies aim to reduce vulnerability to these changing conditions and increase adaptive capacity (World Wildlife Fund, 2009). Governance refers to the interactions by private and public actors and includes the formulation and application of principles to guide their interactions, and care for the institutions that enable them (Munaretto, Siciliano, & Turvani, 2014). The growth in failed attempts to deliver efficient and reliable ecosystem goods and services has led to calls for adaptive governance regimes capable of incorporating the inherent complexity and uncertainty of social-ecological systems (Dietz, Ostrom, & Stern, 2003; Folke, Hahn, Olsson, & Norberg, 2005; Walters, 1997).

We argue that there are four key challenges constraining adaptation at the sub-basin level in the Tisza: policy, fiscal, institutional and capacity. This article begins with an overview of the study area and a description of the methodology. Next is an examination of conceptual domains relating to adaptation in international governance and adaptation in transboundary water management, including multilevel governance. Then the international laws and policies governing transboundary waters and relevant to the Danube basin and the Tisza sub-basin are explored. The final section highlights the challenges to adaptation in the Tisza sub-basin, drawing on expert interviews and document analysis. The article concludes with a brief reflection on the lessons to be drawn from these experiences.

## Overview of study area and methodology

Among the tributaries of the Danube River, the Tisza has the largest catchment area, covering approximately 160,000 km$^2$ (20% of the Danube's catchment area), and serving 14 million inhabitants (Figure 1). The upper portion of the Tisza begins in the Ukrainian Carpathian Mountains, where it moves along the border of Romania, flowing south-west into the flat, middle portion of the great Hungarian Plains, then into the lower Tisza, downstream of the Hungarian–Serbian border, where it joins the Danube River (International Commission for the Protection of the Danube River [ICPDR], 2008). Precipitation is concentrated in the mountainous upper catchment, resulting in some of the most sudden and extreme flooding in Europe, with floods reaching up to 12 m in 24–36 hours (Nagy, Ligetvári, & Schweitzer, 2010).

Extensive runoff, deforestation and channelization have reduced the ability of the catchment area to attenuate the flood wave, leading to sudden water level rise that threatens human lives and extensively damages infrastructure and croplands (ICPDR, 2008). Regional forecasts have indicated that the frequency of extreme floods and serious droughts are expected to increase as a result of climate change (Schneller et al., 2013). After the Netherlands, the Tisza ranks as one of the highest flood-risk areas in Europe, and it is also one of the poorest (Linnerooth-Bayer, Eckenberg, & Vári, 2013).

Extreme flood events occur every 10–12 years on average in the Tisza River sub-basin; however, the trend has been towards increases in all facets of flooding, including peak height, volume, and frequency (Sendzimir, Magnuszewski, Balogh, & Vári, 2007). Climate change projections suggest less precipitation in summer, more precipitation in

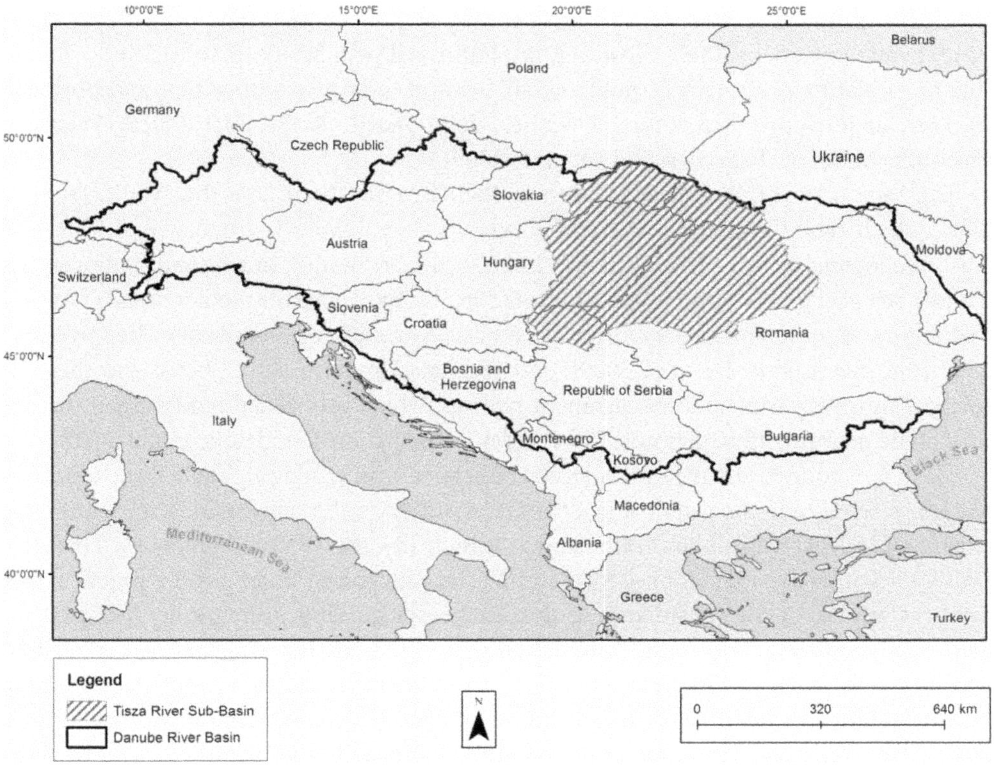


**Figure 1.** The Tisza River sub-basin in the Danube River basin.

winter and spring, and higher temperatures, resulting in earlier snowmelt, which can aggravate three main water-related problems in the Tisza: flooding, inland water stagnation, and drought (Schneller et al., 2013; Werners et al., 2011). Floods are already the most widespread hazard in the Tisza, and, combined with the poor water quality treatment in the region and poor socio-economic conditions (e.g. extreme poverty, social exclusion), they are projected to seriously affect the food supply and increase disease, injury, malnutrition and mortality (Schneller et al., 2013).

The root of the increasing flood stages lies in the original Vasárhelyi Plan of 1870. During this period the Tisza fell within the territory of a larger Hungary, and the Austrian and Hungarian ruling parties developed the Tisza to meet socio-political demands for grain production and export, habitation and flood protection (Sendzimir et al., 2008). The Tisza was shortened by 400 km and deepened to hasten water flow and facilitate navigation and transport. In recent history, Hungary has opted for river engineering methods that created large networks of levees and approximately 3000 km of embankments. These remain inadequate to protect against increasing flood frequency and discharge, because they require repeated raising; and the need to drain water quickly during floods has led to later scarcity of clean water during drought periods (Linnerooth-Bayer et al., 2013). The 2003 implementation of the Improved Vasárhelyi Development Plan – an elaborate river restoration plan to combine water retention, floodplain rehabilitation, and rural development as a strategy to replace prevailing engineering approaches – holds lessons for adaptation through support of floodplain production systems and environmental protection (Linnerooth-Bayer et al., 2013; Sendzimir et al., 2008). However, the improved plan has been considered a failure due to escalating costs, lack of political will beyond the water authorities, and obstacles to land acquisition (Sendzimir, Flachner, Pahl-Wostl, & Knieper, 2010; Werners, Flachner, Matczak, Falaleeva, & Leemans, 2009).

The four main drivers of adaptation in the Tisza sub-basin are: the cyanide spill of 2001, which raised awareness of transboundary vulnerabilities (particularly of industrial waste) following record-breaking floods in 2001; the promotion of alternative approaches to river science, practice and policy pioneered in Germany and the Netherlands; EU water and natural resource management policy (especially the Water Framework Directive [EC, 1992], the Birds Directive [EC, 2009], and Habitats Directive [EC, 1992]); and the shift towards more experimental management policies, which broadened management targets to include adaptive strategies and stakeholder participation (Sendzimir et al., 2010).

The Tisza countries have experienced challenges with managing their transboundary waters in a nested, consistent and effective manner at the sub-basin level within the existing EU legal and policy frameworks, principally the Water Framework Directive (WFD). Ultimately, adaptation efforts in the Tisza sub-basin highlight the practical and political limitations of subsidiarity. Subsidiarity is a guiding principle for EU governance that requires action – and interaction – between the EU and member states and delimits legislative powers between the EU and member states so that decisions are taken as closely as possible to that of the EU citizen (European Union Member States, 2012). The water legislation governed by the WFD provides an adaptive framework by affording the opportunity to adapt measures into future basin management plans via six-year monitoring and assessment cycles; however, it fails to specifically address climate change, and provides only a cursory mention of sub-basin management (European Community [EC], 2000, Article 13(5)).

## Methodology

The examination of adaptive governance at the Tisza sub-basin level was conducted through a combination of primary data analysis of semi-structured interviews and a literature review of peer-reviewed and secondary data, including an analysis of laws, policies and institutions in the Danube basin and the Tisza sub-basin. This qualitative approach followed methods described by Patton, Sawicki, and Clark (2013), who suggest that the mixed use of interviews, review of reports, and data analysis overcomes the potential for bias. Between the eight-month period of January to August 2013, 71 interviews were conducted in various locations throughout Europe. The interviews took place with experts working at the International Commission for the Protection of the Danube River, in the expert groups of the International Commission for the Protection of the Danube River (Tisza Group, River Basin Management, and Standing Working Group), and respondents working at the ministries, water management directorates, and non-governmental organizations in the Tisza countries, as well as with experts working within the European Commission, the United Nations, and the United Nations Economic Commission for Europe involved in the Danube basin and the Tisza sub-basin. Given the respondents' public roles, the interviews are intentionally left anonymous to foster candidness in responses (Table 1). The questions focused on how

**Table 1.** Professional affiliations of experts interviewed and corresponding acronyms.

| Role | Acronym |
|---|---|
| European Commission official | EC |
| Ecosystem and biodiversity | EB |
| Danube River basin | DRB |
| Tisza sub-basin | TSB |
| Regional water manager | RWM |
| Flood protection | FP |
| Government official | GO |
| United Nations Economic Commission for Europe officer | UNECE |
| Freshwater biologist | FB |
| Regional water director | RWD |
| Global Environment Facility / United Nations Development programme officer | GEF/UNDP |

Danube basin and Tisza sub-basin policies and laws were implemented in practice, as well as the perceptions of the experts regarding the interplay of implementation as it concerned adaptive governance and the role of climate change adaptation in the Tisza sub-basin.[1]

## Governance of adaptation in international waters

Globalization has profoundly altered the theory of government and the nation-state, as well as the roles and functions of the political actors regulating public affairs and promoting development. Politics has become more polycentric, with the nation-state as one of many levels in a complex system of overlapping and sometimes competing agencies of governance (Finger, Tamiotti, & Allouche, 2006).

### *Governance of international waters*

There is an ongoing shift from government to governance, in which formal and informal institutions and individuals, public and private, are involved in a continuing process of cooperation and accommodation in order to actively balance conflicting or diverse interests (Dellapenna & Gupta, 2009). Government is no longer the sole decision-making authority, exerting sovereign control over civil society; instead, ideas of multilevel governance contribute to policy development and implementation through the participation of a variety of actors in diverse settings (Pahl-Wostl, 2009). Governance is a reflection of shared goals and behaviours that may not be derived from formal prescriptive responsibility; nor does it necessarily require police power to ensure compliance (Rosenau, 1992).

Increasingly, governance includes the role of non-state actors, as well as the private sector and public participation. Participatory approaches in natural resource management reflect the emergence of new modes of governance and knowledge generation in times of increasing uncertainty and complexity (Pahl-Wostl et al., 2007). Increasingly, these participatory approaches are finding application in the management of international waters (Bruch, Janksy, Nakayama, & Salewicz, 2005). Due to the resource-intensive nature of participatory processes, they can result in decreased efficiency, even while they tend to increase compliance and overall effectiveness (Pahl-Wostl, 2009).

Multilevel governance refers to governance practices at the local, state/provincial, regional, national and international levels that work in a connected manner, within a single, integrated command structure, and where decisions made at one level impact other levels (Finger et al., 2006; Kuhlmann, 2001; Marshall, 2008). Multilevel governance in polycentric systems suggests a nested hierarchy of decision making, and authority at multiple levels (Akamani & Wilson, 2011). While commentators have theorized that redundancies in these systems present an advanced ability to adapt to changing environments and therefore higher resiliency, multilevel governance has noted inefficiencies with respect to climate change adaptation due to ineffective vertical interplay stemming from the large number of decision points and actors (Newig & Fritsch, 2009; Pahl-Wostl, 2009). Improved coordination is the intended outcome, but the cost of coordination, reaching agreement and enforcing such an agreement can be quite high, and if coordination fails, a duplication of efforts and additional costs can ensue (Huitema et al., 2009).

The principle of subsidiarity motivates and underpins the process of assigning responsibilities across governance levels. In areas where the EU does not have exclusive competence, the principle of subsidiarity seeks to protect the capacity of the member states to take decisions and action, and authorizes intervention by the EU only when the objectives of an action cannot be sufficiently achieved by a member state (European Parliament [EP], 2004). The European Community has adopted subsidiarity as a central organizing principle. The 2007 Treaty of Lisbon expanded its application to include proportionality where competencies between the EU and member states are defined, and control mechanisms were introduced in order to monitor its application (EC, 2010). Marshall (2008) notes that while the principle of subsidiarity may open up interpretations in use that can be beneficial for promoting experimentation and learning across governance systems in how tasks are assigned to various levels, it is important to detail the criteria by which the nesting of subunits at higher levels should occur. Governments can underestimate the capacities of subunits to address particular problems, thereby inappropriately retaining responsibilities that should be undertaken by lower levels of government.

### *Basin- and sub-basin-level governance*

The 1992 Dublin Conference on Water and the Environment, and Agenda 21, have operationalized theory into policy by directing that participatory approaches in water governance be carried out at the basin or sub-basin level; however, challenges remain in reconciling the boundaries of this resource with its respective institutions (International Conference on Water and the Environment, 1992; United Nations Conference on Environment and Development, 1992). The scale of governance of river basins is associated with spatial area as well as the degrees of government authority, not only vertically (international, national, regional, state, local), but also horizontally (scope of activity and authority) (Griggs, 2015). Ekstrom and Young (2009) theorize that failures in spatial fit, as applied to environmental resource management, occur when an institution fails to account for the nature, functionality and dynamics of the ecosystem it influences. While theories of subsidiary suggest a downscaling process to more local-level actors, particularly for natural resource governance, global climate change necessitates an upscaling of policy.

While a river basin approach may seem simple, problems and opportunities within a basin are multiple and overlapping, and vary from the local to the regional, thus increasing the geographic scale of institutional arrangements (Huitema et al., 2009; Moss, 2012). At the same time, too many autonomous centres of decision making without clear institutional roles or set modalities for interaction or coordination can also constrain the implementation of government policies, such as those for climate change mitigation and adaptation (Nanni, 2012). In this regard, the WFD represents an ambitious attempt to arrange water resources based upon the principle of basin-level management (Moss, 2012). The integrated approach is promoted through water pollution control and principles and practices at the basin level that aim to achieve 'good status' for surface and ground water by 2015 (EC, 2000). In a 2003 WFD guidance document, the European Community recognized that "by creating a spatial unit for water management based solely on river basins, spatial conflicts could occur with policy sectors structured on administrative and political boundaries" (EC, 2003b, p. 17). However, the river basin can often be too large a unit to manage effectively, and in some of the world's largest rivers – such as the Volga, the Amazon and the Danube – failure to properly integrate sub-basin management weakens vertical coordination in multilevel governance (United Nations Environment Programme [UNEP], 2014b). In addition to basin size, other factors that limit effective management are interjurisdictional conflicts among governments, the presence of groundwater basins that affect surface water management, and other water management areas (e.g. hydropower) requiring different management options, all of which exist in the Tisza (Caponera, 2007). While water managers argue that not all problems require a basin-level approach, alternative levels of management are not often considered because the river basin is considered the optimal spatial unit for managing water (Grünewald, 2008; von Keitz & Kessler, 2008). And arguments have been made nationally in support of selective forms of basin-level management when specific issues are considered advantageous, particularly in the case of upstream–downstream relations, flooding, drought and low water levels (Moss, 2012). However, formal recognition and legitimacy must be provided for lower levels of basin management, as without it member states may undermine existing governance structures and power relations between stakeholders, and create difficult-to-overcome institutional gaps between basin countries (Del Moral & Do Ó, 2014). Uncertainty remains in regard to spatial fit at the sub-basin level, as well as how to manage adaptation when neither level is specifically prescribed by EU law.

### *Adaptive governance*

Given the complex nature of water resources and the socio-economic factors affecting water use, much uncertainty exists regarding long-term water needs and the availability of water resources (Gleick, 2000). Adaptive management provides a framework for governing water resources in a manner that can account for these uncertainties. Adaptive governance involves the devolution of management rights and power sharing in order to promote participation; however, for participation to occur, collaborative networks must exist (Folke et al., 2005). Thus, adaptive governance can be conceived as the synthesis of collaborative management and adaptive management (Huntjens, Pahl-Wostl, & Grin, 2010).

Few institutional frameworks are developed specifically for adaptive governance, although components of adaptive frameworks are often present (Bruch, 2009; Troell & Swanson, 2014). Finding a balance between bottom-up and top-down governance is an important element for adaptive management in river basins and large-scale, complex systems (Huntjens et al., 2010). Adaptive governance requires secure, adequate and flexible funding. While a variety of funding mechanisms and informal networks are available, providing dedicated resources to train, support, incentivize and institutionalize capacities into practice is still challenging (Wyborn, 2015). A more decentralized approach would, for example, attempt to integrate best practices from smaller financed pilot projects and scale them up to catalyze change at multiple levels and across sectors, with the intention of learning at larger scales and for longer periods of time (Barchiesi, Sanchez, Cross, Pérez de Madrid, & Onencan, 2014).

## *Policy frameworks*

Adaptation in international river basins is governed by a range of global, regional and national laws, policies and soft-law instruments. In the Tisza sub-basin, this includes the 1992 UN Framework Convention on Climate Change (UNFCCC), the 1992 Helsinki Watercourses Convention, the 1997 UN Watercourses Convention, the WFD and the EU Strategy on Adaptation to Climate Change, as well as adaptation policies at the level of the Danube Basin.

### *United Nations*

The UNFCCC presents several guiding principles for the international community to utilize in preventing and adapting to climate change. Under Article 4, clear commitments on adaptation to the adverse impacts of climate change are listed, including how to formulate and implement national programmes to facilitate adequate adaptation to climate change (Article 4(1)(b)).[2] While the UNFCCC provides guidance on how to address climate change adaptation, the ability to assess institutions, and the role of institutions in relation to their adaptive capacity, are not addressed (Gupta et al., 2010).

While the UNFCCC is most effective at guiding adaptation at the national level (Gupta et al., 2010), the UN Economic Commission for Europe (UNECE) is a regional commission that oversees the implementation of regional agreements, including the 1992 Convention on the Protection and Use of Transboundary Watercourses and International Lakes (the Helsinki Watercourses Convention) (UNECE, 2010). The Helsinki Watercourses Convention – formerly a regional, now a universal framework – is open to the participation of a variety of countries within and outside Europe, including Asia. The convention aims to protect and ensure the quantity, quality and sustainable use of transboundary water resources by facilitating cooperation among shared watercourses (UNECE, 2010). Though the convention does not explicitly mention climate adaptation, it provides a framework for transboundary cooperation and the development of adaptation strategies, and requires parties to enter into bilateral and multilateral agreements to eliminate contradictions with the principles of the convention. This includes provisions for consultation, research and development, monitoring and assessment, and the establishment of institutions for cooperation and management of shared watercourses. At the 2009 Meeting of the Parties, member states adopted the

Guidance on Water and Adaptation to Climate Change, which provides non-binding guidance on how to perform risk assessments, measure vulnerability, and design and implement adaptation strategies. The guidance is reinforced by pilot projects that are undertaken by UNECE to strengthen the capacity of developing countries to create basin-wide adaptation strategies.

The 1997 UN Convention on the Law of the Non-navigational Uses of International Watercourses (UN Watercourses Convention) entered into force in August 2014. The convention is significant mainly for the codification of three principles of customary international law: equitable and reasonable utilization; prevention of significant harm; and prior notification of planned measures (McCaffrey, 2014). A significant role of the convention is the encouragement of watercourse states to enter into watercourse agreements and establish a framework of general principles to guide the behaviour of the states (Sands, 2007). Interestingly, not all riparians of the Tisza have signed or ratified the convention – only Hungary has (United Nations [UN], 1997). It was reported in multiple interviews that experts from Hungary were interested in seeking specific sub-basin management arrangements for the Tisza sub-basin, through the development of a sub-basin commission or another negotiated framework (interviews DRB #11, GEF/UNDP #54, RWD #49).

While the UNECE Watercourses Convention was intended as a regional instrument for Europe and was eventually opened for accession to states beyond the European region, the UN Watercourses Convention and the UNECE Watercourses Convention remain compatible, and the treaties have essentially the same object and purpose: cooperative use, management and protection of shared freshwater resources (McCaffrey, 2014). Article 3 of the UN Watercourses Convention also expressly respects pre-existing basin agreements, and therefore will not disrupt the interpretation or implementation of regional or basin agreements (UNEP, 2014a).

### *EU directives and policies on adaptation*

In 2000 the EU adopted the WFD, which establishes an adaptive framework in water policy and aims at achieving 'good status' for European waters. The directive shifts away from national, control-specific directives (e.g. water pollution, groundwater, agricultural management) and towards basin-level governance (Dellapenna & Gupta, 2009). This is carried out through a six-year cyclical process via cooperation among nationally identified competent authorities from each basin country, and is considered adaptive by allowing the opportunity to incorporate information into each new basin management cycle (EC, 2000). Specifically, as part of the WFD the EU member states must: (1) identify the individual river basins within their national territory and assign them to international river basin districts (RBDs, e.g. the Danube Basin district); (2) characterize RBDs in terms of pressures, impacts and economic uses of water, including a register of protected areas within each RBD; (3) carry out calibration of the ecological status classification systems by 2006; (4) make operational the monitoring of water status by 2006; (5) identify a programme of measures for achieving the environmental objectives of the WFD; (6) produce and publish river basin management plans for each RBD, including the designation of heavily modified water bodies, by 2009; (7) implement water pricing policies that enhance the sustainability of water resources by 2010; (8) make the programme of measures operational by 2012; and (9) implement the programme of measures and achieve the environmental objectives by 2015 (EC, 2003a). However, the WFD has been criticized for failing to explicitly reference climate change, calling into question its ability to adequately address climate change issues (Nanni, 2012).

In 2013, the EU released its Strategy on Adaptation to Climate Change, which reinforced the recommendations of the UNFCCC to create national adaptation strategies and risk management plans (European Commission, 2013). In recognition of the cross-sectoral nature of climate change adaptation and the need to integrate these activities across multiple levels and projects, the EU dedicated a portion of its cohesion policy funding mechanism to member states that want to improve their adaptation measures, as long as they have developed the requisite national or regional risk assessments outlined by the UNFCCC (European Commission, 2013). The cohesion policy is the EU's principal investment policy and has the primary responsibility of mainstreaming the environment into EU programmes and projects by strengthening vertical and horizontal approaches to management. To meet the criteria for funding under this mechanism, a country's gross domestic product must be less than 75% of the Community average – this removes almost all the territory of the Tisza except Romania and some portions of Hungary. However, the EU has endorsed a macro-regional strategy for three specific regions that could benefit from strengthened cooperation and economic and social cohesion: the Danube Region, the Baltic Sea Region, and the Adriatic and Ionian Region (European Commission, 2014b). Thus, the EU's adaptation strategy does not align adaptation activities with those at the basin level, nor does it align with other EU directives. It is up to each member state to decide whether and how they will adapt to climate change, and the process for applying for adaptation funding from the EU requires meeting guidelines from a global authority (the UNFCCC).

From a policy and legal perspective, adaptation at the basin level is therefore directed at the national level through the UNFCCC and the WFD, but not at the sub-basin level; any activities that occur at the sub-basin level exceed what is required by law.

### *Danube basin-level adaptation policy*

In 1998, predating the adoption of the 2000 WFD, the International Commission for the Protection of the Danube River (ICPDR) was established as a transnational body to implement the Danube River Protection Convention (DRPC). The DRPC is the primary legal instrument governing transboundary water management and cooperation in the Danube basin, and ensures the sustainable and equitable management and use of the Danube River by all countries sharing the basin, including non-EU countries; the EU is also a member of the DRPC (ICPDR, 2008). Because the WFD is an EU mechanism, it is only binding on member states of the EU. Through the DRPC, the Danube countries appointed the ICPDR as the coordinating body for the WFD, and EU and non-EU countries have agreed to manage their portion of the Danube basin according to WFD regulations (ICPDR, 2008). Therefore, in the event of a dispute related to the DRPC, rules for arbitration and the involvement of the UN's International Court of Justice have been stipulated (ICPDR, 1994).

In the 2010 Danube Declaration, the EU and high-level representatives from Danube countries committed to reinforcing sustainable, transboundary management, with a particular emphasis on developing a climate change adaptation strategy for the basin, and to organizing a conference with relevant financial institutions and donors to draw attention to the financial constraints some countries in the Danube Basin face and to identify mechanisms for the financing of projects (ICPDR, 2010). One important step in improving transboundary management of the Danube was the development of the 2013 Strategy on Adaptation to Climate Change for the Danube River Basin (ICPDR, 2013). Although the strategy was envisioned as a method for integrating climate change adaptation into the Danube River Basin Management Plan (DRBMP) of 2015 and future iterations, it highlights the need for more comprehensive planning at the sub-basin and national levels (ICPDR, 2013). The strategy also notes that while conducting a basin-wide vulnerability assessment would be beneficial, following EU Common Implementation Strategy guidelines, the compilation of existing local and national vulnerability assessments throughout the basin was chosen instead (ICPDR, 2013). The ICPDR's adaptation strategy illustrates some of the challenges in planning, balancing and prioritizing its limited resources for adaptation (Barchiesi et al., 2014).

In the first DRBMP, released in 2009, climate change was addressed as an issue of basin-wide importance, and the identification of future pressures on the aquatic environment was considered a priority (ICPDR, 2009a). In the 2015 iteration of the DRBMP, the Danube countries acknowledged the uncertainties related to climate change stresses on water resources (ICPDR, 2015). They also addressed the difficulty in decoupling climate stressors from anthropogenic stressors and therefore integrated climate change into the DRBMP. While there are no specific climate change adaptation measures, the DRBMP indicates that, by noting the effects of climate change on water-related sectors, cooperation can take place among various inter-sectoral activities, e.g. flood risk management, inland navigation, hydropower and agriculture (ICPDR, 2015).

## Adaptation innovations in the Tisza

Building on the theories of adaptive governance discussed earlier, and the functional adaptive management strategies in the Danube Basin, this section explores the source of adaptation innovations in the Tisza, including memoranda of understanding, the Integrated Tisza Basin Management Plan, and the Tisza Sub-basin Flood Management Plan. While the WFD does not explicitly prescribe management at the sub-basin level, the Tisza sub-basin has developed several noteworthy innovations. The ability to coordinate and implement these actions over the long term is essential for consistent and sustainable management.

### *Memoranda of understanding*

At the sub-basin level, the WFD allows the development of supplemental actions. In this regard, the Tisza countries have forged a memorandum of understanding (MOU) to manage activities at the sub-basin level. From this MOU, an ad hoc Tisza Group was formed in 2004 to coordinate the activities of the MOU.

The MOU is distinct from legally binding treaties, although MOUs may be viewed – in the words of the UN's International Law Commission – as "treaties in simplified form" (International Law Commission, 1966). They can provide signatories the benefits of entering into an agreement without having to meet the formalities associated with negotiating, ratifying and amending a treaty. Unlike its sister sub-basin the Sava, however, the Tisza has been managing its activities without a formal basin commission, and without formal treaties beyond those shared bilaterally with riparian states (Table 2).

The 2004 MOU that created the Tisza Group also included a provision to develop a sub-basin management plan by 2009. In 2010, a ministerial statement, 'Towards the Development and Implementation of a River Basin Management Plan for the Tisza River Basin', was signed by the Tisza countries; this statement sought to ensure the ultimate completion of the first Tisza sub-basin management plan in 2011 (ICPDR, 2010).

In 2011 the Tisza Group signed a new MOU, 'Strengthening of Tisza River Basin Cooperation', in addition to setting a 2012 deadline for developing a case study for climate change impacts on the Tisza sub-basin, as well as a 2015 deadline for the updated Tisza Basin Management Plan. The MOU also promotes regional cooperation on the protection of mountain resources, with the Framework Convention on the Protection and Sustainable Development of the Carpathians (Carpathian Convention, 2006). While participation and cooperation with the Carpathian Convention are ongoing, the first tasks of the MOU have not been accomplished; reasons for this will be discussed below in the upcoming section on challenges to sub-basin adaptation.

### *The Integrated Tisza River Basin Management Plan (ITRBMP)*

While the delay in ultimately creating the ITRBMP reflects common challenges to sub-basin management – namely fiscal and capacity challenges – interest in advancing activities at this level was reported in multiple interviews (interviews FP #33, GEF/UNDP #54, GO #36, RWD #47, TSB #25). Funding for the ITRBMP was ultimately provided by the governments of Tisza basin countries, EU funding, and the UN Development Programme/Global Environment Facility. This split the activities of the ITRBMP into two separate modes: first, to develop a sub-basin management plan for the Tisza based on funding from the Tisza countries and the EU; and second, to implement a set of pilot projects focused on mainstreaming wetlands and floodplains restoration into national policy (ICPDR, 2011).

In contrast to the DRBMP, the ITRBMP accounted for rivers with a catchment size over 1000 km$^2$ (instead of over 4000 km$^2$), natural lakes over 10 km$^2$ (instead of 100 km$^2$), main canals, and groundwater bodies over 1,000 km$^2$ and of basin-wide

**Table 2.** Water-related bilateral agreements among Tisza countries (adapted from ICPDR, 2009a).

| Countries | Transboundary watercourses | Disasters/emergencies | Environmental protection |
|---|---|---|---|
| Hungary–Romania | 1986 | | 2000 |
| Hungary–Slovakia | | No date | 1999 |
| Hungary–Ukraine | 1997 | 1998 | 1993 |
| Ukraine–Romania | 1997 | | |
| Ukraine–Slovakia | 1995 | 2000 | |

importance. Recognizing that many of the problems of water quality and quantity do not appear in isolation, and that climate change can present major challenges to the sub-basin, the Tisza countries identified the pressures and potential impacts to the region that could occur from each of the issues and that currently affect two or more Tisza countries. These include hydromorphological pressures from flood protection measures, accidental pollution from flooding, loss of wetlands, solid waste, groundwater depletion due to over-abstraction, and increased irrigation and surface water abstraction (ICPDR, 2011). Furthermore, each country reported the extent to which it had a climate change strategy in place (whether one was present or not), as well as any activities that were taking place in the sub-basin at the local level (ICPDR, 2008).

Finally, the ITRBMP explicitly addressed horizontal measures related to the integration of water quality and quantity (ICPDR, 2011, p. 111). Here the Tisza countries recognized that local action alone is not sufficient to effectively implement the ITRBMP, and requested that consistent action be taken by the Tisza Group as well as through bilateral commissions addressing water management in the Tisza, including under the work of the regional Carpathian Convention. The ITRBMP recommended that inter-ministerial or inter-sectoral committees be established to help coordinate implementation, and to develop an overall communication strategy for the Tisza sub-basin in order to target different levels of authority and include aspects of climate change for long-term sub-basin management. Compensation schemes and incentives must be considered in long-term management of sub-basin activities, particularly in cases where flood protection and water retention could be requested as land management alternatives.

### *Sub-basin-level Tisza flood action plan*

In acknowledgement of the number of fatalities caused by flooding, the large numbers of people displaced, and the extensive damage caused both to the environment and to the economy, the EU adopted the 2007 Directive on the Assessment and Management of Flood Risks (Flood Directive, FD) (EC, 2007). The terminology of the FD is unique in that, unlike the WFD, it provides for situations where the units of management can be something other than river basins – they can also be coastal areas (European Commission, 2014a). However, European Commission guidance documents on the WFD and FD indicate that a sub-basin cannot be a unit of management for the purposes of the FD because it has to include all surface water flows that terminate at the sea (European Commission, 2014a, p. 12). While the FD expressly states that this cannot be a sub-basin, the flood events in 2002 and again in 2007 moved the Flood Protection Expert Group of the ICPDR to develop a sub-basin-level flood action plan for the Tisza River basin in 2009 (ICPDR, 2009b; interview GEF/UNDP #54). A potentially problematic aspect of the plan, however, is that unlike in the Danube basin, and as directed by the FD, the Tisza plan indicates that it will only be updated as appropriate, or as determined by the bilateral river commissions, which means that it is not necessarily linked to broader Danube basin governance, the cyclical basin management cycle of the Danube, or the adaptive management process of the Tisza (ICPDR, 2009b).

## Challenges to sub-basin adaptation

The Tisza Group was created to manage the pressures on water resources, including adapting to climate change, but its activities have largely faltered as a result of the lack of legal and policy frameworks, inadequate funding, and institutional and capacity challenges. These are considered now.

### *Policy challenges*

Without a policy framework to guide sub-basin management in the EU WFD, the Tisza lacks formal vertical integration into the broader Danube basin management process. This presents complications for subsidiarity in climate change adaptation at the sub-basin level, given that adaptation is often implemented at the domestic level through national policies. While the monitoring and assessment present in the river basin management plan cycle may help detect climate change–related impacts on water resources, such basin-level efforts often do not translate into national-level adaption policies or action (Nanni, 2012). Hungary, Romania, Slovakia and Ukraine have regularly submitted Annex I National Communication Reports, containing information on national vulnerability and adaptation to climate change per the UNFCCC, and Serbia developed its first non–Annex I National Communication Report in 2010, indicating the financial and capacity needs for developing the first National Action Plan for Adaptation. Such action plans reported to the UNFCCC are not mandated under the WFD, but they are tied to EU funding mechanisms, which provides an impetus to develop them. While any ongoing adaptation activities would be included in the submitted national reports, none of these national reports contain information on projects related to the Tisza. However, if there is a lack of inclusion by institutions in vertical governance at the sub-basin level, a balance is not found between centralized and decentralized control, and adaptive management is less likely to occur. Interviews also indicated that government documents on national water policy discuss the use of water, but not the need for cooperation among transboundary bodies governing water and adaptation – and without country-level interest in furthering sub-basin level activities, action will continue as directed by the WFD and national water law (interviews GO #36 and #48, RWD #47). Given the interest expressed by Tisza countries in strengthening horizontal measures in the ITRBMP, this represents additional policy challenges at the national level in sub-basin management.

### *Fiscal challenges*

While the EU Common Implementation Strategy helps develop a Europe-wide understanding for taking a basin-level approach to management, and provides technical guidance in implementing the WFD, there is no specification of funding schemes for river basin management in the Common Implementation Strategy or the WFD, which leaves the question of funding projects for climate adaptation, or other water-related projects across river basins, entirely up to the member states (Moss, 2012). Each EU fiscal cycle provides the opportunity for basins to decide what projects they want to develop, and a variety of funding mechanisms are available based on the type of project.

Funding for adaptation to climate change in Europe is also offered through a variety of instruments, and is aligned with the six-year fiscal cycle of the corresponding WFD river basin management plans. To fund the administration of the ICPDR, the countries are charged an annual fee assessed from taxes, which is usually an equal amount across the basin, except for times of country transition or other ad hoc exceptions. These fees also partially fund pilot projects for the Danube basin, as agreed by the Danube countries – the fees do not drive activities at the sub-basin level.

For projects being undertaken by the ICPDR, funding is applied for from a variety of sources (Table 3). As indicated previously, basin-level activities are specifically prescribed under the WFD, and are therefore given priority. Interviews in Hungary and Serbia revealed that perceptions regarding funding insecurities led to preferences for more local pilot projects within the Tisza sub-basin (interviews GEF/UNDP #54, GO #36 and #48, RWD #47, RWM #26 and #27). Concerns that projects developed at the sub-basin level would not be completed, or that funding would run out and that local citizens would not be reimbursed for their investments, was reported during interviews, and examples were provided of situations where this had occurred in Romania, Hungary, Serbia and Ukraine (interview, GO #36 and #48, RWD #47, TSB #25, UNECE #39).

### *Institutional challenges*

The ICPDR has a consultative and advisory function, but the basin states are the ultimate decision-making authorities. The activities of each expert group are led by a chair nominated by the countries represented in each group. The position is supposed to change every two years, and the ICPDR cannot recommend who can be seated at these meetings – this is the role of the basin states (interview DRB #12). The activities of the Tisza Group were initially led by a representative of the European Commission, but over time the commission has lessened its role in the workings of the sub-basin and has looked to the countries to take the lead on the activities they want to see advance. Interviews indicated that commission participation as chairpersons was a driver of activities, but that as the commission assumed a reduced role, the effectiveness of the group also lessened (interviews DRB #12, EC #14, TSB 25). Suggestions for strengthening effectiveness in the Tisza Group have been to nominate someone who is active at multiple levels within one government, who is familiar with the directives and laws of the countries, and who can report across multiple levels, including the ICPDR (interviews DRB #11 and #12, EC #4, RWM #27).

**Table 3.** Sources of funding for the Danube basin (adapted from ICPDR, 2009a).

| |
|---|
| National commitments from basin countries |
| EU funding mechanisms (e.g. Common Agricultural Policy, Cohesion Fund, Structural and Life Funds, EU Neighbourhood Fund) |
| International funding institutions (e.g. European Investment Bank, European Bank for Reconstruction and Development, World Bank) |
| Water pricing policies (e.g. application of polluter-pays principle) |
| Other external organizations (e.g. Global Environment Facility, Environment and Security Initiative, World Wildlife Fund) |

Differing visions about what the Tisza Group should do and how it should operate, compounded by perceptions of counterparts in other states as difficult and driving their own agendas, have effectively reduced the productivity of the Tisza Group (interviews DRB #11 and #12, GO #36, TSB #25). The Tisza Group MOU was updated in 2011 and included the intention to develop a case study for climate change impacts in the Tisza by the end of 2012. The study has yet to occur, and there are no current plans to complete it (ICPDR, 2011). Additional activities for the Tisza Group include a 2015 date for the development of the updated ITRBMP (ICPDR, 2011), but this too has yet to occur.[3] Interviewees described how countries often know what goals they are trying to achieve nationally in order to meet the suite of EU water quality directives, but this does not always translate into sub-basin cooperation (interviews DRB #12, GO #36, TSB #25). Furthermore, problems of political divisions within the countries or gaps in knowledge can lead to issues of trust and delays in financing projects that countries had originally agreed to support. This has specifically been the case with projects relating to climate change adaptation in the Tisza sub-basin, where a strong interest in advancing projects on adaptation has been voiced by multiple people at multiple levels, but where concerns over the lack of coordination and direction have led to cessation of activities (interviews DRB #12, EC #4, GO #36 and 49, RWD #47, TSB #25, UNECE #34).

### *Capacity challenges*

While the WFD requires each country to assign a competent authority in river basin management, the responsibilities of national ministries and agencies, their institutional roles and modalities for coordination of water resources management, and the interaction with local authorities, water users and NGOs, which would normally be defined in new legislation, are left to national governments (Nanni, 2012). And, because there are only a limited number of national experts in each country capable of participating in meetings where knowledge of both hydrologic systems and water governance is necessary, it is not uncommon for the same experts to attend multiple meetings.

The consistency in attendance would seem important, given the large number of activities from the UN, EU, Danube, Tisza, national, regional and local levels; however, some experts reported feeling overextended (interviews EB# 7, EC #4, FB #42, GO #36 and 51). Given the vast differences between local and regional climate conditions, experts have requested that training be made available to better understand the variability in decisions and levels for which they are managing adaptation projects (interviews GO #36, RWD #47). Uncertainty related to climate change has been considered an excuse for not taking action, or for not updating climate change into national policies (interviews GO #36, RWD #47). Ministries in Hungary have worked to determine the scale of possible impacts of climate change, but due to the lack of agreement on climate change scenarios, for example, this has not been written into Hungarian water policy (Huntjens et al., 2010). The author also observed limitations in the ability of country representatives to be present at the Standing Working Group meeting when the ICPDR and high-level representatives meet to decide on objectives for the year ahead and provide political guidance for the entirety of the basin.

Experts interviewed also noted that the generation and exchange of information and data allows compatibility of perceptions and fosters communication among parties, which – over time – aids in building commitment towards common goals, helps address difficulties in a cooperative, technical and ultimately more effective manner, and builds trust. Furthermore, this also allows coordinated action to be decentralized to the local level through smaller pilot projects where communities and local government can better see tangible results (interviews GEF/UNDP #54, GO #48, RWD #47, RWM #27, UNECE #39 and #61).

## Conclusions

While it has been increasingly recognized that the effects of climate change on water resources are best addressed in a river basin context (Bruch & Troell, 2011; Nanni, 2012), the failure, or inability, to properly integrate sub-basin management weakens vertical coordination in multilevel governance (UNEP, 2014b). The WFD supports horizontal coordination by linking water quality, water quantity and environmental integrity. However, multilevel governance posits that a tiered structure of river basin and sub-basin organizations are in place and operating effectively – a tenet of effective decentralization and subsidiarity that the EU WFD discounts, at least when it comes to sub-basin management.

Experiences in the Tisza indicate that there are many challenges to governance at the sub-basin level, and that these challenges have constrained the ability to undertake effective sub-basin measures to adapt. Notwithstanding these challenges, the Tisza Group has shown the potential to develop incremental measures to adapt to climate change, especially in response to specific flood events and industrial accidents.

In acknowledgement of financial resource constraints, there are four options that could improve vertical integration of adaptive governance. First, a sub-basin commission could be created. A sub-basin commission could provide regional management of the Tisza sub-basin and would require an agreement of the Tisza member states that would allow the sub-basin states to highlight the pressures in need of the most attention (flooding and adaptation to climate change). Second, the Tisza Group could be housed at and have actions coordinated through the ICPDR. In this regard, the Tisza Group could hold a permanent position in the ICPDR, similar to the expert groups. Previous forms of the Tisza Group have worked on an ad hoc basis though memoranda of understanding developed to address particular activities. Third, existing or new bilateral treaties could be used for sub-basin management. While bilateral treaties exist among many of the Tisza countries (Table 2), many are outdated or cover only specific arrangements (e.g. disasters/emergencies, environmental protection). These would need to be individually developed or updated with each neighbouring sub-basin country and define the objectives to be achieved by the bilateral parties. Finally, a framework for managing the Tisza could be developed. This option allows for the development of a sub-basin river organization, without the formal creation of a sub-basin commission. More flexibility is possible here, since these types of organizations can exist through councils, committees, or agencies through the establishment of a mandate for the organization (what it is expected to do), its authority (including formal and informal actors), and its capacity (resources and financing). All four options face policy, financial and resource constraints similar to those witnessed to date, and thus are unlikely to improve adaptive management at the sub-basin level unless more resources are provided. Alternatively, it will be necessary to rethink the assertion that adaptation needs to occur at all levels.

This is not unique to the Tisza sub-basin. Of the more than 260 transboundary basins in the world, many have transboundary sub-basins, and many are in the developing world. These countries have fewer resources, and are likely for the foreseeable future to have fewer resources for sub-basin management. Therefore, more attention should be paid to conceptual and operational frameworks governing adaptation in transboundary sub-basins where resources are limited.

## Notes

1. Questions relevant to adaptation and multilevel governance included: (1) Are there any policies that directly or indirectly address decentralization of water governance in the Tisza sub-basin? (2) What gaps exist between policy and practice in multilevel governance? (3) What are some of the constraints and opportunities for adaptation in the Tisza? (4) What are the trends in adaptation, and what level of governance are they coming from?
2. All sub-articles in 4(1) include other adaptation-related commitments.
3. While the 2004 and 2011 MOUs are non-binding, they have provided a framework for establishing the Tisza Group and developing the Tisza sub-basin management plan and eventually the flood management plan. Despite the MOU to develop an adaptation strategy for the Tisza, to date this has not been done.

## Acknowledgements

We thank the Southern Illinois University IGERT Program in Watershed Science and Policy and associated colleagues for their support. We also thank the International Commission for the Protection of the Danube River for assisting in obtaining data, and for hosting Shanna during the spring of 2013 while she conducted her research.

## Funding

This material is based upon work supported by the United States National Science Foundation [grant no. 0903510]. Any opinions expressed here are those of the authors and do not necessarily reflect the views of the National Science Foundation.

## References

Akamani, K., & Wilson, P. I. (2011). Towards the adaptive governance of transboundary water resources. *Conservation Letters*, *4*, 409–416. doi:10.1111/j.1755-263X.2011.00188.x

Barchiesi, S., Sanchez, J. C., Cross, K., Pérez de Madrid, M., & Onencan, A. M. (2014). *Adaptation planning: Views towards resilience and up-scaling success to enhance transboundary water governance* (IUCN Environmental Policy and Law Paper No. 75). Gland, Switzerland: IUCN. https://portals.iucn.org/library/sites/library/files/documents/IUCN-EPLP-no.075.pdf

Becker, G. (2005). *Transboundary river basin management regimes: The Tisza Basin case study* (Background Report to Deliverable 1.3.1). Amsterdam: NeWater.

Bruch, C. (2009). Adaptive water management: Strengthening laws to cope with uncertainty. In A. K. Biswas, C. Torjada, & R. Izquierdo (Eds.), *Water management beyond 2020* (pp. 89–113). Berlin: Springer.

Bruch, C., Janksy, L., Nakayama, M., & Salewicz, K. (Eds.). (2005). *Public participation in the governance of international freshwater resources.* Tokyo: UNU Press.

Bruch, C., & Troell, J. (2011). Legalizing adaptation: Water law in a changing climate. *Water International, 36*(7), 828–845.

Caponera, D. A. (2007). *Principles of water law and administration: National and international.* London: Taylor & Francis.

Carpathian Convention. (2006, January). Framework convention on the protection and sustainable development of the carpathians. *Kyiv.*

Csagoly, P. (2000). *The cyanide spill at Baia Mare, Romania: Before, during and after.* Szentendre: Regional Environment Centre.

Del Moral, L., & Do Ó, A. (2014). Water governance and scalar politics across multiple-boundary river basins: States, catchments, and regional powers in the Iberian Peninsula. *Water International, 39*(3), 333–347. doi:10.1080/02508060.2013.878816

Dellapenna, J. W., & Gupta, J. (2009). The evolution of Global Water Law. In J. W. Dellapenna & J. Gupta (Eds.), *The evolution of the law and politics of water* (pp. 3–19). Delft, Netherlands: Springer. Retrieved from http://link.springer.com/chapter/10.1007/978-1-4020-9867-3_1

Dietz, T., Ostrom, E., & Stern, P. C. (2003). The struggle to govern the commons. *Science, 302,* 1907–1912. doi:10.1126/science.1091015

Ekstrom, J. A., & Young, O. R. (2009). Evaluating functional fit between a set of institutions and an ecosystem. *Ecology and Society, 14*(2), 16.

European Commission. (2013). Communication from the Commission to the European Parliament, the Council, the European Economic and Social Committee and the Committee of the Regions: An EU strategy on adaptation to climate change. *COM(2013)216FINAL.*

European Commission. (2014a). *Links between the Floods Directive (FD 2007/60/EC) and Water Framework Directive (WFD 2000/60/EC)* (Technical Report – 2014–078). Luxembourg: European Union.

European Commission. (2014b, February 9). Macro-regional strategies. Retrieved from http://ec.europa.eu/regional_policy/index.cfm/en/policy/cooperation/macro-regional-strategies/

European Community. (1992). Directive 92/43/EEC of 21 May 1992 on the conservation of natural habitats and of wild flora and fauna. *Official Journal of the European Communities, L206,* 007–50.

European Community. (2000). Directive 2000/60/EC of the European Parliament and of the Council of 23 October 2000 establishing a framework for Community action in the field of water policy. *Official Journal of the European Communities, L327,* 01–73.

European Community. (2003a). Common implementation strategy for the Water Framework Directive (2000/60/EC). Guidance document No. 1. *Official Journal of the European Communities, L2985,* 1–270.

European Community. (2003b). Planning process, common implementation strategy for the Water Framework Directive 2006/60/EC. Guidance document No.11. *Official Journal of the European Communities, L2985,* 1–78.

European Community. (2007). Directive 2007/60/EC on the assessment and management of flood risks. *Official Journal of the European Communities, L288,* 27–34.

European Community. (2009). Directive 2009/147/EC of the European Parliament and of the Council of 30 November 2009 on the Conservation of wild birds. *Official Journal of the European Communities, L20,* 07–25.

European Community. (2010). Consolidated versions of the treaty on European Union and the treaty on the functioning of the European Union. *Official Journal of the European Communities 2010/C83/01 and 2010/C83/02,* pp. 01–403.

European Parliament. (2004). The principle of subsidiarity. *European Parliament Fact Sheets.* Retrieved from http://www.europarl.europa.eu/facts_2004/1_2_2_en.htm

European Union Member States. (2012). Consolidated version of the treaty on European Union and the treaty on the functioning of the European Union. *Official Journal of the European Union, C326*, 1–390.

Finger, M., Tamiotti, L., & Allouche, J. (Eds). (2006). *The multi-governance of water: Four case studies*. Albany: State University of New York Press.

Folke, C., Hahn, T., Olsson, P., and Norberg, J. (2005). Adaptive governance of social-ecological systems. *Annual Review of Environmental Resources 30*, 441–473. doi: 10.1146/annurev.energy.30.050504.144511.

Gleick, P. (2000). The changing water paradigm: A look at twenty-first century water resources development. *Water International, 25*(1), 127–138. doi:10.1080/02508060008686804

Griggs, N. S. (2015). Misalignment of watershed and jurisdictional boundaries: The importance of scale. *Water Policy, 17*(6), 1079–1092.

Grünewald, U. (2008). Voraussetzungen für eine erfolgreiche Flussgebietsbewirtschaftung: Klare einzugsgebietsbezogene Ursachen-Wirkungs-Analysen und klares einzugsgebietsbezogenes Handeln [Conditions for successful river basin management: Catchment-based cause-effect analysis and clear catchment-related action]. *Korrespondenz Wasserwirtschaft, 1*(8), 423–426.

Gupta, J., Termeer, C., Klostermann, J., Meijerink, S., Van den Brink, M., Jong, P., ... Bergsma, E. (2010). The adaptive capacity wheel: A method to assess the inherent characteristics of institutions to enable the adaptive capacity of society. *Environmental Science & Policy, 13*, 459–471. doi:10.1016/j.envsci.2010.05.006

Huitema, D., Mosert, E., Egas, W., Moellenkamp, S., Pahl-Wostl, C., & Yalcin, R. (2009). Adaptive water governance: assessing the institutional prescriptions of adaptive (Co-) management from a governance perspective and defining a research agenda. *Ecology and Society, 14* (1), 26.

Huntjens, P., Pahl-Wostl, C., & Grin, J. (2010). Climate change adaptation in European river basins. *Regional Environmental Change, 10*(4), 263–284. doi:10.1007/s10113-009-0108-6

International Commission for the Protection of the Danube River. (1994). *Convention on cooperation for the protection and sustainable use of the danube river (Danube River Protection Convention)*. Vienna: Author. Retrieved from https://www.icpdr.org/main/sites/default/files/DRPC%20English%20ver.pdf

International Commission for the Protection of the Danube River. (2008). *Analysis of the Tisza River Basin 2007*. Vienna: Author. Retrieved from http://www.icpdr.org/main/sites/default/files/Tisza_RB_Analysis_2007.pdf

International Commission for the Protection of the Danube River. (2009a). *Danube river district management plan*. Vienna: Author. Retrieved from http://www.icpdr.org/main/sites/default/files/DRBM_Plan_2009.pdf

International Commission for the Protection of the Danube River. (2009b). *Sub-basin level flood action plan Tisza River Basin*. Vienna: Author. http://www.icpdr.org/main/sites/default/files/FAP09_Tisza.pdf

International Commission for the Protection of the Danube River. (2010, February 16). *Danube declaration adopted at the ministerial meeting*. Vienna: Ministerial Declaration.

International Commission for the Protection of the Danube River. (2011). *Integrated Tisza River Basin management plan*. Vienna: Author. Retrieved from http://www.icpdr.org/main/sites/default/files/Uploaded%20-%20ITRBM%20PLan%20-%20Jan%202011_V2GWcomprev%20Okt2011.pdf

International Commission for the Protection of the Danube River. (2013). *ICPDR strategy on adaptation to climate change*. Vienna: Author. Retrieved from http://www.icpdr.org/main/sites/default/files/nodes/documents/icpdr_climate-adaptation-strategy.pdf

International Commission for the Protection of the Danube River. (2015). *Danube River Basin district management plan- update 2015*. Vienna: Author. Retrieved from http://www.icpdr.org/main/sites/default/files/nodes/documents/drbm_plan_-_update_2015_-_report_-_dec_2014.pdf

International Conference on Water and the Environment. (1992). The Dublin statement on water and sustainable development. Retrieved from: http://www.un-documents.net/h2o-dub.htm

International Law Commission. (1966). Draft articles on the Law of Treaties with commentaries. *Yearbook of the International Law Commission, Vol. II.* A/CN.4/SER.A/1966/Add.1. New York: United Nations.

Kuhlmann, S. (2001). Future governance of innovation policy in Europe - Three scenarios. *Research Policy, 30*, 953–976.

Linnerooth-Bayer, J., Eckenberg, L., & Vári, A. (2013). Catastrophe models and policy processes: Managing flood risk in the Hungarian Tisza River Basin - An introduction. In A. Amendola, J. Linnerooth-Bayer, L. Ekenberg, & A. Vári (Eds.), *Integrated catastrophe risk modeling: Supporting policy processes* (pp. 171–179). Dordrecht: Springer Science.

Marshall, G. R. (2008). Nesting, subsidiarity, and community-based environmental governance beyond the local level. *International Journal of the Commons, 2*(1), 75–97. doi:10.18352/ijc.50

McCaffrey, S. C. (2014, May 25). The entry into force of the 1997 watercourses convention. *International Water Law Project Blog*. Retrieved from http://www.internationalwaterlaw.org/blog/2014/05/25/dr-stephen-mccaffrey-the-entry-into-force-of-the-1997-watercourses-convention/

Moss, T. (2012). Spatial fit, from panacea to practice: Implementing the EU Water Framework Directive. *Ecology and Society, 17*(3), 2. doi:10.5751/ES-04821-170302

Munaretto, S., Siciliano, G., & Turvani, M. E. (2014). Integrating adaptive governance and participatory multicriteria methods: A framework for climate adaptation governance. *Ecology & Society, 19*(2), 74. doi:10.5751/ES-06381-190274

Nagy, I., Ligetvári, F., & Schweitzer, F. (2010). Tisza river valley: Future prospects. *Hungarian Geographical Bulletin, 59*(4), 361–370.

Nanni, M. (2012). Legislation as a tool in support of adaptive water management in response to climate change. *Water International, 37*(6), 628–639. doi:10.1080/02508060.2012.714966

Newig, J., & Fritsch, O. (2009). Environmental governance: Participatory, multilevel - and effective? *Environmental Policy and Governance, 19*, 197–214. doi:10.1002/eet.v19:3

Pahl-Wostl, C. (2009). A conceptual framework for analysing adaptive capacity and multi-level learning processes in resource governance regimes. *Global Environmental Change, 19*, 354–365. doi:10.1016/j.gloenvcha.2009.06.001

Pahl-Wostl, C., Sendzimir, J., Jeffrey, P., Aerts, J., Berkamp, G., & Cross, K. (2007). Managing change toward adaptive water management through social learning. *Ecology and Society, 12*(2), 30.

Patton, C., Sawicki, D., & Clark, J. (2013). *Basic methods of policy analysis and planning*. Geneva: Pearson Education.

Rosenau, J. N. (1992). Governance, order, and change in world politics. In J. N. Rosenau & E. O. Czempiel (Eds.), *Governance without government: Order and change in world politics* (pp. 1–29). Cambridge: Cambridge University Press.

Sands, P. (2007). The development of the 1997 convention on the non-navigational uses of international watercourses by the international law commission. In D. K. Anton, J.I. Charney, P. Sands, T. Shoenbaum, & M. Young (Eds.), *International Environmental Law: Cases, Materials, Problems* (pp. 800–801). New Jersey, USA: LexisNexis.

Schneller, K., Bálint, G., Chicos, A., Csete, M., Dzurdzenik, J., Göncz, A., ... Pálvolgyi, T. (2013). Climate change impacts on the Hungarian, Romanian, and Slovak Territories of the Tisza catchment Area. In P. Schmidt-Thomé & S. Greiving (Eds.), *European climate vulnerabilities and adaptation: A spatial planning perspective* (pp. 205–229). Chichester, West Sussex: Wiley Blackwell.

Sendzimir, J., Flachner, Z., Pahl-Wostl, C., & Knieper, C. (2010). Stalled regime transition in the upper Tisza River Basin: The dynamics of linked action situations. *Environmental Science & Policy, 13*, 604–619. doi:10.1016/j.envsci.2010.09.005

Sendzimir, J., Magnuszewski, P., Balogh, P., & Vári, A. (2007). Anticipatory modeling of biocomplexity in the Tisza River Basin: First steps to establish a participatory adaptive framework. *Environmental Modelling & Software, 22*, 599–609. doi:10.1016/j.envsoft.2005.12.032

Sendzimir, J., Magnuszewski, P., Flachner, Z., Balogh, P., Molnar, G., Sarvari, A., & Nagy, Z. (2008). Assessing the resilience of a river management regime: Informal learning in a shadow network in the Tisza River Basin. *Ecology and Society, 13*(1), 11.

Troell, J., & Swanson, G. (2014). *Adaptive water governance and the principles of International Water Law* (IUCN Environmental Policy and Law Paper No. 75). Gland: IUCN. https://portals.iucn.org/library/sites/library/files/documents/IUCN-EPLP-no.075.pdf

United Nations. (1997). Convention on the Law of the non-navigational uses of international watercourses. *Doc. A/51/869.* Retrieved from: https://treaties.un.org/Pages/ViewDetails.aspx?src=IND&mtdsg_no=XXVII-12&chapter=27&lang=en#EndDec

United Nations Conference on Environment and Development. (1992). Agenda 21: Programme for action for sustainable development. *A/Conf.151/26(1992).* Retrieved from: http://sustainabledevelopment.un.org/index.php?page=view&nr=23&type=400&menu=35

United Nations Economic Commission for Europe (UNECE). (2010). *Guidance on water and adaptation to climate change.* Geneva: UN.

United Nations Environment Programme (UNEP). (2014a, November 26–28). *Freshwater law and governance: Global and regional perspectives for sustainability.* Paper presented at the First International Environment Forum for Basin Organizations, Nairobi, Kenya.

United Nations Environment Programme (UNEP). (2014b). *Towards integrated water resources management: International experiences in development of river basin organizations.* Khartoum: Author. Retrieved from http://www.unep.org/disastersandconflicts/portals/155/countries/Sudan/pdf/SudanWRM.pdf

von Keitz, S., & Kessler, P. (2008). Grenzen des Flußgebietsmanagements. Folgt die Wasserwirtshaft dem Falschen Ansatz? [Limits of water basin management: Is water management following the right approach?] *Korrespondenz Wasserwitschaft, 1*(7), 354–360.

Walters, C. (1997). Challenges in adaptive management of riparian and coastal ecosystems. *Conservation Ecology, 1*(2), 1.

Werners, S. E., Flachner, Z., Matczak, P., Falaleeva, M., & Leemans, R. (2009). Exploring earth system governance: A case study of floodplain management along the Tisza river in Hungary. *Global Environmental Change, 19*(4), 503–511. doi:10.1016/j.gloenvcha.2009.07.003

Werners, S. E., West, J., Leemans, R., Tàbara, J. D., Dai, X., Flachner, Z., … Trombi, G. (2011). Opportunities and constraints for climate adaptation in regional water and land use planning. In W. L. Filho (Ed.), *The economic, social and political elements of climate change* (pp. 669–692). Berlin: Springer.

World Wildlife Fund. (2009). *Shifting course: Climate adaptation for water management institutions.* Washington, DC: Author.

Wyborn, C. A. (2015). Connecting knowledge with action through coproductive capacities: Adaptive governance and connectivity conservation. *Ecology & Society, 20*(1), 11. doi:10.5751/ES-06510-200111

# Index

Notes: Page numbers in *italics* refer to figures
Page numbers in **bold** refer to tables
Page numbers with 'n' denote notes

For Product Safety Concerns and Information please contact our EU
representative GPSR@taylorandfrancis.com
Taylor & Francis Verlag GmbH, Kaufingerstraße 24, 80331 München, Germany

www.ingramcontent.com/pod-product-compliance
Ingram Content Group UK Ltd.
Pitfield, Milton Keynes, MK11 3LW, UK
UKHW050754180926
475701UK00007B/1047

* 9 7 8 0 3 6 7 5 8 9 6 6 0 *